Vanadium Dioxide-Based Thermochromic Smart Windows

Vanadium Dioxide-Based Thermochromic Smart Windows

edited by
Yi Long
Yanfeng Gao

CRC Press
Taylor & Francis Group
Boca Raton London New York

CRC Press is an imprint of the
Taylor & Francis Group, an **informa** business

Vanadium Dioxide-Based Thermochromic Smart Windows

First published 2021 by Jenny Stanford Publishing Pte. Ltd.

Published 2023 by CRC Press
Taylor & Francis Group
6000 Broken Sound Parkway NW, Suite 300
Boca Raton, FL 33487-2742

ISBN 13: 978-981-4877-06-0 (hbk)

Visit the Taylor & Francis Web site at
http://www.taylorandfrancis.com

and the CRC Press Web site at
http://www.crcpress.com

British Library Cataloguing-in-Publication Data
A catalogue record for this book is available from the British Library.

To my PoPo, Ms. Li Meiying

Contents

Preface

In 2018, the Intergovernmental Panel on Climate Change (IPCC) reduced the global warming allowance from 2 to 1.5°C to alarm the world about the great urgency of climate change. This emphasized the significance of energy conservation and carbon emission reduction. The building sector consumes approximately 40% of the total energy, while heating, ventilation, and air-conditioning (HVAC) consume half of the energy in buildings. To address this issue, improving building energy efficiency is critical in energy conservation.

Windows are considered as the least energy-efficient part of the building envelope. In hot seasons, most of the window-directed solar energy is converted into heat, which leads to a high demand for cooling, while in winter, windows are responsible for 30% of energy loss. The most studied energy-saving windows are focused on chromogenic technologies, including electrochromic, photochromic, and thermochromic. Among these, thermochromic materials are considered the most cost-effective, with rational stimulus and zero energy input.

Vanadium dioxide (VO_2)-based thermochromic smart windows are the most widely studied category. In this book, I have had the honour of inviting top scientists from all over the world to cover a wide range of topics focused on VO_2-based smart windows, including history, mechanisms, fabrication, strategies, and perspectives. This is the first book dedicated to this specific area. We do hope that it would serve as a textbook for young researchers who are interested in solving the biggest problems of window energy consumption and, more importantly, inspire fresh ideas in and facilitate the commercialization of this particular technology.

Thermochromic smart windows is a small community trying to solve a big problem. This community has been working very hard to conduct fundamental scientific study, carry out engineering work to keep improving performance, and deliver commercial products to society. We hope this technology, in very near future, to become one of the options of cutting down energy consumption in buildings.

Editing and writing this book was not an easy task. I would personally like to show my appreciation for my dearest colleagues: Prof. Yanfeng Gao, Prof. Claes G. Granqvist, Prof. Ping Jin, Prof. Yuanyuan Cui, Prof. Xiudi Xiao, Dr. Xun Cao, Prof. Ming Li, Prof. Xuchuan Jiang, Prof. Dimitra Vernardou, Prof. Cheng Chang, Prof. Litao Kang, and Prof. Haibo Jin. Special thanks to my students, Prof. Ning Wang, Dr. Yujie Ke, Prof. Yang Zhou, Dr. Shancheng Wang, Alvis Vu Tuan Duc, and ChengZhi Zhou. We grow up together.

I would like to show my appreciation to my funding support, the National Research Foundation, Prime Minister's Office, Singapore, under its Campus for Research Excellence and Technological Enterprise (CREATE) program. They generously sponsored me to start this research in 2011, and the sweet fruits have ripened for harvesting. My gratitude is beyond words. I am grateful to my program leaders, Profs. Pooi See Lee and Shlomo Magdassi. Their consistent support and tolerance helped me tremendously.

Last but not the least, my family: SF, Zhi Xing, and Hong Yi. My love to you is unconditional.

Yi Long
June 2020

Chapter 1

Thermochromic VO_2 for Energy-Efficient Glazing: An Introduction

Claes G. Granqvist
Department of Materials Science and Engineering, Ångström Laboratory, Uppsala University, SE-75121 Uppsala, Sweden
claes-goran.granqvist@angstrom.uu.se

Vanadium dioxide, VO_2, has numerous potential applications in forthcoming technologies. The largest one, at least in terms of surface area, is in glazing (i.e., windows and glass facades) for energy-efficient buildings. These applications are based on the fact that VO_2 has thermochromic properties, and thin films and nanoparticle composites can transmit less solar energy above a certain "critical" temperature in the vicinity of room temperature than below this temperature, which implies that the demand on energy-guzzling air cooling can be lowered. The transmittance of visible light, on the other hand, is not influenced by temperature to a corresponding degree. This chapter outlines the basic properties of VO_2 and discusses how this material can be modified in order to be of practical interest for glazing. The roles of multilayering, doping, nanostructuring, etc., are introduced. In-depth discussions of these and other aspects are given in subsequent chapters.

Vanadium Dioxide-Based Thermochromic Smart Windows
Edited by Yi Long and Yanfeng Gao

ISBN 978-981-4877-06-0 (Hardcover), 978-1-003-17731-9 (eBook)
www.jennystanford.com

1.1 The Big Challenge

Environmental challenges serve as prime drivers for new technology [1–3], and the business case for clean energy is growing [4]. The quest for new energy technology is fueled by the increasing quantity of greenhouse gases in the atmosphere. Due to the greenhouse gases, the amount of CO_2 has risen dramatically from ~315 parts per million (ppm) at the end of the 1950s to about 410 ppm in 2018, while the growth rate has almost tripled [5]. The CO_2 injection originates largely from energy conversion, specifically the burning of coal, oil, and gas. Another greenhouse gas, which is presently emerging as an environmental threat, is hydrofluorocarbon emitted from air-conditioning units [6, 7], whose numbers are growing rapidly, particularly in developing countries. The increasing amount of CO_2 is strongly believed to be of significance for life on Earth and leads to global warming and rising sea levels [8], and there are many secondary and perilous influences of climate change connected with socioeconomic effects and increased risk for violent conflict [9–11], health issues [12, 13], and modifications of the geosphere [14] and biosphere [15–17]. Self-reinforcing feedback loops represent special threats and might lead to a transition to a "hothouse Earth" with temperatures much higher than during any interglacial period during the last millennium [18]. Furthermore, it should be noted that the global population is growing and is expected to be some 50% larger in the year 2100 than it is today [19]; this population is increasingly living in megacities [20], which act as "urban heat islands" and tend to have average temperatures several degrees Celsius higher than the temperatures of the surrounding countryside [21]. The human influence is strong enough to distinguish the present geological era (the Anthropocene) from the one prior to the Industrial Revolution (the Holocene) [22].

The gloomy prospects outlined above underscore that radical steps must be taken to decarbonize the energy sector, and this realization leads directly to a focus on buildings which account for 30% to 40% of the global use of primary energy [23]. Energy efficiency in the built environment is sometimes overlooked as an opportunity for CO_2 abatement [24], but, in fact, there are many emerging "green" technologies—often based on nanomaterials—that can be put to work [1, 25–27]. Particular attention should be given to glazing (i.e., windows and glass facades), which very often allow undesired energy transfer with excessive inflow or outflow of heat and associated need for energy-demanding cooling and heating of buildings. It is then patently clear that *glazing with adjustable throughput of solar energy and visible light* can lower the energy demand. Such glazing is often called "smart," "intelligent," or "adaptive" and uses a class of materials known as "chromogenic" [28, 29]. These materials can be of several different types and can alter their properties upon temperature change (thermochromic materials, as discussed in the present book), exposure to electrical voltage or current (electrochromic materials), irradiation by ultraviolet light (photochromic materials), and exposure to oxidizing/reducing conditions (gasochromic materials).

The present chapter introduces thermochromic vanadium dioxide (VO_2)-based materials for applications in energy-efficient glazing. This work began more than thirty years ago with research at the Lawrence Berkeley National Laboratory in the USA [30, 31] and through concurrent efforts by the author's team in Sweden [32, 33]. Research on thermochromic VO_2-based materials has increased sharply during recent years, as can be seen from a database such as Web of Science; looking for citations to articles with "thermochromic" or "thermochromism" together with "VO_2" or "vanadium dioxide" in their titles yielded about 100 entries/year in 2005, more than 1000 entries/year in 2013, and more than 2000 entries/year in 2018. Obviously, this introductory chapter covers only a small fraction of the relevant scientific literature.

Glazing based on VO_2, or any other inorganic thermochromic material, does not yet exist as a commercial product, but the field of thermochromics has advanced greatly during recent years, and it is the contention of the author and many others that practical implementation of thermochromic glazing may be in the offing [34–43]. It should also be observed that thermochromics is of interest not only for glazing but also, for example, for roof-type ceramic tiles [44] and for a vast number of other applications.

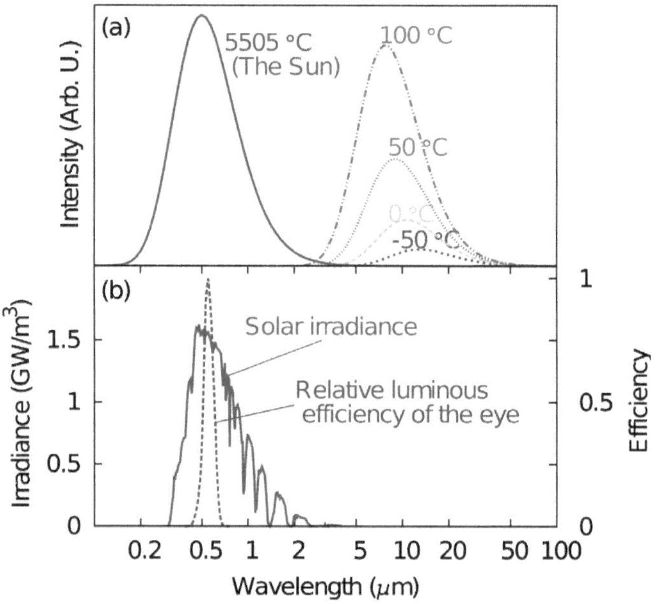

Figure 1.1 (a) Blackbody spectra for the shown temperatures (including the Sun's surface temperature). (b) Typical spectral solar irradiance for clear weather and at sea level and the relative sensitivity of the human eye.

1.2 Thermal, Solar, and Luminous Radiation

Thermochromics is about the control of luminous (visible) and solar light. Figure 1.1 reports key features of the radiation around us. *Thermal radiation*, introduced in Fig. 1.1a, is governed by blackbody curves (shown for four values of temperature in the range of $-50°C < \tau < 100°C$) multiplied by an emittance less than unity; it is evident that this radiation lies at $\lambda > 2$ μm, where λ denotes wavelength. *Solar irradiation* incident on Earth's atmospheric envelope can be represented by blackbody-like radiation emerging from the Sun's surface at around 5500°C; this radiation falls at $0.2 < \lambda < 3$ μm. At ground level and during clear weather, solar irradiation is typically the one shown in Fig. 1.1b; distinct minima in the bell-shaped curve are dominated by absorption in water vapor, and short-wavelength absorption is due to ozone. *Visible light* is represented by the luminous efficiency of the light-adapted human eye, which corresponds to a bell-shaped curve for $0.4 < \lambda < 0.7$ µm; its peak is at $\lambda = 0.55$ µm. It should be noted that distinct wavelength ranges—overlapping or not—characterize the various types of radiation.

Quantitative data on luminous (lum) and solar (sol) transmittance, denoted by T_{lum} and T_{sol}, respectively, are obtained from

$$T_{\text{lum,sol}} = \int d\lambda \varphi_{\text{lum,sol}}(\lambda) T(\lambda) / \int d\lambda \varphi_{\text{lum,sol}}(\lambda), \qquad (1.1)$$

where $T(\lambda)$ is the spectral transmittance, φ_{lum} is the eye's sensitivity [45], and φ_{sol} is the solar irradiance when the sun stands 37° above the horizon (known as the "air mass 1.5" spectrum) [46]. Analogous formulas apply to reflectance R. Furthermore, it is convenient introduce a quantitative measure for solar energy modulation by

$$\Delta T_{\text{sol}} = T_{\text{sol}}(\tau < \tau_c) - T_{\text{sol}}(\tau > \tau_c), \qquad (1.2)$$

where τ_c is a "critical" temperature at which the optical properties of the thermochromic material are changed, as further discussed below.

1.3 Thermochromic Materials, Notably Vanadium Dioxide

Thermochromic properties can be found in inorganic and organic materials as well as among composites. The inorganics may be most important for glazing and are considered first.

Inorganic thermochromic materials are conventionally and conveniently introduced via their electrical conductivity, as seen from the compilation of data in Fig. 1.2 [31]. The conductivity changes abruptly, sometimes by orders of magnitude, at rather well-defined temperatures τ_c as the result of reversible alterations of the crystal structure. Vanadium dioxide is especially interesting owing to the fact that τ_c occurs at about 68°C—which is not hugely different from room temperature—as first observed sixty years ago [47]. The change takes place between a low-temperature ($\tau < \tau_c$) structure with semiconducting properties and a high-temperature ($\tau > \tau_c$) structure with metallic-like electrical conductivity.

The optical properties of VO_2 are radically changed at τ_c as well. This feature—i.e., thermochromism—is illustrated in Fig. 1.3 with

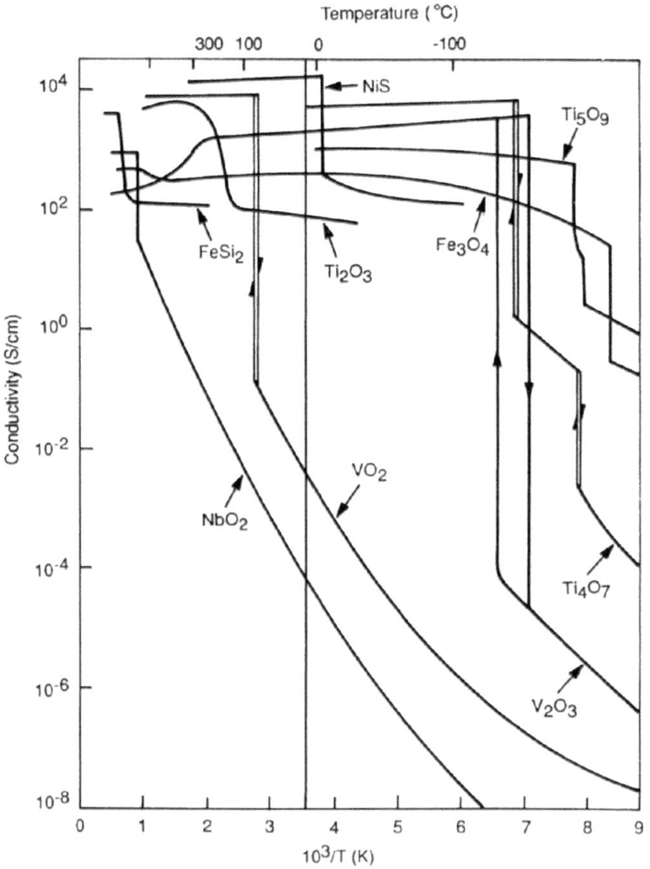

Figure 1.2 Electrical conductivity versus temperature (upper horizontal axis) and reciprocal temperature (lower horizontal axis) for metal-based compounds. The vertical line indicates room temperature. Vertical arrows signify hysteretic transitions. Reprinted from [31], Copyright © 1986 Elsevier.

regard to a 50 nm thick film of VO_2 and a corresponding nanoparticle composite containing the same amount of VO_2. The panels on the left show $T(\lambda)$ (upper) and $R(\lambda)$ (lower) in the wavelength range for solar irradiation and demonstrate that $T(\lambda)$ is greater in the semiconducting state at $\tau < \tau_c$ than in the metallic-like state at $\tau > \tau_c$ as long as λ is larger than about 1 μm. The difference in $T(\lambda)$ between low and high temperatures is enhanced at larger

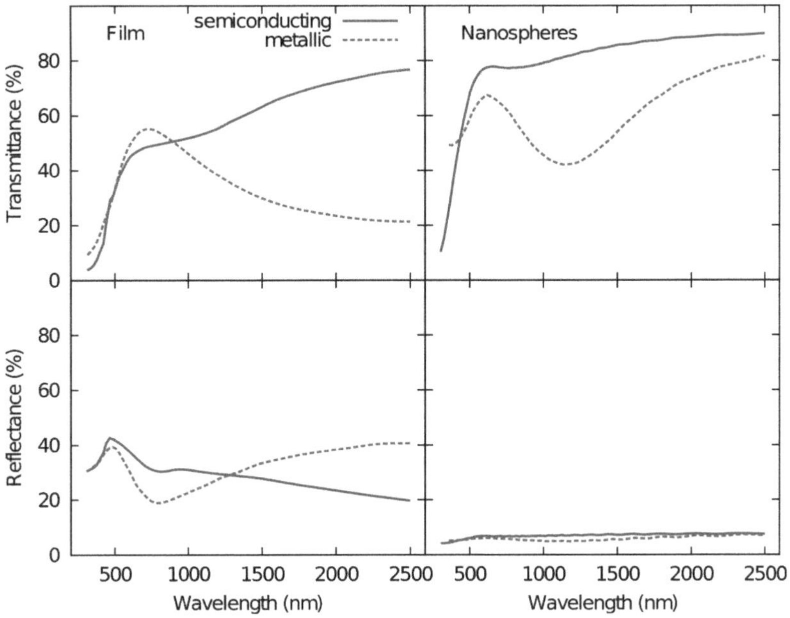

Figure 1.3 Spectral transmittance (upper panels) and reflectance (lower panels) of a 50 nm thick VO_2 film (left-hand panels) and of a layer comprising a dilute dispersion of VO_2 nanospheres, with an equivalent VO_2 thickness of 50 nm, in a medium mimicking transparent glass and polymer (right-hand panels).

wavelengths. Clearly this kind of variation of $T(\lambda)$ is the desired one for energy-efficient fenestration, and glazing with a thermochromic thin film of VO_2 admits more energy at a low temperature than at a high temperature, as noted many times in the literature [23, 48–50]. In addition, Fig. 1.3 shows that $R(\lambda)$ increases monotonically toward longer wavelengths, as expected for a metal. The right-hand panels in Fig. 1.3 pertain to a VO_2-based nanoparticle composite; it shows even larger thermochromism than the corresponding VO_2 film and demonstrates that nanostructuring can be of huge importance, as will be discussed later.

What is the physics behind the large property change at τ_c? Insights into this issue can be obtained from the atomic structure of VO_2, which is monoclinic at $\tau < \tau_c$ ($P2_1/c$ in crystallographic notation) and tetragonal ($P4_2/mnm$) at $\tau > \tau_c$. The two phases are usually referred to as monoclinic (M_1) and rutile (R). There are also more than ten additional metastable VO_2 polymorphs [51], none of which is thermochromic (however, one may note VO_2(B), which is an interesting material for batteries and supercapacitors).

The M_1 and R phases of VO_2 have subtle—yet extremely important—differences, and M_1 is characterized by V^{4+} cations along the rutile *c* axis forming homopolar bonds and undergoing structural twist [52, 53], as illustrated in Fig. 1.4 [54]. The decreased symmetry, as well as the formation of a bonding-antibonding pair, leads to the splitting of the vanadium d band, which can be seen in Fig. 1.5. The fact that the Fermi level E_F falls in between the two parts of the $d_{\|}$ band at $\tau < \tau_c$, while E_F lies inside this band at $\tau > \tau_c$, accounts for the metal–insulator transition manifested in the electrical and optical properties. The crystallographic change shown in Fig. 1.4 is characteristic of a structure-driven transition (known as Peierls-type). However, there is also ample evidence in favor of strong electron–electron correlations being a driving force behind

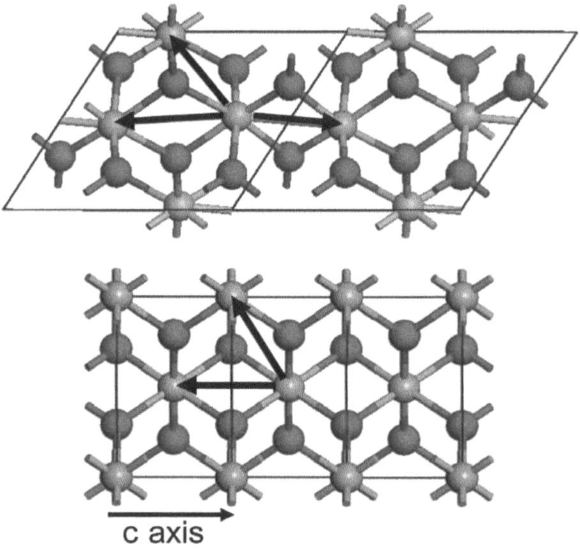

Figure 1.4 Low-temperature monoclinic (M_1) structure (upper panel) and high-temperature tetragonal (R) structure (lower panel) of VO_2. Essential changes of the V^{4+} ions (gray symbols) are indicated by arrows. Reprinted from [54], Copyright © 2009 American Physical Society.

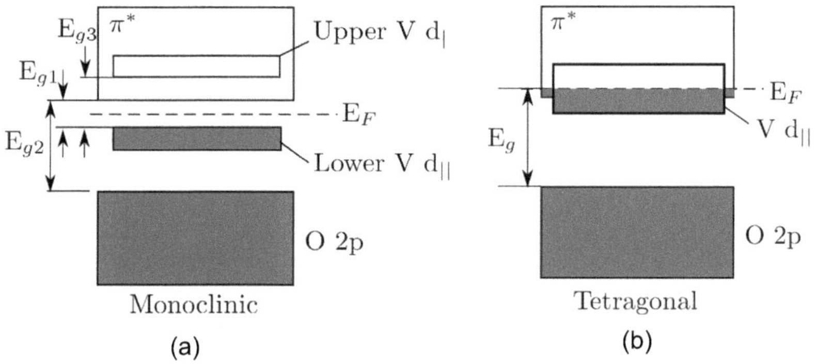

Figure 1.5 Schematic drawing of the band structure of VO_2 in a low-temperature monoclinic (M_1) structure (a) and a high-temperature tetragonal (R) structure (b). The bands are represented conventionally. E_g denotes bandgaps, and E_F is Fermi energy.

the property change at τ_c (Mott transition). A voluminous scientific literature discusses Peierls versus Mott, but a more Solomonic picture is emerging in which both mechanisms are at work for VO_2 [55–57].

Thermochromic properties are common among organic materials but are often confined to rather narrow wavelength ranges and therefore are more suitable for controlling T_{lum} than T_{sol}. Color-neutral thermochromic layers and laminates are well known [58]. Other options are provided by thermotropic "cloud gels" that may be positioned between glass panes [59]. These gels consist of polymers, and "clouding" (i.e., transition to a diffusely scattering state) can take place at a "critical" temperature as the outcome of reversible thermochemical dissolution and thermally induced changes in the lengths of the polymer molecules.

1.4 Thin Films and Nanoparticles of VO_2: Some Challenges and Opportunities

Thin films and nanoparticles (nanorods, nanobelt, etc.) of VO_2 can be prepared by many techniques, but the fabrication is often fraught with challenges. Vanadium has a high affinity for oxygen and is able

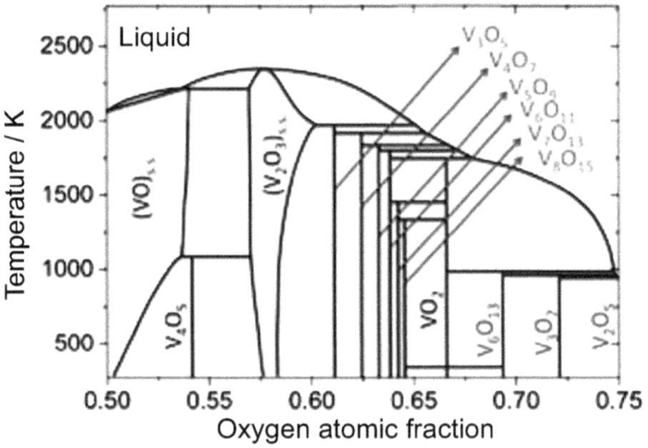

Figure 1.6 Phase diagram for the VO-$VO_{2.5}$ system. Reprinted from [60], Copyright © 2012 Elsevier.

to form compounds containing V^{5+}, V^{4+}, V^{3+}, or V^{2+}. It is then not surprising that the vanadium-oxygen phase diagram is complicated [60]. Figure 1.6 indicates some twenty different phases and shows that the VO_2 phase space is narrow with regard to the oxygen fraction. A reaction route involving SO_2 rather than O_2 may, in fact, be more convenient but is yet not fully explored [61]. Further complexity stems from the fact that the various oxides can have several different crystal structures, as noted above.

Details of the thin film or nanoparticle preparation techniques influence the properties of the ensuing material, which is the subject of a large scientific and technical literature and will be discussed elsewhere in this book. Here, the main focus is on magnetron sputter deposition, which—along with spray pyrolysis (chemical vapor deposition)—is a suitable technique for large-area manufacturing and has proven commercial potential. Magnetron sputtering can be implemented with in-line thin film deposition onto float glass [62] and, alternatively, with roll-to-roll deposition [63, 64] onto flexible substrates of temperature-stable polymers and ultrathin glass. Techniques such as these are able to coat areas on the scale of square kilometers per year, which may be demanded for the full-scale production of thermochromic glazing. A glimpse of the potential market can be seen from the fact that the production

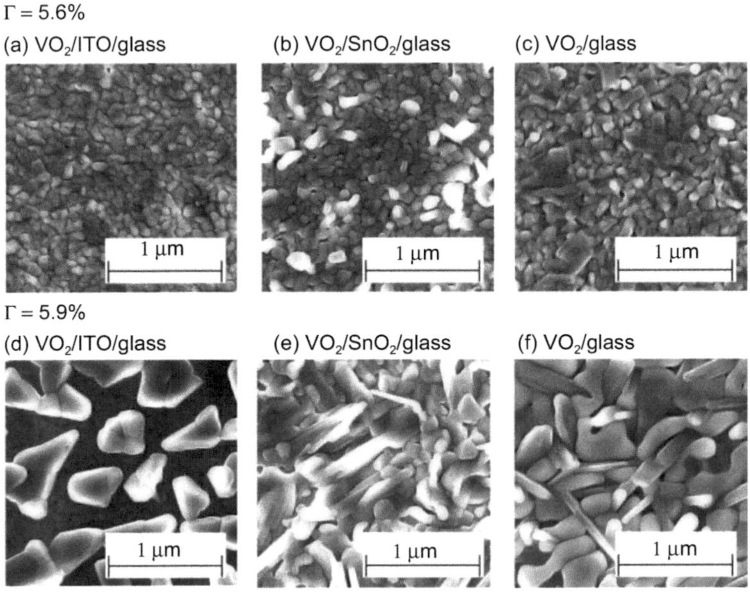

Figure 1.7 Scanning electron micrographs of VO_2 deposits prepared by reactive dc magnetron sputtering onto the following substrates: electrically conducting In_2O_3:Sn (ITO) on glass (a, d), electrically nonconducting SnO_2 on glass (b, e), and bare glass (c, f). Sputtering was carried out in a mixture of oxygen and argon, with the oxygen fraction denoted by Γ. The lateral size is indicated by scale bars. Reprinted from [67], Copyright © 2015 AIP Publishing.

of float glass—most of which is used for glazing—in 2018 was a remarkable 10^{10} square meters per year and growing [65]! Toxicity of VO_2 is an important aspect, especially for large-scale manufacturing [66].

Thin film growth is normally affected by the underlying substrate. This aspect is elaborated with regard to sputter-deposited VO_2 films in Fig. 1.7, where it is evident that the deposit can consist of a dense arrangement of fine regular grains (Fig. 1.7a), well separated large particles (Fig. 1.7d), or a tortuous arrangement of irregular structural units (Figs. 1.7e and 1.7f) depending on whether the deposition is onto glass, SnO_2, or In_2O_3:Sn (known as ITO) [67].

Thin films of thermochromic VO_2 can be prepared by oxidizing a substoichiometric oxide or by reducing an overstoichiometric

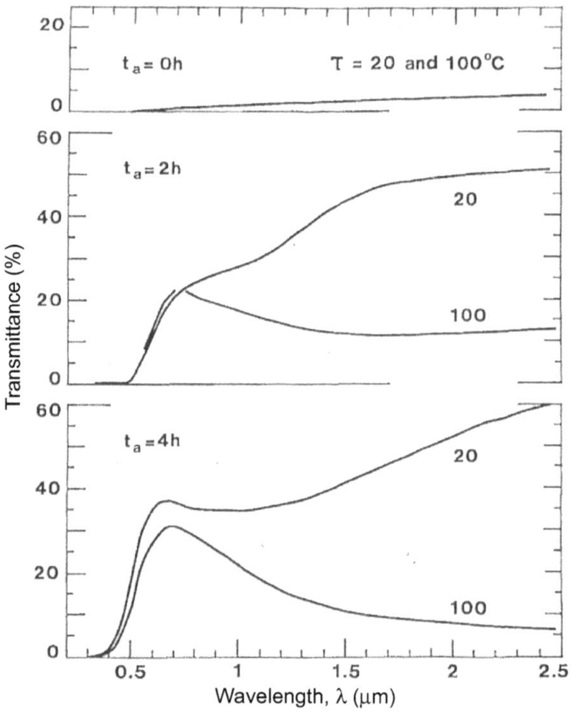

Figure 1.8 Spectral transmittance of a vanadium-based film after deposition and annealing at the temperatures t_a. Data are shown for $\tau < \tau_c$ and at $\tau > \tau_c$. Reprinted with permission from [69], Copyright © 1991 The Optical Society.

oxide, as has been demonstrated many times, which clearly relieves some of the requirements for process control during thin film manufacturing. Rapid thermal treatment can be used to improve process control [68]. Figure 1.8 displays result from an early study in which an approximately 120 nm thick metallic vanadium film was annealed at about 400°C in air for 2 and 4 h [69]. This treatment led to a film with clear-cut thermochromism and exhibiting electrical and optical properties consistent with those in Figs. 1.2 and 1.3. The film thickness was increased by a factor around 2.3, as expected from the transformation from V to VO_2.

Successive transformation of vanadium-based nanoparticles to thermochromic VO_2 is an interesting possibility. One process,

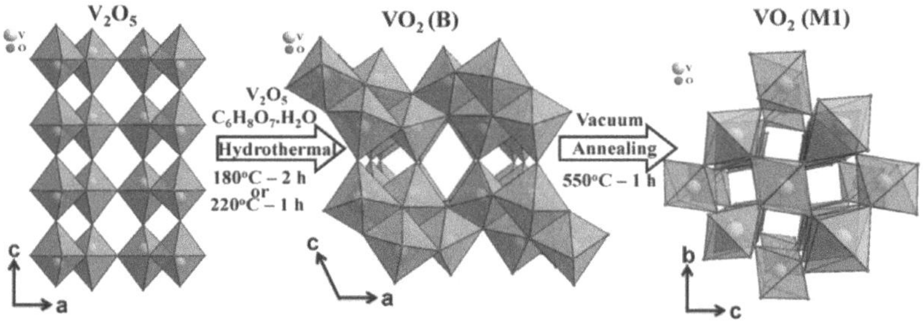

Figure 1.9 Successive transformation of V_2O_5 to $VO_2(B)$ by the hydrothermal process and then of $VO_2(B)$ to $VO_2(M_1)$ by vacuum annealing. Reprinted from [70], Copyright © 2014 Elsevier.

illustrated in Fig. 1.9, uses V_2O_5 and citric acid as precursors in a single-step hydrothermal process yielding $VO_2(B)$ nanoplatelets, which are then transformed to $VO_2(M_1)$ by vacuum annealing for 1 h [51]. In another process, $VO_2(B)$ nanobelts were transformed to thermochromic W-doped $VO_2(M_1)$ via a one-step hydrothermal route in the presence of tungstic acid [70].

1.5 Toward Practical VO_2-Based Thermochromic Glazing: A Multistep Approach

Single thin films of pure VO_2 have been in focus thus far in this introductory chapter. Their thermochromic properties are not good enough for widespread applications, however, and additional steps must be taken to meet a number of challenges [71], as discussed next.

1.5.1 Achieving Long-Term Durability

The first of these challenges is that VO_2 is not thermodynamically stable—which is evident from the phase diagram in Fig. 1.6—and postdeposition oxidation is expected to eventually lead to nonthermochromic V_2O_5 [72]. Figure 1.10 shows data for an 80 nm thick sputter-deposited VO_2 film in an as-prepared thermochromic state and after heating it in dry air at 300°C for 1 h, which has

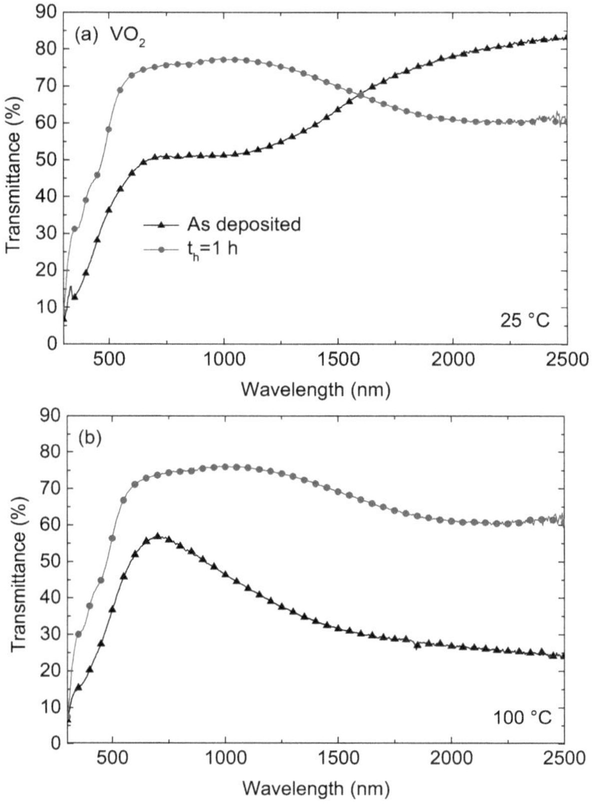

Figure 1.10 Spectral transmittance of a VO_2 film in an as-deposited state and after annealing at 300°C for 1 h. Data were recorded at $\tau < \tau_c$ (a) and $\tau > \tau_c$ (b). Reprinted from [73], Copyright © 2014 Elsevier.

clearly yielded nonthermochromic properties [73]. The remedy to this undesired transformation is to apply a top coating that is dense enough to prevent further oxidation of VO_2, and Fig. 1.11 reports spectral data for the same type of film as in Fig. 1.10 but after coating it with 10 or 30 nm of sputter-deposited Al_2O_3. Now treatments at 300°C in dry air for time periods of up to 30 h were unable to alter the thermochromism substantially, except for the thinnest Al_2O_3 film deposited after treatment for the longest time possible, in which case some minor decline in the thermochromism was noted. Protection by Al_2O_3 is also efficient in humid air [73, 74]. Other

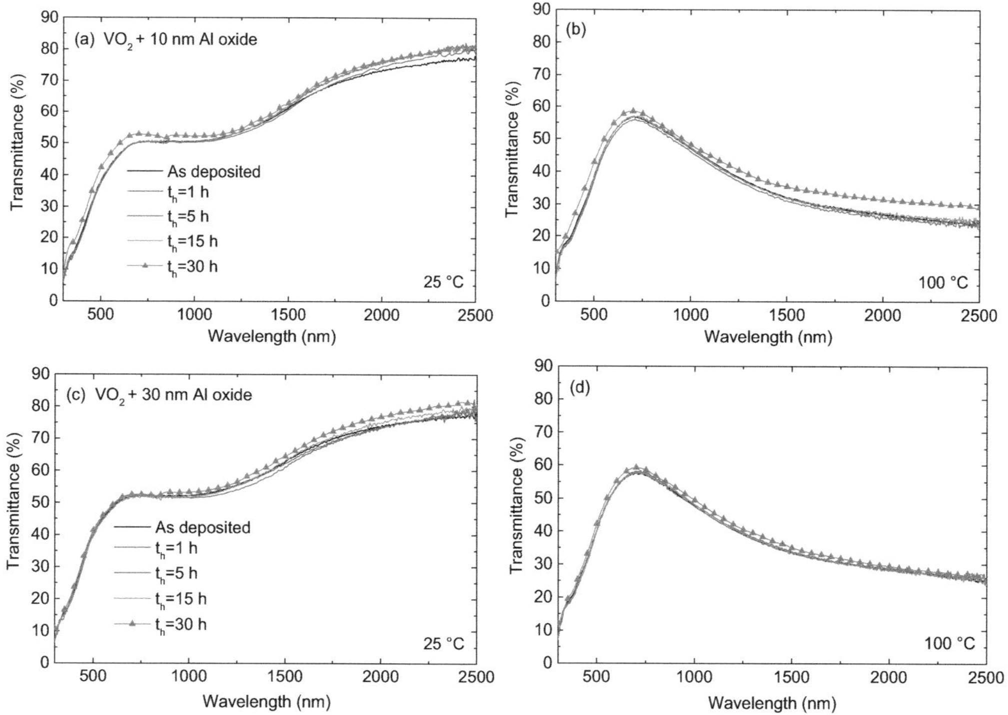

Figure 1.11 Spectral transmittance of VO_2 films, coated with 10 nm (a, b) and 30 nm (c, d) of Al_2O_3, in an as-deposited state and after annealing at 300°C for the shown duration t_h. Data were recorded at $\tau < \tau_c$ (a, c) and $\tau > \tau_c$ (b, d). Reprinted from [73], Copyright © 2014 Elsevier.

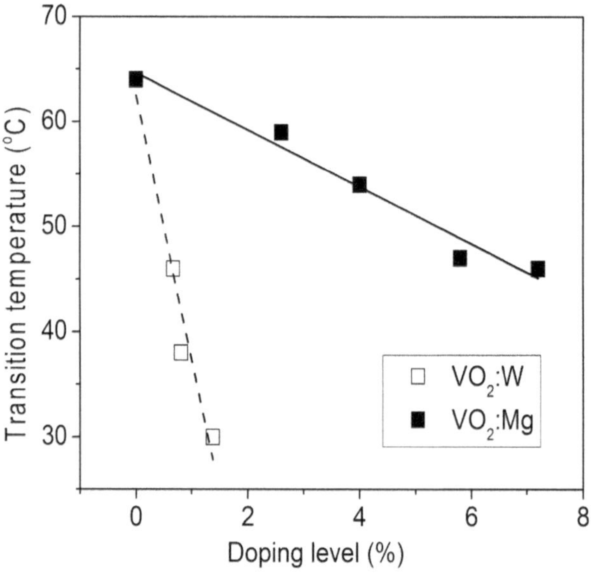

Figure 1.12 "Critical" transition temperature τ_c between semiconducting and metallic-like states in sputter-deposited VO_2 films doped with tungsten or magnesium. Reprinted from [79], Copyright © 2009 AIP Publishing.

oxidation-preventing measures include thin films of SiO_2 [75] as well as ZnO or alumina shells on VO_2 nanoparticles [76, 77].

1.5.2 Having the Thermochromic Shift at Room Temperature

Another challenge, which can be met in a straightforward manner, is the fact that τ_c of VO_2 is well above room temperature, specifically at 68°C for a bulk-like material [47]. However, it has been known for a long time that doping—i.e., replacing some of the vanadium atoms with another species—can increase or decrease the "critical" temperature, and tungsten is particularly efficient for lowering τ_c [52, 78]. Figure 1.12, based on electrical measurements, reports that about 2 at.% of tungsten is able to displace τ_c to approximately room temperature [79], and other data show the same result as long as the VO_2 is homogeneous and well crystallized [71]. Infrared emissivity in the 7.5–14 μm wavelength range is shown in Fig. 1.13 for sol-

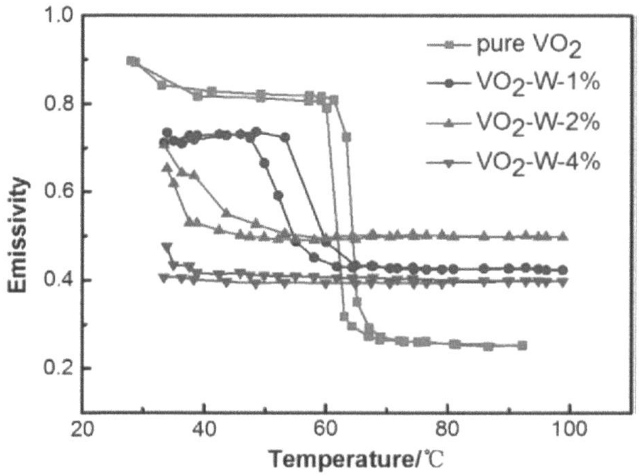

Figure 1.13 Emittance versus temperature for sol-gel-deposited VO_2 films doped with the shown amounts of tungsten. Some hysteresis is apparent in the thermochromic transitions. Reprinted from [80], Copyright © 2016 Elsevier.

gel-deposited films and gives direct proof that τ_c is decreased by the addition of tungsten and also indicates that the magnitude of the property change is diminished as the amount of tungsten is increased [80]. Normally, the dopant is incorporated by some codeposition technique, but postsynthetic diffusive annealing is another alternative, which was used recently for doping with boron [81].

1.5.3 Enhancing the Luminous Transmittance

Glazing should normally have luminous transmittance of a high value in order to provide sufficient daylighting and indoors–outdoors visual contact. This requirement is challenging for thermochromic VO_2 films, and T_{lum} is no larger than about 40% even if the film is as thin as 50 nm (cf Fig. 1.3). One straightforward method to boost the transmittance is via antireflection coatings, which can also serve to create desired color rendering and prevent further oxidation. There are many studies of multilayer structures with VO_2 films below other oxide thin films, or in between such films, with refractive indices and film thicknesses adjusted to maximize T_{lum} [75, 82, 83]. If the top layer is of TiO_2, it may provide photocatalytic,

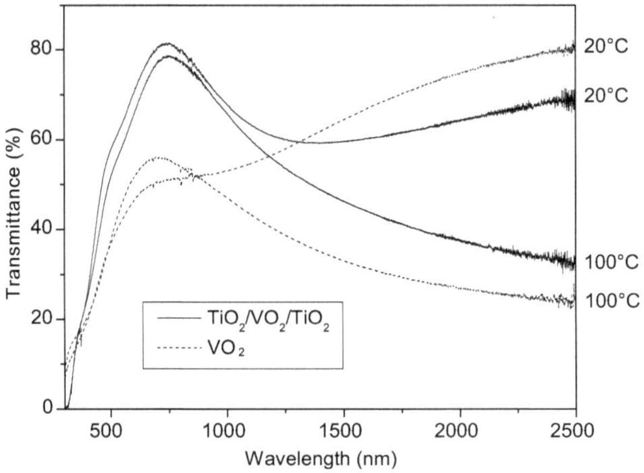

Figure 1.14 Spectral transmittance of a VO_2 film and of such a film located between two 75 nm thick TiO_2 films to prevent reflection. The data were recorded at $\tau < \tau_c$ and $\tau > \tau_c$. Reprinted from [82], Copyright © 2009 Elsevier.

self-cleaning, and antifogging effects [84], whereas ZnO top layers can give antibacterial and biocompatible functionality [85].

Figure 1.14 shows one example of a 50 nm thick sputter-deposited VO_2 film and such a film located between two 75 nm thick TiO_2 films [82]. The midluminous transmittance is seen to increase from about 40% to about 60% as a result of the TiO_2 films, which is a substantial improvement. Interesting angular properties can be achieved with a larger number of layers, as demonstrated for two VO_2 films positioned between three TiO_2 films [86]. Surface engineering is a viable option for enhancing the properties of VO_2 thin films and, for example, "bioinspired" cone-shaped surface structuring can be employed for antireflection treatment [87].

The fact that $T(\lambda)$ as well as $R(\lambda)$ are low at $\lambda < 500$ nm (cf Fig. 1.3) shows that optical absorption is high, and obviously antireflection treatments cannot do anything about this. To enhance T_{lum} radically, it is necessary to lower the short-wavelength absorption, which implies that the dominating optical bandgap of VO_2 (cf Fig. 1.5) should be widened. A particularly efficient dopant for this purpose is magnesium, as found both experimentally [88, 89] and computationally [90], and other band-widening agents

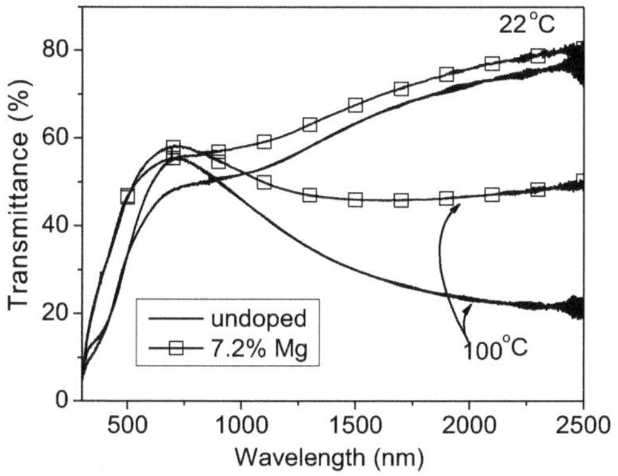

Figure 1.15 Spectral transmittance of 50 nm thick magnesium-doped and undoped VO_2 films recorded at $\tau < \tau_c$ and $\tau > \tau_c$. Reprinted from [79], Copyright © 2009 AIP Publishing.

are certain alkaline earth metals [89], zinc [91], and terbium [92]. Codoping with tungsten and a second element can combine increased T_{lum} with lowered τ_c. Replacing some of the oxygen in VO_2 by fluorine also increases T_{lum}, as has been known for many years [93, 94]. Figure 1.15, from the original report on bandgap widening by magnesium doping [79], shows $T(\lambda)$ for sputter-deposited 50 nm thick films of doped and pure VO_2 at $\tau < \tau_c$ and $\tau > \tau_c$ and demonstrates that the dopant displaces the high-transmittance region toward shorter wavelengths at the same time as it has the unwanted effect of diminishing ΔT_{sol}. Figure 1.12 demonstrates that τ_c drops as the magnesium content is enhanced, although not as rapidly as on doping with tungsten. Figure 1.16 indicates that T_{lum} is enhanced from 39% to 51% as the magnesium content is increased from 0% to 7.2%, which is significant.

1.5.4 Boosting the Solar Energy Modulation

The change in $T(\lambda)$ at τ_c, shown in the left-hand panels of Fig. 1.3, takes place predominantly at wavelengths for which solar irradiance is weak, and the associated undesirably small magnitude of ΔT_{sol} (see Eq. 1.2) for VO_2 films has been a major obstacle for developing practical thermochromic glazing. However, the discovery via computation [95, 96] that a layer consisting of

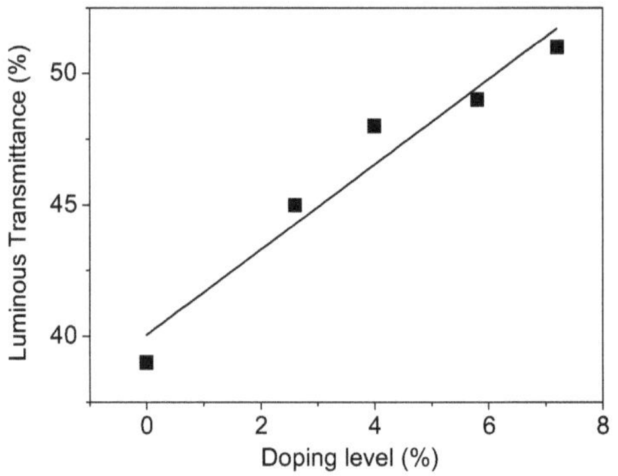

Figure 1.16 Luminous transmittance as a function of magnesium content for 50 nm thick VO_2-based films recorded at $\tau < \tau_c$. The line is drawn to guide the eye. Reprinted from [79], Copyright © 2009 AIP Publishing.

VO_2 nanoparticles—rather than a continuous VO_2 thin film—can enhance ΔT_{sol} radically was a game changer. The sizes should not be much larger than 20 nm in order to avoid undesirable optical scattering from nanoparticles surrounded by air [97] or from freely exposed grains in a discontinuous thin film (cf Fig. 1.7) [98], whereas larger sizes can be tolerated if their embedding material has a refractive index similar to that of VO_2. Such "nanothermochromism" [95] is illustrated in the right-hand panels of Fig. 1.3, which show $T(\lambda)$ and $R(\lambda)$ for a 5 μm thick layer of a transparent material representing typical glass and polymers (with a refractive index of 1.5) with 1 vol.% of well dispersed VO_2 nanospheres. The chosen parameters yield an "equivalent" VO_2 thickness of 50 nm, implying that the nanosphere data can be easily compared with the results of the continuous VO_2 film.

It should be noted that VO_2 films and VO_2-based nanoparticle composites differ in terms of a number of important features: (i) the nanosphere-containing composite has a much larger optical transmittance than the corresponding thin film, (ii) metallic-like nanoparticles display a pronounced transmittance minimum at $0.7 < \lambda < 1.5$ μm, which strongly curtails the transmittance of intense solar radiation, while T_{lum} is influenced only marginally, and (iii) the nanospheres are optically absorbing rather than reflecting. The deep minimum in $T(\lambda)$—corresponding to an absorption maximum—for

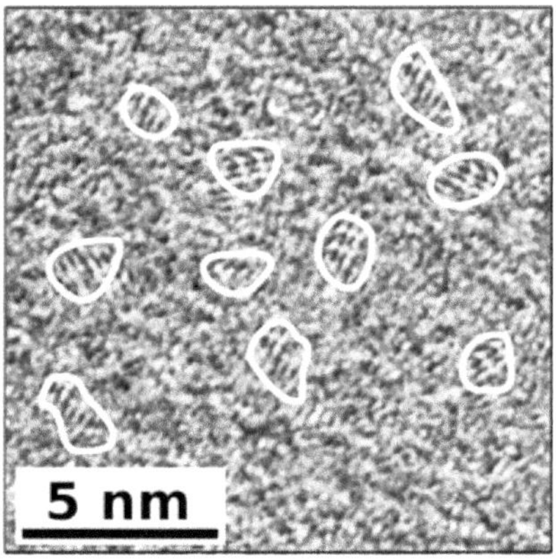

Figure 1.17 High-resolution transmission electron micrograph of a nanocrystalline VO_2-Al_2O_3 composite thin film; encircled areas indicate VO_2. Reprinted from [101], Copyright © 2008 Elsevier.

infrared solar irradiation stems from plasmon resonance (collective oscillations of free electrons) in VO_2 at $\tau > \tau_c$ [99] and lies in a wavelength range that is almost ideal for controlling solar energy throughput without affecting the luminous properties (cf Fig. 1.1).

Nanoparticles and nanorods based on VO_2 can be prepared by a large number of chemical and physical techniques, and many parameters affect the specific nanofeatures [67, 100]. One example is reported in Fig. 1.17, which shows that tiny VO_2-based crystallites can be occluded in a matrix of Al_2O_3 by cosputtering onto a substrate at about 400°C [101]. Another example is given in Fig. 1.18, which shows a VO_2-SiO_2 composite made by sputter deposition; rounded vanadium-rich nanoparticles 100–300 nm in size are apparent [102]. Chemical methods are commonly used to make nanomaterials; direct and scalable synthesis of VO_2 nanoparticles [103] and nanowires [104] can be accomplished by hydrothermal techniques, and templating techniques give additional process flexibility [105].

1.5.5 Performance Limits for Thermochromic Glazing

The results presented above, as well, as in the literature, show that a variety of combinations of T_{lum} and ΔT_{sol} can be obtained in

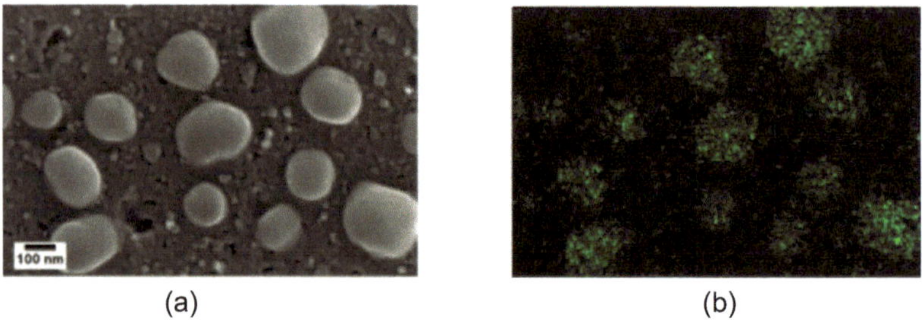

Figure 1.18 Scanning electron micrograph (a) and energy-dispersive X-ray data (b) for a sputter-deposited VO_2-SiO_2 film; the green dots signify vanadium. Reprinted from [102], Copyright © 2016 IEEE.

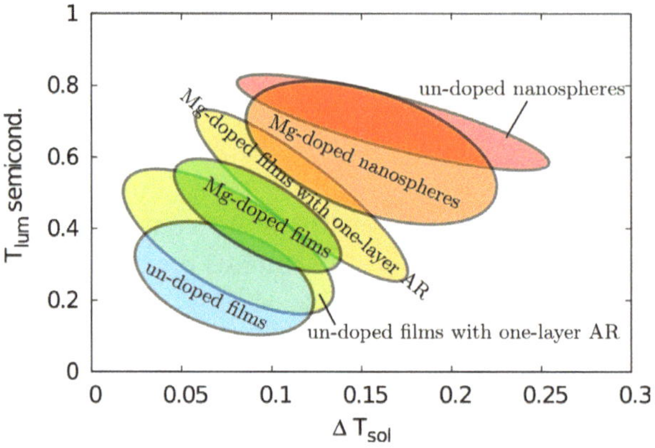

Figure 1.19 Schematic performance limits of VO_2-based materials in thermochromic glazing. AR denotes antireflection. Reprinted from [35], Copyright © 2014 AIP Publishing.

VO_2-based materials. Figure 1.19 aims at summarizing today's state-of-the-art materials [35]; the data pertain specifically to $T_{lum}(\tau < \tau_c)$, but the luminous transmittance is rather temperature independent.

Thin films of pure VO_2 have the lowest values of T_{lum} and ΔT_{sol}, but the optical performance can be improved by antireflection layers, and especially good properties were obtained with multilayers of TiO_2 and VO_2 (cf Fig. 1.14). Magnesium-containing VO_2 films have even better properties, which can be improved still further by antireflection coatings. However, superior performance can be reached with VO_2 nanoparticles, which can give a luminous transmittance of about 60% together with a thermochromic solar energy modulation larger than 20%. Magnesium-containing nanoparticles do not provide any better results than pure VO_2 nanoparticles, since the magnesium lowers ΔT_{sol} to an extent that is not balanced by bandgap widening. A slight improvement in T_{lum} and ΔT_{sol} may be feasible if the nanoparticles have optimized shape and orientation [95] and with core–shell nanostructures [96].

1.6 Some Conclusions and Comments

This introductory chapter has outlined the basics of thermochromism in vanadium dioxide and presented some key developments during a more than thirty-year-long journey from the idea of inorganics-based thermochromic fenestration to devices that in 2018 appeared to be ready for the market. The material of choice has been VO_2 the entire time and, perhaps disappointingly, no serious contender is on the radar till today. It was emphasized that pure VO_2 is not good enough for major glazing-related applications, but (i) protective surface layers can prevent its transformation to nonthermochromic V_2O_5, (ii) doping—especially with tungsten—can lead to thermochromic properties change at room temperature, (iii) antireflection coatings as well as doping can increase the luminous transmittance, and (iv) nanostructuring is able to boost the solar energy modulation range and the luminous transmittance. The majority of the work on VO_2-based materials is strictly experimental, but theory support—specifically hybrid functional (first-principles) calculation—has been used extensively to study the effects of doping and of nonstoichiometry.

It should be observed that the very significant developments in VO_2-based materials are not in themselves sufficient for large-scale applications, which must use low-cost high-throughput manufacturing. Two principally different routes can be distinguished: one employs in-line multilayer thin film deposition onto glass and the other is founded on the fabrication of VO_2-based nanoparticles and their subsequent incorporation in a transparent polymer matrix. The nanoparticle approach is compatible with efficient roll-to-roll fabrication of flexible thermochromic foils [106–108], which can be used as laminates for sheet glass [109].

References

1. Smith, G. B. and Granqvist, C. G. (2010). *Green Nanotechnology: Solutions for Sustainability and Energy in the Built Environment* (CRC Press, Boca Raton, FL, USA).
2. Ginley, D. S. and Cahen, D. (2012). *Fundamentals of Materials for Energy and Environmental Sustainability* (Cambridge University Press, New York, NY, USA).
3. Granqvist, C. G. and Niklasson, G. A. (2018). Solar energy materials for thermal applications: a primer, *Sol. Energy Mater. Sol. Cells*, **180**, pp. 213–226.
4. Obama, B. (2017). The irreversible momentum of clean energy: private sector efforts help drive decoupling of emissions and economic growth, *Science*, **355**, pp. 126–129.
5. US Department of Commerce, National Oceanic and Atmospheric Administration, Earthscan System Research Laboratory, Global Monitoring Division (2018). http://www.esrl.noaa.gov/gmd/ccgg/trends/index.html
6. Hawken, P. (2017). *Drawdown* (Penguin Books, New York, NY, USA).
7. Reese, A. (2018). Slow coolant phaseout could worsen warming, *Science*, **359**, p. 1084.
8. Stocker, T. F., Qin, D., Plattner, G.-K., Tignor, M. M. B., Allen, S. K., Boschung, J., Nauels, A., Xia, Y., Bex, V. and Midgley, P. M. (2013). *Climate Change 2013: The Physical Science Basis. Intergovernmental Panel on Climate Change, Working Group I Contribution to the Fifth Assessment Report on the Intergovermental Panel on Climate Change* (Cambridge University Press, New York, NY, USA).
9. Scheffran, J., Brzoska, M., Kominek, J., Link, P. M. and Schilling, J. (2012). *Climate Change, Human Security and Violent Conflict: Challenges for Social Stability* (Springer, Berlin, Germany).
10. Burke, M., Hsiang, S. M. and Miguel, E. (2015). Global non-linear effect of temperature on economic production, *Nature*, **527**, pp. 235–239.
11. Missirian, A. and Schlenker, W. (2017). Asylum applications respond to temperature fluctuations, *Science*, **358**, pp. 1610–1614.

12. Matthews, T. K. R., Wilby, R. L. and Murphy, C. (2017). Communicating the deadly consequences of global warming for human heat stress, *Proc. Natl. Acad. Sci. U.S.A.*, **114**, pp. 3861–3866.

13. Mora, C., Dousset, B., Caldwell, I. R., Powell, F. E., Geronimo, R. C., Bielecki, C. R., Counsell, C. W. W., Dietrich, B. S., Johnston, E. T., Louis, L. V., et al. (2017). Global risk of deadly heat, *Nat. Clim. Change*, **7**, pp. 501–506.

14. Schädel, C., Bader, M. K.-F., Schuur, E. A. G., Biasi, C., Bracho, R., Čapek, P., De Baets, S., Diáková, K., Ernakovic, J., Estop-Aragones, C., et al. (2016). Potential carbon emissions dominated by carbon dioxide from thawed permafrost soils, *Nat. Clim. Change*, **6**, 950–953.

15. Wheeler, T. and von Braun, J. (2013). Climate change impacts global food security, *Science*, **341**, pp. 508–513.

16. Henson, S. A., Beaulieu, C., Ilyina, T., John, J. G., Long, M., Séférian, R., Tjiputra, J. and Sarmiento, J. L. (2017). Rapid emergence of climate change in environmental drivers of marine ecosystems, *Nat. Commun.*, **8**, pp. 14682-1–14682-9.

17. Hughes, T. P., Kerry, J. T., Álvarez-Noriega, M., Álvarez-Romero, J. G., Anderson, K. D., Baird, A. H., Babcock, R. C., Beger, M., Bellwood, D. R., Berkelmans, R., et al. (2017). Global warming and recurrent mass bleaching of corals, *Nature*, **543**, pp. 373–377.

18. Steffen, W., Rockström, J., Richardson, K., Lenton, T. M., Folke, C., Liverman, D., Summerhayes, C. P., Barnosky, A. D., Cornell, S. E., Crucifix, M., et al. (2018). Trajectories of the Earth system in the Anthropocene, *Proc. Natl. Acad. Sci. U.S.A.*, **15**(33), pp. 8252–8259.

19. United Nations, Department of Economic and Social Affairs (2017). *World Population Prospects: The 2017 Revision* (United Nations, New York, NY, USA).

20. United Nations, Department of Economic and Social Affairs (2018). *World Urbanization Prospects: The 2018 Revision* (United Nations, New York, NY, USA).

21. Akbari, H., Cartalis, C., Kolokotsa, D., Muscio, A., Pisello, A. L., Rossi, F., Santamouris, M., Synnefa, A., Wong, N. H. and Zinzi, M. (2016). Local climate change and urban heat island mitigation techniques: the state of the art, *J. Civ. Eng. Manage.*, **22**(1), pp. 1–16.

22. Waters, C. N., Zalasiewicz, J., Summerhayes, C., Barnosky, A. D., Poirier, C., Galuszka, A., Cearreta, A., Edgeworth, M., Ellis, E. C., Ellis, M., et al. (2016). The Anthropocene is functionally and stratigraphically distinct from the Holocene, *Science*, **351**, pp. aad2622-1–aad2622-10.

23. United Nations Environmental Programme (2009). *Buildings and Climate Change: Summary for Decision-Makers* (UNEP Sustainable Buildings and Climate Initiative, Paris, France).
24. Richter, B., Goldston, D., Crabtree, G., Glicksman, L., Goldstein, D., Greene, D., Kammen, D., Levine, M., Lubell, M., Savitz, M., et al. (2008). How America can look within to achieve energy security and reduce global warming, *Rev. Mod. Phys.*, **80**, pp. S1–S107.
25. Pacheco-Torgal, F., Diamanti, M. V., Nazari A. and Granqvist, C. G. (eds.) (2013). *Nanotechnology in Eco-Efficient Construction* (Woodhead, Cambridge, UK).
26. Pacheco-Torgal, F., Labrincha, J. A., Cabeza, L. F. and Granqvist, C. G. (eds.) (2015). *Eco-Efficient Materials for Mitigating Building Cooling Needs: Design, Properties and Applications* (Woodhead, Cambridge, UK).
27. Pacheco-Torgal, F., Kalaiselvam, S., Granqvist, C. G. and Ivanov, V. (eds.) (2016). *Nano and Biotech Based Materials for Energy Building Efficiency* (Springer, Cham, Switzerland).
28. Granqvist, C. G. (1990). Chromogenic materials for transmittance control of large-area windows, *Crit. Rev. Solid State Phys. Mater. Sci.*, **16**, pp. 291–308.
29. Lampert, C. M. and Granqvist, C. G. (1990). *Large-Area Chromogenics: Materials and Devices for Transmittance Control*, SPIE Institutes for Advanced Optical Technologies, Vol. 4 (SPIE Optical Engineering Press, Bellingham, WA, USA).
30. Lee, J. C., Jorgenson, G. V. and Lin, R. J. (1986). Thermochromic materials research for optical switching, *Proc. Soc. Photo-Opt. Instrum. Eng.*, **692**, pp. 2–7.
31. Jorgenson, G. V. and Lee, J. C. (1986). Doped vanadium oxide for optical switching films, *Sol. Energy Mater.*, **14**, pp. 205–214.
32. Babulanam, S. M., Eriksson, T. S., Niklasson, G. A. and Granqvist, C. G. (1986). Thermochromic VO_2 films for energy-efficient windows, *Proc. Soc. Photo-Opt. Instrum. Eng.*, **692**, pp. 8–18.
33. Babulanam, S. M., Eriksson, T. S., Niklasson, G. A. and Granqvist, C. G. (1987). Thermochromic VO_2 films for energy-efficient windows, *Sol. Energy Mater.*, **16**, pp. 347–363.
34. Gao, Y., Luo, H., Zhang, Z., Kang, L., Chen, Z., Du, J., Kanehira, M. and Cao, C. (2012). Nanoceramic VO_2 thermochromic smart glass: a review on progress in solution processing, *Nano Energy*, **1**, pp. 221–246.

35. Li, S.-Y., Niklasson, G. A. and Granqvist, C. G. (2014), Thermochromic undoped and Mg-doped VO_2 thin films and nanoparticles: optical properties and performance limits for energy efficient windows, *J. Appl. Phys.*, **115**, pp. 053513-1–053513-10.

36. Warwick, M. E. A. and Binions, R. (2014). Advances in thermochromic vanadium dioxide films, *J. Mater. Chem. A*, **2**, pp. 3275–3292.

37. Ji, Y.-X., Boman, M., Niklasson, G. A. and Granqvist, C. G. (2016). Thermochromics for energy-efficient buildings: thin surface coatings and nanoparticle composites, in *Nano and Biotech Based Materials for Energy Building Efficiency*, Pacheco-Torgal, F., Buratti, C., Kalaiselvam, S., Granqvist, C. G. and Ivanov, V., eds. (Springer, Cham, Switzerland), pp. 71–96.

38. Wang, S., Liu, M., Kong, L., Long, Y., Jiang, X. and Yu, A. (2016). Recent progress in VO_2 smart coatings: strategies to improve the thermochromic properties, *Prog. Mater. Sci.*, **81**, pp. 1–54.

39. Li, M., Magdassi, S., Gao, Y. and Long, Y. (2017). Hydrothermal synthesis of VO_2 polymorphs: advantages, challenges and prospects for the application of energy efficient smart windows, *Small*, pp. 1701147-1–701147-25.

40. Granqvist, C. G. and Niklasson, G. A. (2017). Thermochromic oxide-based thin films and nanoparticle composites for energy-efficient glazing, *Buildings*, **7**, pp. 3-1–3-20.

41. Chang, T.-C., Cao, X., Bao, S.-H., Ji, S.-D., Luo, H.-J. and Jin, P. (2018). Review on thermochromic vanadium dioxide based smart coatings: from lab to commercial application, *Adv. Manuf.*, **6**, pp. 1–19.

42. Cui, Y., Ke, Y., Liu, C., Chen, Z., Wang, N., Zhang, L., Zhoi, Y., Wang, S., Gao, Y. and Long, Y. (2018). Thermochromic VO_2 for energy-efficient smart windows, *Joule*, **2**(9), pp. 1707–1746.

43. Xu, F., Cao, X., Luo, H. and Jin P. (2018). Recent advances in VO_2-based thermochromic composites for smart windows, *J. Mater. Chem. C*, **6**, pp. 1903–1919.

44. Gonçalves, A., Resende, J., Marques, A. C., Pinto, J. V., Nunes, D., Marie, A., Goncalves, R., Pereira, L., Martins, R. and Fortunato, E. (2016). Smart optically active VO_2 nanostructured layers applied in roof-type ceramic tiles for energy efficiency, *Sol. Energy Mater. Sol. Cells*, **150**, pp. 1–9.

45. Wyszecki, G. and Stiles, W. S. (2000). *Color Science: Concepts and Methods, Quantitative Data and Formulae*, 2nd ed. (Wiley, New York, NY, USA).

46. ASTM (2008). G173-03 standard tables of reference solar spectral irradiances: direct normal and hemispherical on a 37° tilted surface, in *Annual Book of ASTM Standards* (American Society for Testing and Materials, Philadelphia, PA, USA), Vol. 14.04, http://rredc.nrel.gov/solar/spectra/am1.5.
47. Morin, F. J. (1959). Oxides which show a metal-to-insulator transition at the Neel temperature, *Phys. Rev. Lett.*, **3**, pp. 34–36.
48. Saeli, M., Piccirillo, C., Parkin, I. P., Binions, R. and Ridley, I. (2010). Energy modelling studies of thermochromic glazing, *Energy Build.*, **42**, pp. 166–1673.
49. Saeli, M., Piccirillo, C., Parkin, I. P., Ridley, I. and Binions, R. (2010). Nano-composite thermochromic thin films and their application in energy-efficient glazing, *Sol. Energy Mater. Sol. Cells*, **94**, pp. 141–151.
50. Hoffmann, S., Lee, E. S. and Clavero, C. (2014). Examination of the technical potential of near-infrared switching thermochromic windows for commercial building applications, *Sol. Energy Mater. Sol. Cells*, **123**, pp. 66–80.
51. Rao Popuri, S., Miclau, M., Artemenko, A., Labrugere, C., Villesuzanne, A. and Pollet, M. (2013). Rapid hydrothermal synthesis of VO_2(B) and its conversion to thermochromic VO_2(M1), *Inorg. Chem.*, **52**, pp. 4780–4785.
52. Goodenough, J. B. (1971). The two components of the crystallographic transition in VO_2, *J. Solid State Chem.*, **3**, pp. 490–500.
53. Eyert, V. (2011). The metal-insulator transition of VO_2: a band theoretical approach, *Ann. Phys. (Lpz.)*, **11**, pp. 650–702.
54. Booth, J. M. and Casey, P. S. (2009). Anisotropic structure deformation in the VO_2 metal-insulator transition, *Phys. Rev. Lett.*, **103**, pp. 086402-1–086402-4.
55. Koethe, T. C., Hu, Z., Haverkort, M. W., Schüßler-Langeheine, C., Venturini, F., Brookes, N. B., Tjernberg, O., Reichelt, W., Hsieh, H. H., Lin, H.-J., Chen, C. T. and Tjeng, L. H. (2006). Transfer of spectral weight and symmetry across the metal-insulator transition in VO_2, *Phys. Rev. Lett.*, **97**, pp. 116402-1–116402-4.
56. Tomczak, J. M., Aryasetiawan, F. and Biermann, S. (2008). Effective bandstructure in the insulating phase versus strong dynamical correlations in metallic VO_2, *Phys. Rev. B*, **78**, 115103-1–115103-6.
57. Weber, C., O'Regan, D. D., Hine, N. D. M., Payne, M. C., Kotliar, G. and Littlewood, P. B. (2012). Vanadium dioxide: a Peierls–Mott insulator stable against disorder, *Phys. Rev. Lett.*, **108**, pp. 256402-1–256402-5.

58. Seeboth, A. and Lötzsch, D. (2014). *Thermochromic and Thermotropic Materials* (Pan Stanford Publishing, Singapore).
59. Lee, H. Y., Cai, Y., Bi, S., Liang, Y. N., Song, Y. and Hu, X. M. (2017). A dual-response nano-composite toward climate-adaptable solar modulation for energy-saving smart windows, *ACS Appl. Mater. Interfaces*, **9**, pp. 6054–6063.
60. Kang, Y.-B. (2012). Critical evaluation and thermodynamic optimization of the VO–$VO_{2.5}$ system, *J. Eur. Ceram. Soc.*, **32**, pp. 3187–3198.
61. Ji, Y.-X., Boman, M., Niklasson, G. A. and Granqvist, C. G. (2016). Thermochromic VO_2 films made by thermal oxidation of vanadium in SO_2, *Sol. Energy Mater. Sol. Cells*, **144**, pp. 713–716.
62. Gläser, H. J. (2000). *Large Area Coating* (von Ardenne Anlagentechnik, Dresden, Germany).
63. Bishop C. A. (2015). *Vacuum Deposition onto Webs, Films and Foils*, 3rd ed. (William Andrew, Waltham, MA, USA).
64. Bishop, C.A. (2015). *Roll-to-Roll Vacuum Deposition of Barrier Coatings*, 2nd ed. (Wiley, Hoboken, NJ, USA).
65. Freedonia (2018). *Global Flat Glass: Demand and Sales Forecasts, Market Share, Market Size, Market Leaders*, 13th ed. (The Freedonia Group, Cleveland, OH, USA).
66. Zhou, H., Li, J., Bao, S., Wang, D., Liu, X. and Jin, P. (2015). The potential cytotoxicity and mechanism of VO_2 thin films for intelligent thermochromic windows, *RSC Adv.*, **5**, pp. 106315–106324.
67. Montero, J., Ji, Y.-X., Li, S.-Y., Niklasson, G. A. and Granqvist, C. G. (2015). Sputter deposition of thermochromic VO_2 films on In_2O_3:Sn, SnO_2, and glass: structure and composition versus oxygen partial pressure, *J. Vac. Sci. Technol. B*, **33**, pp. 031805-1–031805-7.
68. Ba, C. O. F., Fortin, V., Bah, S. T., Vallée, R. and Pandurang, A. (2016). Formation of VO_2 by rapid thermal annealing and cooling of sputtered vanadium thin films, *J. Vac. Sci. Technol. A*, **34**, pp. 031505-1–031505-8.
69. Jiang, S.-J., Ye, C.-B., Khan, M. S. R. and Granqvist, C. G. (1991). Evolution of thermochromism during oxidation of evaporated vanadium films, *Appl. Opt.*, **30**, pp. 847–851.
70. Zhang, Y., Zhang, X., Huang, Y., Huang, C., Niu, F., Meng, C. and Tan, X. (2014). One-step hydrothermal conversion of VO_2(B) into W-doped VO_2(M) and its phase transition and optical switching properties, *Solid State Commun.*, **180**, pp. 24–27.

71. Li, S.-Y., Niklasson, G. A. and Granqvist, C. G. (2012). Thermochromic fenestration with VO_2-based materials: three challenges and how they can be met, *Thin Solid Films*, **520**, pp. 3823–3828.
72. Talledo A. and Granqvist, C. G. (1995). Electrochromic vanadium-pentoxide-based films: structural, electrochemical, and optical properties, *J. Appl. Phys.*, **77**, pp. 4655–4666.
73. Ji, Y.-X., Li, S.-Y., Niklasson, G. A. and Granqvist, C. G. (2014). Durability of thermochromic VO_2 thin films under heating and humidity: effect of Al oxide top coatings, *Thin Solid Films*, **562**, pp. 568–573.
74. Wang, X., Cao, Y., Yang, C., Yan, L. and Li, Y. (2016). Vanadium dioxide film protected with an atomic-layer-deposited Al_2O_3 thin film, *J. Vac. Sci. Technol. A*, **34**, pp. 01A106-1–01A106-7.
75. Chang, T., Cao, X., Dedon, L. R., Long, S., Huang, A., Shao, Z., Li, N., Luo, H. and Jin, P. (2018). Optical design and stability study for ultrahigh-performance and long-lived vanadium dioxide-based thermochromic coatings, *Nano Energy*, **44**, pp. 256–264.
76. Chen, Y., Zeng, X., Zhu, J., Li, R., Yao, H., Cao, X., Ji, S. and Jin, P. (2017). High performance and enhanced durability of thermochromic films using VO_2@ZnO core–shell nanoparticles, *ACS Appl. Mater. Interfaces*, **9**, pp. 27784–27791.
77. Tong, K., Li, R., Zhu, J., Yao, H., Zhou, H., Zeng, X., Ji, S. and Jin P. (2017). Preparation of VO_2/Al-O core–shell structure with enhanced weathering resistance for smart window, *Ceram. Int.*, **43**, pp. 4055–4061.
78. Greenberg, C. B. (1983). Undoped and doped VO_2 films grown from $VO(OC_3H_7)_3$, *Thin Solid Films*, **110**, pp. 7 –82.
79. Mlyuka, N. R., Niklasson, G. A. and Granqvist, C. G. (2009). Mg doping of thermochromic VO_2 films enhances the optical transmittance and decreases the metal-insulator transition temperature, *Appl. Phys. Lett.*, **95**, pp. 171909-1–171909-3.
80. Liu, D., Cheng, H., Xing, X., Zhang, C. and Zheng, E. (2016). Thermochromic properties of W-doped VO_2 thin films deposited by aqueous sol–gel method for adaptive infrared stealth application, *Infrared Phys. Technol.*, **77**, pp. 339–343.
81. Alivio, T. E., Sellers, D. G., Asayesh-Ardakani, H., Braham, E. J., Horrocks, G. A., Pelcher, K. E., Villareal, R., Zuin, L., Shamberger, P. J., Arróyave, et al. (2017). Postsynthetic route for modifying the metal-insulator transition of VO_2 by interstitial dopant incorporation, *Chem. Mater.*, **29**, pp. 5401–5412.

82. Mlyuka, N. R., Niklasson, G. A. and Granqvist, C. G. (2009). Thermochromic multilayer films of VO_2 and TiO_2 with enhanced transmittance, *Sol. Energy Mater. Sol. Cells*, **93**, pp. 1685–1687.
83. Chen, Z., Gao, Y., Kang, L., Du, J., Zhang, Z., Luo, H., Miao, H. and Tan, G. (2011). VO_2-based double-layered films for smart windows: optical design, all-solution preparation and improved properties, *Sol. Energy Mater. Solar Cells*, **95**, pp. 2677–2684.
84. Powell, M. J., Quesada-Cabrera, R., Taylor, A., Teixeira, D., Papakonstantinou, I., Palgrave, R. G., Sankar, G. and Parkin, I. P. (2016). Intelligent multifunctional $VO_2/SiO_2/TiO_2$ coatings for self-cleaning, energy-saving window panels, *Chem. Mater.*, **28**, pp. 1369–1376.
85. Zhou, H., Li, J., Bao, S., Li, J., Liu, X. and Jin, P. (2016). Use of ZnO as antireflective, protective, antibacterial, and biocompatible multifunction nanolayer of thermochromic VO_2 nanofilm for intelligent windows, *Appl. Surf. Sci.*, **363**, pp. 532–542.
86. Mlyuka, N. R., Niklasson, G. A. and Granqvist, C. G. (2009). Thermochromic VO_2-based multilayer films with enhanced luminous transmittance and solar modulation, *Phys. Status Solidi A*, **206**, pp. 2155–2160.
87. Taylor, A., Parkin, I., Noor, N., Tummeltshammer, C., Brown, M. S. and Papakonstantinou, I. (2013). A bioinspired solution for spectrally selective thermochromic VO_2 coated intelligent glazing, *Opt. Express*, **21**, pp. A750–A764.
88. Li, S.-Y., Mlyuka, N. R., Primetzhofer, D., Hallén, A., Possnert, G., Niklasson, G. A. and Granqvist, C. G. (2013). Bandgap widening in thermochromic Mg-doped VO_2 thin films: quantitative data based on optical absorption, *Appl. Phys. Lett.*, **103**, pp. 161907-1–161907-4.
89. Dietrich, M. K., Kramm, B. G., Becker, M., Meyer, B. K., Polity, A. and Klar, P. J. (2015). Influence of doping with alkaline earth metals on the optical properties of thermochromic VO_2, *J. Appl. Phys.*, **117**, pp. 185301-1–185301-8.
90. Hu, S., Li, S.-Y., Ahuya, R., Granqvist, C. G., Hermansson, K., Niklasson, G. A. and Scheicher, R. H. (2012). Optical properties of Mg-doped VO_2: absorption measurements and hybrid functional calculations, *Appl. Phys. Lett.*, **101**, pp. 201902-1–201902-4.
91. Jiang, M., Bao, S., Cao, X., Li, Y., Li, S., Zhou, H., Luo, H. and Jin, P. (2014). Improved luminous transmittance and diminished yellow color in VO_2 energy efficient smart thin films by Zn doping, *Ceram. Int.*, **40**, pp. 6331–6334.

92. Wang, N., Duchamp, M., Dunin-Borokowski, R. E., Liu, S., Zeng, X. T., Cao, X. and Long, Y. (2016). Terbium-doped VO_2 thin films: reduced phase transition temperature and largely enhanced luminous transmittance, *Langmuir*, **32**, pp. 759–764.
93. Khan, K. A. and Granqvist, C. G. (1989). Thermochromic sputter-deposited vanadium oxyfluoride coatings with low luminous absorptance, *Appl. Phys. Lett.*, **55**, pp. 4–6.
94. Burkhardt, W., Christmann, T., Franke, S., Kriegseis, W., Meister, D., Meyer, B. K., Niessner, W., Schalch, D. and Scharmann, A. (2002). Tungsten and fluorine co-doping of VO_2 films, *Thin Solid Films*, **402**, pp. 226–231.
95. Li, S.-Y., Niklasson, G. A. and Granqvist, C. G. (2010). Nanothermochromics: calculations for VO_2 nanoparticles in dielectric hosts show much improved luminous transmittance and solar energy transmittance modulation, *J. Appl. Phys.*, **108**, pp. 063525-1–063525-8.
96. Li, S.-Y., Niklasson, G. A. and Granqvist, C. G. (2011). Nanothermochromics with VO_2-based core–shell structures: calculated luminous and solar optical properties, *J. Appl. Phys.*, **109**, pp. 113515-1–113515-5.
97. Laaksonen, K., Li, S.-Y., Puisto, S. R., Rostedt, N. K. J., Ala-Nissila, T., Granqvist, C. G., Nieminen, R. M. and Niklasson, G. A. (2014). Nanoparticles of TiO_2 and VO_2 in dielectric media: conditions for low optical scattering and comparison between effective medium and four-flux theories, *Sol. Energy Mater. Sol. Cells*, **130**, pp. 132–137.
98. Montero, J., Ji, Y.-X., Granqvist, C. G. and Niklasson, G. A. (2016). Thermochromic light scattering from particulate VO_2 layers, *J. Appl. Phys.*, **119**, pp. 085302-1–085302-6.
99. Lopez, R., Haynes, T. E., Boatner, L. A., Feldman, L. C. and Haglund, Jr., R. F. (2002). Temperature-controlled surface plasmon resonance in VO_2 nanorods, *Opt. Lett.*, **27**, pp. 1327–1329.
100. Li, S.-Y., Namura, K., Suzuki, M., Niklasson, G. A. and Granqvist, C. G. (2013). Thermochromic VO_2 nanorods made by sputter deposition: growth conditions and optical modelling, *J. Appl. Phys.*, **114**, pp. 033516-1–033516-11.
101. Romanyuk, A., Steiner, R., Marot, L., Spassov, V. and Oelhafen, P. (2008). nc-VO_2/Al_2O_3 nanocomposite films prepared by dual target magnetron sputtering, *Thin Solid Films*, **516**, pp. 8513–8516.

102. Ji, Y.-X., Niklasson, G. A. and Granqvist, C. G. (2016). Direct formation of thermochromic composite films of VO_2 nanoparticles in SiO_2 hosts, *2016 IEEE 16th International Conference on Nanotechnology* (*IEEE-NANO*), pp. 823–825.

103. Powell, M. J., Marchand, P., Denis, C. J., Bear, J. C., Darr, J. A. and Parkin, I. P. (2015). Direct and continuous synthesis of VO_2 nanoparticles, *Nanoscale*, **7**, pp. 18686–18693.

104. Horrocks, G. A., Singh, S., Likely, M. F., Sambandamurthy, G. and Banerjee, S. (2014). Scalable hydrothermal synthesis of free-standing VO_2 nanowires in the M_1 phase, *ACS Appl. Mater. Interfaces*, **6**, pp. 15726–15732.

105. Wu, S., Tian, S., Liu, B., Tao, H., Zhao, X., Palgrave, R. G., Sankar, G. and Parkin, I. P. (2018). Facile synthesis of mesoporous VO_2 nanocrystals by a cotton-template method and their enhanced thermochromic properties, *Sol. Energy Mater. Sol. Cells*, **176**, pp. 427–434.

106. Kim, H., Kim, Y., Kim, K. S., Jeong, H. Y., Jang, A.-R., Han, S. H., Yoon, D. H., Suh, K. S., Shin, H. S., Kim, T. Y. and Yang, W. S. (2013). Flexible thermochromic window based on hybridized VO_2/graphene, *ACS Nano*, **7**, pp. 5769–5776.

107. Kim, Y., Yu, S., Park, J., Yoon, D., Dayaghi, A. M., Kim, K. J., Ahn, J. S. and Son, J. (2018). High-throughput roll-to-roll fabrication of flexible thermochromic coatings for smart windows with VO_2 nanoparticles, *J. Mater. Chem. C*, **6**, pp. 3451–3458.

108. Fleer, N. A., Pelcher, K. E., Zou, J., Nieto, K., Douglas, L. D., Sellers, D. G. and Banerjee, S. (2017). Hybrid nanocomposite films comprising dispersed VO_2 nanocrystals: a scalable aqueous-phase route to thermochromic fenestration, *ACS Appl. Mater. Interfaces*, **9**, pp. 38887–38900.

109. Srirodpai, O., Wootthikanokkhan, J., Nawalertpanya, S., Yuwawech, K. and Meeyoo, V. (2017). Preparation, characterization and thermochromic properties of EVA/VO_2 laminate films for smart window applications and energy efficiency in building, *Materials*, **10**, pp. 53-1–53-21.

Chapter 2

Effect of Doping on the Thermochromic Performance of VO_2

Ning Wang,[a] Yuanyuan Cui,[b] Yanfeng Gao,[b] and Yi Long[c,d]

[a]*Shenzhen Institutes of Advanced Technology, Chinese Academy of Sciences, 1068, Xueyuan Avenue, Shenzhen 518055, China*
[b]*School of Materials Science and Engineering, Shanghai University, Shanghai 200444, China*
[c]*School of Materials Science and Engineering, Nanyang Technological University, 50 Nanyang Avenue, Singapore 639798, Singapore*
[d]*Singapore-HUJ Alliance for Research and Enterprise (SHARE), Nanomaterials for Energy and Energy-Water Nexus (NEW), Campus for Research Excellence and Technological Enterprise (CREATE), Singapore 138602, Singapore*
ning.wang@siat.ac.cn, cui-yy@shu.edu.cn

Doping with cations/anions is an effective approach to improve the thermochromic performance of VO_2 by tuning the bandgap, the charge carrier density, and the strain modulation. In this chapter, we aim to elaborate the roles of the dopants in tailoring the thermochromic performance of VO_2. Some of the dopants discussed in this chapter are H (hydrogen), Mg (magnesium), W (tungsten), F (fluorine), N (nitrogen), and the Sb (antimony). The doping effect will be reviewed in depth in terms of both experimental and calculation aspects.

Vanadium Dioxide-Based Thermochromic Smart Windows
Edited by Yi Long and Yanfeng Gao

ISBN 978-981-4877-06-0 (Hardcover), 978-1-003-17731-9 (eBook)
www.jennystanford.com

2.1 Introduction

Doping is a method to introduce cations/anions to replace the V/O atomic sites in the VO_2 lattice and has been widely used to the tailor optical and electronic properties of VO_2 [1–7]. Doping can be accomplished via the sol-gel [3, 8, 9], pulse laser deposition [10–12], chemical vapor deposition [13–15], and magnetic sputtering approaches [16–18]. Doping VO_2 with metal cations (M^{n+}) where $n > 4$ could introduce extra free electron (e^-) carriers (negative doping) and where $n < 4$ could generate extra hole (h^+) carriers (positive doping), which could be used to tune the electronic band structure of VO_2 [19]. In addition to the charge carriers, the extra local strain arising from the larger/smaller ionic radius of dopants in the atomic substitution could also contribute to property tailoring [20–22]. In this chapter, the role of dopants in tailoring the thermochromic performance of VO_2 will be illustrated both by experimental investigations and theoretical simulations, which will help the authors to understand the doping effect in depth.

2.2 Experimental Investigations

2.2.1 Mg Doping

Compared with V^{4+} (ionic radius 0.58 Å) in VO_2, Mg^{2+} has a lower valence state and a larger ionic radius 0.72 Å, which benefits doping with Mg with extra h^+ carriers and compressive strain. As depicted in the Clausius–Clapeyron equation (Eq. 2.1) [22], the phase transition temperature T_C can be modulated with uniaxial extra stress applied along the c axis. As reported by Cao et al. [22], the applied 2.2% compressive uniaxial strain could be enough for reducing the T_C of VO_2 to room temperature and such a phenomenon is also observed in a joule heating experiment (Fig. 2.1). With respect to the h^+ carriers, the itinerant or mobile h^+ carriers could reduce the T_C, and the localized h^+ carriers trapped by the discrete V^{5+} in the positive doping could increase the T_C. In $V_{1-x}Mg_xO_2$ with the formal valence formula $V^{4+}_{1-3x}V^{5+}_{2x}Mg^{2+}_xO_2$, each Mg^{2+} dopant is

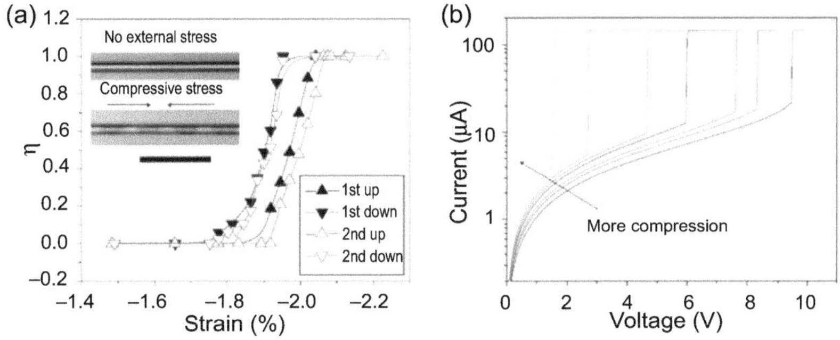

Figure 2.1 (a) Uniaxial compression reversibly induces a metal–insulator transition (MIT) at room temperature in clamped VO_2 beams. The inset shows representative optical images of metal–insulator domains along a stressed beam. (b) Room-temperature I–V characteristic of a VO_2 beam under different axial compressions, showing an MIT induced by joule heating, with both the threshold voltage and current being significantly reduced by the external compression. The experiment was carried out under ambient conditions. The scale bar in (a) is 10 μm [22].

assumed to generate one mobile and localized h^+ carrier (trapped by the discrete V^{5+}) [19]. Therefore, the larger ionic radius and the extra mobile h^+ carrier in Mg doping could reduce the T_C, and the T_C reducing rate is ~3 K/at.% [23–25]. In addition to the T_C reducing capability, Mg doping could increase the visible transmittance via bandgap widening [24, 25], where the Fermi level is shifted downward due to the additional h^+ doping level besides the valence band.

$$\mathrm{d}T_C/\mathrm{d}\sigma = (\varepsilon_0/T_C^0)/\Delta H \tag{2.1}$$

2.2.2 W Doping

W-doped VO_2 ($V_{1-x}W_xO_2$) has the formal valence formula $V^{4+}_{1-3x}V^{3+}_{2x}W^{6+}_xO_2$, where the W^{6+} cation has the ionic radius 0.6 Å, slightly larger than that of V^{4+}, and each W^{6+} substitution could result in two extra electrons that are either mobile polarons or trapped by the discrete V^{3+} cation. The strain effect could be neglected for W doping since there is only a little difference between the ionic radii of V^{4+} and W^{6+}. Irrespective of the localized or delocalized nature, the extra electrons induced by W doping always result in a sharp drop in T_C. As observed in experiments, the T_C reducing efficiency for W doping could reach 20–26 K/at.%, which depends on the fabrication methods of W-doped VO_2 crystalline structures [18, 23, 26–37]. In tailoring the electronic structure of VO_2, W doping could reduce the bandgap by moving up the Fermi level as the electron doping level is formed besides the conduction band.

2.2.3 Sb Doping

As for Sb doping, Sb^{3+} and Sb^{5+} have the ionic radii 0.76 Å and 0.6 Å, respectively. As reported by Gao et al. [38], Sb^{3+} doping resulted in ultrafine VO_2 nanoparticles (size 8–30 nm) in a one-step hydrothermal synthesis, but Sb^{5+} doping enhanced the growth of VO_2 nanoparticles with a larger size, 200–300 nm, which was attributed to the positively and negatively charged surface of the VO_2 nanoparticles doped with Sb^{3+}/Sb^{5+}. Actually, the extra strain induced by Sb^{3+} doping in the lattice with a larger ionic radius should be another reason for the ultrafine particle size of VO_2, and the group believed that the nucleation rate of VO_2 could be largely improved by extra strain in the materials. In addition, Sb^{3+} doping showed a T_C reducing rate of $\sim$1.4 K/at.%, which should be mainly ascribed to the extra compressive strain in the lattice arising from the atomic substitution of V^{4+} (0.58 Å) by Sb^{3+} (0.76 Å).

2.2.4 F Doping

F^- as an effective anion dopant, F^- doping has been widely investigated for the performance improvement of VO_2. F-doped VO_2 with the formal valence formula $V^{4+}_{1-y}V^{3+}_{y}O^{2-}_{2-y}F^{-}_{y}$ occupies the extra electron carrier that is able to reduce the T_C for the phase transition. It is still not clear which kind of carriers (localized or delocalized) should be assigned for the extra electrons in F doping, but improved thermochromic performance has been observed in F-doped VO_2 nanostructures [39–42]. As reported by Dai et al. [39], with 2.93 at.% F doping, the T_C could be reduced to 35°C, and a large solar-heat shielding ability (35.1%), an excellent solar modulating ability

(10.7%), as well as a decent visible transmittance (48.7%), could be achieved at the same time. The large T_C reducing rate, 11 K/at.%, for F doping should be due to the totally or partially delocalized nature of the generated electron carriers, which needs to be clarified in future investigations.

2.2.5 N Doping

Compared with F^- doping, N^{3-} doping could result in extra hole carriers with the formal valence formula $V^{4+}_{1-y}V^{5+}_{y}O^{2-}_{2-y}N^{3-}_{y}$. Due to the localized nature of N doping in VO_2 [6, 43], the performance enhancement of VO_2 via N doping was rarely investigated. Recently, a low T_C reducing rate of 1.4 K/at.% for N doping was reported by Zhang et al., where a low-energy N^{2+} ion sputtering method was used to prepare N-doped VO_2 thin films [6]. With X-ray adsorption near-edge structure spectroscopy and photon electronic spectroscopy characterizations, the bandgap between the levels $d_{\|}$ and π^* of V 3d was found to be narrowed by the N doping, which was regarded as the reason for the slow T_C reducing. However, the totally or partially delocalized hole carriers may be an alternative reason for this phenomenon, which needs to be proved with more investigations.

2.2.6 H Doping

Different from the atomic substitution in doping with metallic cations, hydrogen doping is done in order to insert protons into the center of the open channels in the lattice of VO_2 [44]. In hydrogen-doped VO_2 [44–49], a strong O–H bond with a short bond length of 0.99–1.03 Å could be formed within the structure and H doping could expand the lattice and increase the electron carrier concentration. As observed in the experiments, the metallic phase of VO_2 could be stabilized at room temperature as the lightest H^+ cations contribute to the extra electrons for the V^{4+} pairs, which could enhance the electron–electron correlation for stabilizing the metallic (rutile) phase [48].

2.3 Simulations of Elemental Doping in VO_2

With the development of fast and inexpensive computers and the availability of modeling and calculation software, simulation of materials has become accessible to scientist and researchers. Elemental doping is a process that intentionally introduces other element(s) into a pure material to modify its properties and performance, and it has been intensively studied by computational simulations. For VO_2, the simulations of elemental doping are frequently conducted under the framework of density functional theories (DFTs), which is one of the powerful strategies to explore its T_C and optical properties. In this section, we will introduce the simulations of elemental doping in VO_2 by answering the following five questions:

1. How does phase transition happen in VO_2?
2. Which method is appropriate for simulating elemental doping in VO_2?
3. What are the criteria for selecting the doping element?
4. How do you run a calculation to simulate the doping of VO_2?
5. How do you analyze the calculation results?

The first question is how phase transition happens in VO_2, or what the mechanism is for phase transition in pure VO_2. This question is vital because it determines at what timescale and length range the simulation would be undertaken. Here, an introduction is given concerning the phase transition mechanism in pure VO_2. A molecular picture based on the relatively simple crystal-field theory is considered, which was initially proposed by Goodenough in 1971 (Fig. 2.2) [19]. In rutile VO_2, a wide π bond and a narrow π^* antibond are formed between the V^{4+} and O^{2-} orbitals and a $d_{\|}$ nonbond is formed between adjacent V^{4+} orbitals along the crystallographic c axis. Although the energy band of VO_2(R) is roughly 2.5 eV, the $d_{\|}$ bands and the unfilled π^* partially overlap, and the Fermi level falls at the point where the π^* and $d_{\|}$ bands overlap, resulting in metallic states. Below the transition temperature, the tilting of the VO_6 octahedra enhances the π overlap between the V^{4+} and O^{2-} orbitals, thereby elevating the antibonding π^* level, whereas

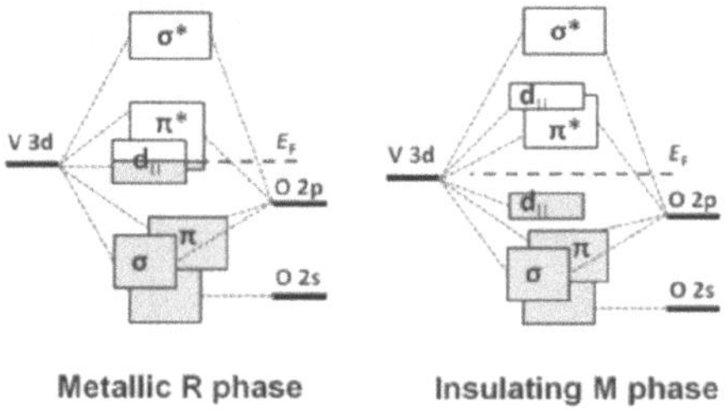

Figure 2.2 Band structures of the metallic R and insulating M phases of VO_2 depicted by molecular orbital diagrams [19].

the $d_{\parallel}$ bonds interact strongly in V–V pairs and then split into $d_{\parallel}$ bonding and antibonding components. A bandgap of ~0.7 eV is formed between the π^* and $d_{\parallel}$ bands, leading to the formation of an insulating phase. Although this model exhibits several discrepancies from the experimental results, it can nevertheless semiqualitatively explain the nature of the phase transition in VO_2.

The debate over the driving force behind the VO_2 phase transition has been ongoing for several decades, regarding whether the electron–electron correlation is strong enough to localize the electrons by forming a Mott–Hubbard insulator (Mott model) or whether structural distortions alone can induce the insulating phase (Peierls model) [50, 51]. Recent measurements and calculations have suggested that both electronical and structural distortion could induce VO_2 phase transition [52]. In brief, at least three distinct timescales emerge in the phase transition. The first step is the rapid dilation of the V–V bond (~307 fs); the second step is a slower, transverse motion of the VO_6 octahedra that locally rearrange to adopt a more rutile-like geometry (~9.2 ps); the third step is a far slower motion (~100 ps) that is ascribed to shear movements at the speed of sound. When phase transition is approached after these three steps, a spontaneous Peierls distortion would be initiated for the cation chains. Finally, the extent of the distortion increases before the Peierls distortion spreads to the orthogonal chains.

The second question is, which method is appropriate to simulate the elemental doping in VO_2? The modeling and simulation of materials is challenging, mainly due to the extreme timescale and length range governing the materials' response. The time may span large scales, from the femtoseconds of atomic vibrations to the decades of use of materials in products. The length scale may range from the angstroms of atoms to the meters of engineered structures. It is well accepted that no single technique will work for all scales. Therefore, a number of methods have been developed, each appropriates for a given range of time and lengths.

Let us consider the VO_2 material in the length range from 1 Å to 100 mm. At the smallest of these scales, the bonding between V and O atoms dominates its behavior. This bonding arises from the underlying electronic structure, and to describe it requires the use of methods of quantum mechanics. At a larger scale, we need to consider the behavior of V and O atoms. At this time, we can use atomic structure methods to describe the bonding; in general these methods are so complicated that we must approximate the bonding with some sort of empirical or approximate function. To understand the behavior of the atoms, we must simulate their behavior, which we can do with various atomistic simulation methods, such as molecular dynamics or the Monte Carlo method. At even larger scales, there are too many atoms for us to consider, so we must find new approaches that focus on the dominant "units." These units may be dislocations, grain boundaries, or some other defect, and the simulations are based on these defects being the fundamental units. The length scale that is dominated by defects is often called the mesoscale [53].

On the basis of the discussions of the phase transition mechanism in the first question, we can deduce that when we simulate elemental doping in VO_2, the timescale is femtoseconds to picoseconds and the range of length is electronic to atomic size. The behaviors of VO_2 are closely related to the types of bonding between the atoms, which represent the distribution of electrons around the nuclei. The ionic bonds between V and O are dominated by the strong electrostatic interactions between the ions. If V is substituted by a certain nonmetallic atom, the covalent bonds between this atom and O will have a localized electronic distribution between atoms. Fundamentally, the properties of doped VO_2 start with its bonding.

A detailed description of bonding in VO_2 requires a calculation of the electronic distributions. The simulation methods that can yield such electronic distributions are called electronic structure methods. Nowadays, numerous codes and software, free or expensive, are widely available and accessible to researchers. These methods go by many names, depending on the approximations used for the wave functions. The Hartree–Fock method was one of the mainstays of electronic structure calculations (especially in quantum chemistry) for decades but has largely been supplanted by an approach based on a DFT formulation of quantum mechanics. For details of DFT, readers can refer to the well-written book by Richard [53].

Also, there are a number of methods within the framework of DFT. Which DFT method is suitable for simulating the elemental doping of VO_2? The premise is that this method can reasonably reproduce the basic physical properties of VO_2(M) and VO_2(R). Zhu and Schwingenschlögl summarized these physical properties as follows [54]: (i) the total energy of the nonmagnetic M phase is lower than that of the nonmagnetic R phase, (ii) the R phase is not magnetic, (iii) the M phase is not magnetic, (iv) the R phase is a metal, and (v) the M phase is a semiconductor. In the past years, various DFT theoretical approaches have been tested in an attempt to uncover the physics of the phase transition in VO_2, such as local density approximation (LDA) [55], generalized gradient approximation (GGA) [54], LDA+U (Hubbard on-site repulsion) methods [56], dynamic mean-field theory (DMFT) [57] and its variants [58], the Green function shielding coulomb potential method (GW) method [59] and its variants [60], and the hybrid functional method (HSE) [50]. However, these methods have problems in reproducing the basic physical properties of VO_2 phases. For example, the LDA method can reproduce the R phase without a bandgap, which indicates that the R phase is a metal, but cannot reproduce the bandgap of the M phase. Thus, the LDA method cannot depict the semiconducting property of M [55]. Although the LDA+U method can reproduce the bandgap of the M phase, it is not in accordance with the experiments when calculating the magnetic properties [56]. Zhu and Schwingenschlögl combined the modified Becke–Johnson potential with the LDA method [54], but later Yuan et al. found that the problems still exist when calculating the magnetic properties of another VO_2 metastable monoclinic phase [61]. The GW method can predict the electronic features of the phase transition only if the parameter-based model variant is employed or self-consistent quasiparticle calculations are performed on the coulomb-hole and screened-exchange (COHSEX) level [59]. Moreover, the HSE method reproduces the metallic and insulating states of the R and M phases, respectively, but not the magnetic properties [50].

Table 2.1 summarizes the advantages and disadvantages of these DFT methods in reproducing the basic physical properties of VO_2, and it can be seen that not a single computational method has been able to completely reproduce the basic physical properties of VO_2. As Yuan et al. discussed in their paper [61], VO_2 is a very sensitive system, in which the atomic and electron structures interact closely with each other, together with the existence of strong electron–electron correlation. Even a slight change in the calculation method will result in a significant difference in the calculation results. Therefore, one should be very cautious in choosing the DFT calculation method.

The third question is what the criteria are in selecting the doping elements. From the theoretical aspect, the electronic phase transition in VO_2 is accompanied by a nearly simultaneous structural phase transition between the VO_2(R) and VO_2(M) phases. If the energy barrier across the phase transition could be lowered by doping, T_C would decrease. Accordingly, in conducting the simulations, the selection of the doping element is commonly based on two aspects: increasing the carrier concentration to accelerate the electronic phase transition and introducing distortion into the atomic structure to assist the structural phase transition.

Firstly, the doping element should increase the carrier concentration of VO_2. When the doping element serves as the donor or acceptor in VO_2, it injects its electrons or holes into the electronic structure of VO_2, which increases the carrier concentration in the system. Because the electronic phase transition of VO_2 is a debatably typical Mott phase transition [64], the increased carrier concentration lowers the energy barrier across the phase transition, thus decreasing the T_C. There are two doping strategies to increase the carrier concentration. One is to insert smaller-sized doping

Table 2.1 Merits and problems of different theoretical approaches in reproducing the physical properties of VO_2

Methods	Criteria[a]					Reference
	(1)	(2)	(3)	(4)	(5)	
LDA	√	–	–	√	×	[55]
LDA/GGA	×	×	×	√	×	[54]
LDA+U	–	×	×	×	√	[56]
DMFT	–	–	–	√	√	[57]
DMFT+V	–	–	–	–	√	[58]
Cluster DMFT	–	–	–	√	√	[62]
Model GW	–	–	–	–	√	[60]
GW	–	–	–	√	×	[59]
LDA+G_0W_0	–	–	–	√	×	[63]
COHSEX+G_0W_0	–	–	–	√	√	[63]
HSE	–	–	–	√	√	[50]
mBJLDA	√	√	√	√	√	[54]

[a]Criteria: (1) The total energy of the nonmagnetic M1 phase is lower than that of the nonmagnetic R phase. (2) The R phase shows no long-range magnetic order. (3) The M1 phase is nonmagnetic. (4) The R phase is metallic. (5) The M1 phase is a semiconductor.
Note: Fulfilled and unfulfilled criteria are shown by √ and ×, respectively, and the – indicates that the criterion is not considered in the reference.
Source: [54]

atoms, such as H [65], Li [66], Na [66], and B [67], into the interstitial sites of VO_2 (Fig. 2.3a). For instance, hydrogen is the lightest atom, which possesses one electron in its outermost orbital. When VO_2 is doped with hydrogen, the H atoms are located at the interstitial sites and inject their electrons into the VO_2 system (Fig. 2.3b) [49, 65, 68]. Gao et al. presented that H efficiently reduces the T_C by 38 K/at.% (Fig. 2.3c) [65]. The other strategy to increase the carrier concentration is to substitute the V sites with a high-valence element (Fig. 2.3d) such as W [27, 69], Nb [70, 71], or Mo [72, 73]. When VO_2 is doped with W atoms, they have been shown to inject some of their electrons into the V 3d valence bands and reduce the T_C by 20–26 K/at.% [69]. He et al. could narrow the optical bandgap from 0.65 eV to 0.54 eV by increasing the W doping concentration (Fig. 2.3e,f) [74].

Secondly, the doping element should introduce structural distortion into VO_2, especially along the V–V chains. VO_2(M) and VO_2(R) belong to the space groups $P2_1/c$ and $P4_2/mnm$, respectively, and

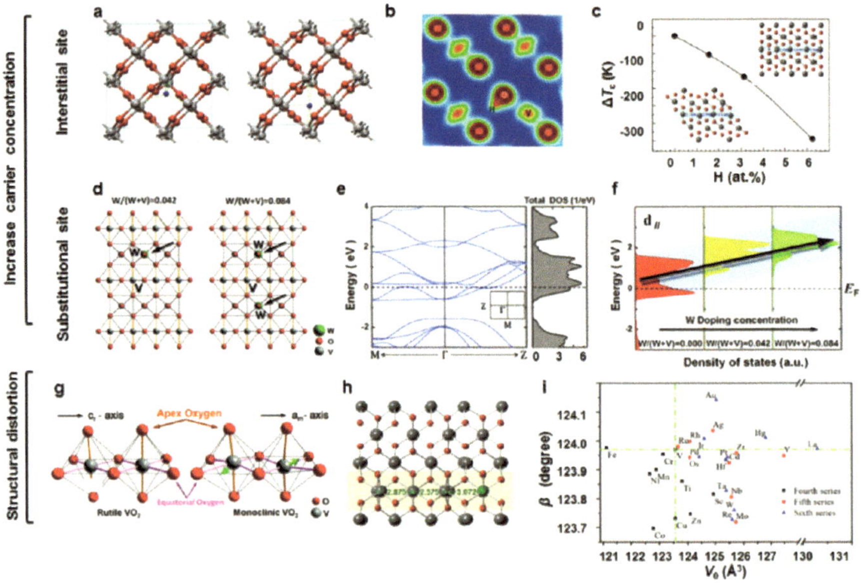

Figure 2.3 The effect of elemental doping on the thermochromic performance of VO_2. (a) The location of the H atom at the tetrahedral (left) and octahedral (right) interstitial site in VO_2, where H, V, and O atoms are indicated by blue, gray, and red spheres, respectively [65]. (b) The electron density of the (0 0 1) plane in H-doped VO_2(R) [65]. (c) The dependence of the transition temperature reduction on the H doping concentration [65]. (d) The location at which a W atom substitutes the V atom in VO_2, where the W, V, and O atoms are indicated by green, gray, and red spheres, respectively [74]. (e) The band structure and total density of states (DOSs) of pure VO_2(R) [74]. (f) The partial DOS of the $d_{\parallel}$ orbital in VO_2 with different W concentrations [74]. (g) Atomic structures for VO_2(R) and VO_2(M) during phase transition; the green arrows indicate the moving directions of V atoms [74]. (h) Sb dopant causes the V–V distances to alternatively vary in VO_2(R). (i) Distribution of transition metal–doped VO_2(M) with respect to the volume and the β angle [78].

their apparent difference lies in the lengths of the V–V bonds, which are constant in VO_2(R) but alternatively varying in VO_2(M) (Fig. 2.3g) [75]. If a dopant causes the V–V distances in VO_2(R) to alternatively change (Fig. 2.3h), thereby resembling that of VO_2(M), the dopant will lower the T_C. For instance, Zhang et al. reported that Be-doped VO_2(R) displays structural distortion around the Be atom, where the V–V chains present dimerization similar to that in VO_2(M) [76]. Through first-principles calculations, they predicted that the reduction of T_C in the Be-doped VO_2 is as large as 58 K/at.% [76]. In addition to the V–V distance, the lattice parameters can also be modified by dopants, which will influence the T_C. Sun et al. proposed that if a dopant can introduce changes in the lattice parameters of VO_2(M) to resemble those of VO_2(R), the dopant will decrease the T_C [77]. They conducted DFT calculations and found that the T_C decreases with the expansion of the lattice and decrease in the β angle of VO_2(M) with transition metal doping (Fig. 2.3i) [77].

Variation in the carrier concentration and structural distortion frequently occurs together, as the doping element either occupies the interstitial sites or substitutes the lattice site of V or O atoms, leading to variations in both the electronic and atomic structures of VO_2.

The fourth question is how to run a calculation to simulate the doping of VO_2. The first step is to construct the supercell models. There are four V atoms and eight O atoms in one VO_2(M) unit cell, whereas there are two V atoms and four O atoms in one VO_2(R) unit cell. Accordingly, the number of atoms in one VO_2(M) unit cell is two times compared to that in one VO_2(R) unit cell. To obtain supercell models with the same number of atoms, a supercell is commonly constructed with $2\times 2\times 4$ primitive unit cells for the VO_2(R) phase and with $2\times 2\times 2$ primitive unit cells for the VO_2(M) phase; thus each supercell consists of 96 atoms. Then the dopant atom is introduced into the supercell. Here comes the question, where is the site of the dopant atom? If the dopant has a size comparable to that of the V atom, such as Ti or W, it would be located at the substitutional site of the V atom. If the dopant has a rather small atomic radius, such as H or Li, it might be doped at the tetrahedral interstitial or octahedral interstitial sites.

The second step is to describe the electronic constraints in the calculations. Here we take an example for the calculations of the energies and electronic structures of the doped VO_2 systems by using the projector-augmented wave (PAW) method in the Vienna Ab Initio Simulation Package (VASP). VASP is a prepackaged code that solves the Schrödinger equation using appropriate approximations for the energy and forces [79, 80]. DFT codes like VASP are powerful tools for scientific research. GGA was employed to account for exchange and correlation in the Perdew–Burke–Ernzerhof form. To overcome the deficiencies of standard exchange-correlation approximations for localized V d states, the rotationally invariant implementation of the GGA+U method was used, in which the effective coulomb repulsion potential $U_{\text{eff}} = U - J$ was chosen to be 3.4 eV. The cut-off energy for the plane-wave basis was 520 eV. The PAW method was applied for all the self-consistent electronic structure calculations [81, 82], with structure optimization based on the calculated forces on each atom. A regular mesh of $2 \times 3 \times 3$ k-points was used to sample the Brillouin zone for the supercell. In all calculations, all atoms were allowed to relax until the Hellmann–Feynman forces acting on them became less than 1×10^{-5} eV/Å.

The third step is to submit the calculations to computers or computing clusters. When the calculation is finished, we will check the main output files to make sure the calculation is completed normally, then go through the energy of each interaction and check the final energy and the change in energy for each interaction.

The fifth question is how to analyze the calculation results. The results commonly include the following aspects: the T_C, electronic structures, atomic configurations, and optical properties. Here, we take an example for the calculations of T_C, electronic structures, and atomic configurations in the H-doped VO_2 systems. Also, we take an example for the calculations of optical properties in Sb-doped VO_2 systems.

Firstly, we can calculate the reduction of T_C. According to the work by Netsianda et al., T_C (T_c) is given by the following expression [83]:

$$T_c = T_{c,0} \frac{\Delta H}{\Delta H_0}, \tag{2.2}$$

where $T_{c,0}$ is the transition temperature of pure VO_2; here the value of 341 K was adopted. ΔH is the enthalpy of phase transition in doped VO_2, and ΔH_0 is the enthalpy of phase transition in pure VO_2. The values of ΔH_0 and ΔH can be obtained from DFT calculations. As there are 64 O atoms, 32 V atoms, and 1 H atom

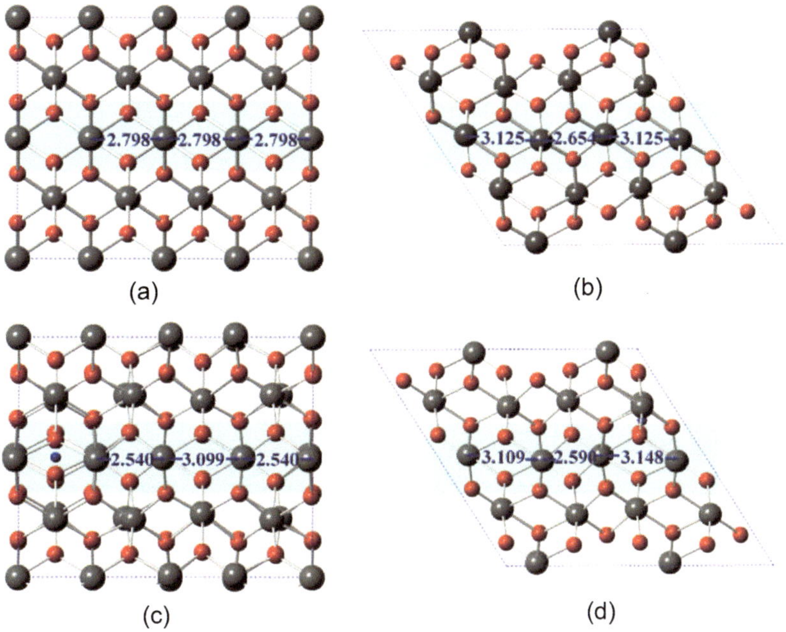

Figure 2.4 The V–V chains along the *c* axis for pure VO_2(R) (a), pure VO_2(M) (b), H-doped VO_2(R) (c), and H-doped VO_2(M) (d). The V, O, and H atoms are indicated by gray, red, and blue spheres, respectively.

in the constructed supercell, the doping concentration is around 3.12%. Applying Eq. 2.2, the value of T_c was calculated to be ~220 K in VO_2 with 3.12% H doping. Recalling that the change in T_c is linearly proportional to the content of the dopant in the compound, we thus have a reduction of ~40 K of 1% H for the doping, suggesting the phase transition could take place at 301 K, which is close to the ambient temperature of 298 K.

Secondly, for atomic configurations around the doped H in VO_2 (Fig. 2.4), it is observed that the crystal structure is distorted around the H atom in VO_2(R), where the V–V chains feature dimerization characteristics like that in the VO_2(M) phase. For instance, the V–V bond length is constantly 2.798 Å in pure VO_2(R) (Fig. 2.4a), while in the H-doped VO_2(R), some V–V bonds were shortened to the range from 2.509 Å to 2.686 Å and the other V–V bonds were lengthened to the range from 2.879 Å to 2.988 Å (Fig. 2.4c). In the case of

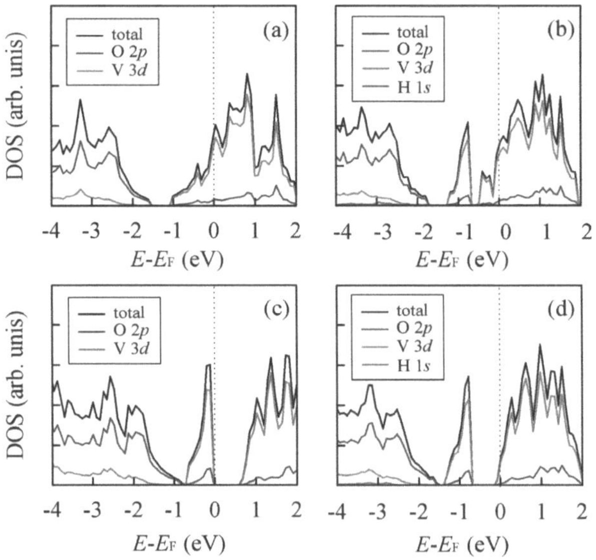

Figure 2.5 The density of states (DOSs) of pure VO_2(R) (a), H-doped VO_2(R) (b), pure VO_2(M) (c), and H-doped VO_2(M) (d). The zero point of the energy axis corresponds to the Fermi level E_F.

VO_2(M), the doped H atom only causes a slight expansion in the local structure. For instance, the long V–V distance in the pure VO_2(M) is 3.125 Å (Fig. 2.4b), while in the H-doped VO_2(M), this distance is slightly shortened to 3.109 Å (Fig. 2.4d), quite close to the value of 3.125 Å in the pure VO_2(M). From the atomic features discussed above, it can be deduced that the crystal structures of VO_2(R) are obviously distorted by the doped H in comparison to that of VO_2(M).

Thirdly, for electronic structures, Fig. 2.5 shows the DOS of pure VO_2 and H-doped VO_2. It is seen from Figs. 2.5a and 2.5b that VO_2(R) and H-doped VO_2(R) exhibit metallic properties as the Fermi level crosses the conduction band. On the contrary, the DOS of pure VO_2(M) in Fig. 2.5c shows that the Fermi level is located at the top of the valence band. It is an intrinsic insulating electronic structure with a bandgap of 0.61 eV, in good agreement with the X-ray photoemission data and other theoretical calculation results. The DOS of H-doped VO_2(M) in Fig. 2.5d shows a shifting up of Fermi level to the edge of the conduction band and a narrowed bandgap of

0.54 eV. Generally, the near-infrared (NIR, 800–1500 nm) absorption in VO_2 film is assigned to the optical transitions across the intrinsic bandgap. Therefore, this narrowed bandgap for H-doped VO_2(M) is expected to result in an increase in the NIR absorption, which needs further experimental verification.

Fourthly, the optical properties of VO_2 can be calculated. Here we take Sb-doped VO_2 as an example. The optical properties are expressed by the dielectric function as $\varepsilon(\omega) = \varepsilon_1(\omega) + i\varepsilon_2(\omega)$ [84]. The imaginary part, $\varepsilon_2(\omega)$, can be described as the details of the real transitions between the unoccupied and occupied states, and the real part, $\varepsilon_1(\omega)$, can be derived through the Kramers–Kronig relationship from the imaginary part. The reflectivity spectra ($R(\omega)$) and absorption coefficient ($\alpha(\omega)$) can be obtained from $\varepsilon_1(\omega)$ and $\varepsilon_2(\omega)$ as follows:

$$R(\omega) = \left| \frac{\sqrt{\varepsilon_1(\omega) + i\varepsilon_2(\omega)} - 1}{\sqrt{\varepsilon_1(\omega) + i\varepsilon_2(\omega)} + 1} \right|^2 \tag{2.3}$$

$$\alpha(\omega) = \frac{\omega}{c} \left[2\sqrt{\varepsilon_1^2(\omega) + \varepsilon_2^2(\omega)} - 2\varepsilon_1(\omega) \right]^{\frac{1}{2}} \tag{2.4}$$

Figure 2.6 displays the absorption and reflectance spectra in the infrared (IR) light range. Figures 2.6a and 2.6b show weak absorption for pure VO_2(M) and strong absorption for pure VO_2(R). It is the same case in the reflection spectra in Figs. 2.6c and 2.6d. These results suggest that IR light could penetrate pure VO_2(M) easily but is significantly blocked by pure VO_2(R), which is in accordance with the experimental results. After Sb doping, both VO_2(R) and VO_2(M) exhibit an increase in the absorption coefficient, as shown in Figs. 2.6a and 2.6b. Meanwhile, Sb-doped VO_2(R) also presents enhanced reflectance in the IR range in Fig. 2.6d. As for Sb-doped VO_2(M) in Fig. 2.6c, the case is slightly complicated since the reflectance shows an obvious increase below 0.4 eV but a decrease above 0.4 eV.

Generally speaking, the relationship between the absorptivity (A), reflectivity (R), and transmissivity (T) in a material can be expressed as

$$A + R + T = 1. \tag{2.5}$$

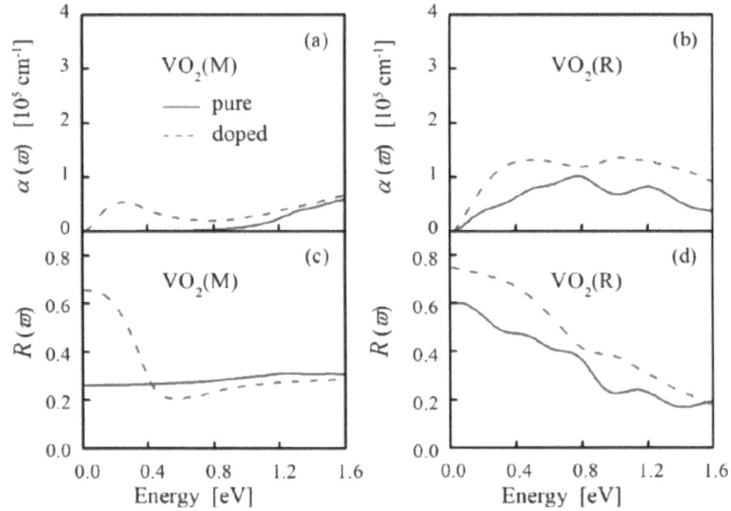

Figure 2.6 Absorption spectra of the infrared light derived from first-principles calculations for VO_2(M) (a) and VO_2(R) (b) and reflectivity spectra for VO_2(M) (c) and VO_2(R) (d). The absorption coefficient $\alpha(\omega)$ is given in 10^5 cm^{-1}.

It can be interpreted from Eq. 2.5 that increased absorptivity and increased reflectivity would result in decreased transmissivity in Sb-doped VO_2(M) when the energy of the IR light is lower than 0.4 eV. However, when the energy of the IR light is above 0.4 eV, the transmissivity is uncertain due to the increased absorptivity but decreased reflectivity. In contrast, when the electrons are radiated by IR light with energy lower than 0.4 eV, the thermal vibration would intensify instead of an electron transition occurring and therefore some of the IR light is absorbed while most of it is reflected, corresponding to the increases in both absorption and reflectivity. In terms of these findings, we may conclude that the switching of the optical properties between VO_2(M) and VO_2(R) is somewhat dependent on the energy of the light. Moreover, the desired switch of the IR light transmittance might degrade to some extent owing to the doping of VO_2 by Sb, especially in the middle-IR and far-IR spectral ranges.

Figure 2.7 presents the absorption and reflectivity spectra in the visible light range. For the absorption spectra in Figs. 2.7a and

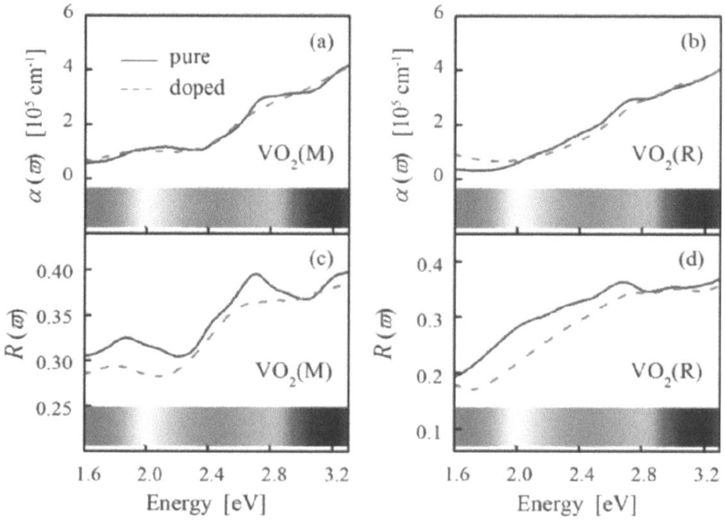

Figure 2.7 Absorption spectra of visible light derived from first-principles calculations for VO_2(M) (a) and VO_2(R) (b) and reflectivity spectra for VO_2(M) (c) and VO_2(R) (d). The absorption coefficient $\alpha(\omega)$ is given in 10^5 cm^{-1}.

2.7b, there is no significant difference between pure and Sb-doped VO_2, indicating unchanged absorption in VO_2 after it is doped with Sb. For the reflectivity spectra in Figs. 2.7c and 2.7d, one can see that the reflectance decreased after Sb doping. As interpreted from Eq. 2.4, unchanged absorptivity and decreased reflectivity would result in increased transmissivity. Therefore, it can be deduced that the introduction of Sb might enhance the transmission of visible light in VO_2.

2.4 Concluding Remarks and Outlook

In summary, the effect of doping on the ability to tailor the thermochromic performance of VO_2 has been demonstrated though experimental findings and theoretical simulations. In the experiments, the localized or delocalized extra electron carriers generated by the cation dopants M^{n+} ($n > 4$) or H^+ insertion could effectively reduce the T_C of VO_2. In M^{n+} ($n < 4$) doping, only the generated mobile hole carriers are able to reduce the T_C, while the localized hole carrier trapped by the discrete V^{5+} could result in an increase in the T_C. Besides the extra carriers, the compressive strain resulting from atomic substitution could also give rise to a decline in the T_C. In theoretical simulations, the specific DFT method mBJLDA could be used to calculate the magnetic and electronic properties of VO_2, while the other DFT simulation approaches, including the LDA, DMFT, GW, and HSE, could not reproduce the magnetic and electronic performances of VO_2 simultaneously. With the optimized calculation approaches, the phase transition, electronic structures, atomic configurations, and optical properties of the doped VO_2 could be well simulated, which can be used to assist in the design of doping experiments. In the near future, it is necessary to explore one or more dopants that have the ability to efficiently reduce the T_C and simultaneously improve the thermochromic performance of VO_2, which will be useful for real application of VO_2 smart materials.

References

1. Li, J., Liu, W., Zhang, X., Chu, P. K., Cheung, K. M. C. and Yeung, K. W. K. (2019). Temperature-responsive tungsten doped vanadium dioxide thin film starves bacteria to death, *Mater. Today*, **22**, pp. 35–49.
2. Xie, B. H., Fu, W. B., Fei, G. T., Xu, S. H., Gao, X. D. and Zhang, L. D. (2018). Preparation and enhanced infrared response properties of ordered W-doped VO_2 nanowire array, *Appl. Surf. Sci.*, **436**, pp. 1061–1066.
3. Seyfouri, M. M. and Binions, R. (2017). Sol-gel approaches to thermochromic vanadium dioxide coating for smart glazing application, *Sol. Energy Mater. Sol. Cells*, **159**, pp. 52–65.
4. Miyazaki, K., Shibuya, K., Suzuki, M., Sakai, K., Fujita, J.-i. and Sawa, A. (2016). Chromium–niobium co-doped vanadium dioxide films: large temperature coefficient of resistance and practically no thermal hysteresis of the metal-insulator transition, *AIP Adv.*, **6**, p. 055012.
5. Dietrich, M. K., Kramm, B. G., Becker, M., Meyer, B. K., Polity, A. and Klar, P. J. (2015). Influence of doping with alkaline earth metals on the optical properties of thermochromic VO_2, *J. Appl. Phys.*, **117**, p. 185301.
6. Zhang, W., Wang, K., Fan, L., Liu, L., Guo, P., Zou, C., Wang, J., Qian, H., Ibrahim, K., Yan, W., Xu, F. and Wu, Z. (2014). Hole carriers doping effect

on the metal-insulator transition of N-incorporated vanadium dioxide thin films, *J. Phys. Chem. C*, **118**, pp. 12837–12844.

7. Li, M., Wu, X., Li, L., Wang, Y., Li, D., Pan, J., Li, S., Sun, L. and Li, G. (2014). Defect-mediated phase transition temperature of VO_2 (M) nanoparticles with excellent thermochromic performance and low threshold voltage, *J. Mater. Chem. A*, **2**, pp. 4520–4523.
8. Warwick, M. E. A. and Binions, R. (2014). Advances in thermochromic vanadium dioxide films, *J. Mater. Chem. A*, **2**, pp. 3275–3292.
9. Wu, J., Huang, W., Shi, Q., Cai, J., Zhao, D., Zhang, Y. and Yan, J. (2013). Effect of annealing temperature on thermochromic properties of vanadium dioxide thin films deposited by organic sol–gel method, *Appl. Surf. Sci.*, **268**, pp. 556–560.
10. Sakai, J., Zaghrioui, M., Phuoc, V. T., Roger, S., Autret-Lambert, C. and Okimura, K. (2013). Pulsed laser-deposited VO_2 thin films on Pt layers, *J. Appl. Phys.*, **113**, pp. 123503–123506.
11. Lysenko, S., Rua, A., Fernandez, F. and Liu, H. (2009). Optical nonlinearity and structural dynamics of VO_2 films, *J. Appl. Phys.*, **105**, p. 043502.
12. Chen, B., Yang, D., Charpentier, P. A. and Zeman, M. (2009). Al^{3+}-doped vanadium dioxide thin films deposited by PLD, *Sol. Energy Mater. Sol. Cells*, **93**, pp. 1550–1554.
13. Powell, M. J., Quesada-Cabrera, R., Taylor, A., Teixeira, D., Papakonstantinou, I., Palgrave, R. G., Sankar, G. and Parkin, I. P. (2016). Intelligent multifunctional $VO_2/SiO_2/TiO_2$ coatings for self-cleaning, energy-saving window panels, *Chem. Mater.*, **28**, pp. 1369–1376.
14. Blackman, C. S., Piccirillo, C., Binions, R. and Parkin, I. P. (2009). Atmospheric pressure chemical vapour deposition of thermochromic tungsten doped vanadium dioxide thin films for use in architectural glazing, *Thin Solid Films*, **517**, pp. 4565–4570.
15. Piccirillo, C., Binions, R. and Parkin, I. P. (2007). Synthesis and functional properties of vanadium oxides: V_2O_3, VO_2, and V_2O_5 deposited on glass by aerosol-assisted CVD, *Chem. Vap. Deposition*, **13**, pp. 145–151.
16. Gagaoudakis, E., Kortidis, I., Michail, G., Tsagaraki, K., Binas, V., Kiriakidis, G. and Aperathitis, E. (2016). Study of low temperature RF-sputtered Mg-doped vanadium dioxide thermochromic films deposited on low-emissivity substrates, *Thin Solid Films*, **601**, pp. 99–105.
17. Zheng, J., Bao, S. and Jin, P. (2015). $TiO_2(R)/VO_2(M)/TiO_2(A)$ multilayer film as smart window: combination of energy-saving, antifogging and self-cleaning functions, *Nano Energy*, **11**, pp. 136–145.

18. Zhou, S., Li, Y., Zhu, H., Sun, R., Zhang, Y., Huang, Y., Li, L., Shen, Y., Zhen, Q., Tong, G. and Fang, B. (2012). Microstructures and thermochromic characteristics of low-cost vanadium–tungsten co-sputtered thin films, *Surf. Coat. Technol.*, **206**, pp. 2922–2926.
19. Goodenough, J. B. (1971). The two components of the crystallographic transition in VO_2, *J. Solid State Chem.*, **3**, pp. 490–500.
20. Wang, N., Duchamp, M., Dunin-Borkowski, R. E., Liu, S., Zeng, X., Cao, X. and Long, Y. (2016). Terbium-Doped VO_2 Thin films: reduced phase transition temperature and largely enhanced luminous transmittance, *Langmuir*, **32**, pp. 759–764.
21. Wang, N., Chew Shun, N. T., Duchamp, M., Dunin-Borkowski, R. E., Li, Z. and Long, Y. (2016). Effect of lanthanum doping on modulating the thermochromic properties of VO_2 thin films, *RSC Adv.*, **6**, pp. 48455–48461.
22. Cao, J., Ertekin, E., Srinivasan, V., Fan, W., Huang, S., Zheng, H., Yim, J. W. L., Khanal, D. R., Ogletree, D. F., Grossmanan, J. C. and Wu, J. (2009). Strain engineering and one-dimensional organization of metal-insulator domains in single-crystal vanadium dioxide beams, *Nat. Nanotechnol.*, **4**, pp. 732–737.
23. Wang, N., Liu, S., Zeng, X. T., Magdassi, S. and Long, Y. (2015). Mg/W-codoped vanadium dioxide thin films with enhanced visible transmittance and low phase transition temperature, *J. Mater. Chem. C*, **3**, pp. 6771–6777.
24. Li, S.-Y., Mlyuka, N. R., Primetzhofer, D., Hallén, A., Possnert, G., Niklasson, G. A. and Granqvist, C. G. (2013). Bandgap widening in thermochromic Mg-doped VO_2 thin films: quantitative data based on optical absorption, *Appl. Phys. Lett.*, **103**, p. 161907.
25. Mlyuka, N. R., Niklasson, G. A. and Granqvist, C. G. (2009). Mg doping of thermochromic VO_2 films enhances the optical transmittance and decreases the metal-insulator transition temperature, *Appl. Phys. Lett.*, **95**, p. 171909.
26. Wang, N., Duchamp, M., Xue, C., Dunin-Borkowski, R. E., Liu, G. and Long, Y. (2016). Single-crystalline W-doped VO_2 nanobeams with highly reversible electrical and plasmonic responses near room temperature, *Adv. Mater. Interfaces*, **3**, p. 1600164.
27. Hu, L. T., Tao, H. Z., Chen, G. H., Pan, R. K., Wan, M. N., Xiong, D. H. and Zhao, X. J. (2016). Porous W-doped VO_2 films with simultaneously enhanced visible transparency and thermochromic properties, *J. Sol-Gel Sci. Technol.*, **77**, pp. 85–93.

28. Gonçalves, A., Resende, J., Marques, A. C., Pinto, J. V., Nunes, D., Marie, A., Goncalves, R., Pereira, L., Martins, R. and Fortunato, E. (2016). Smart optically active VO_2 nanostructured layers applied in roof-type ceramic tiles for energy efficiency, *Sol. Energy Mater. Sol. Cells*, **150**, pp. 1–9.
29. Zhu, J., Zhou, Y., Wang, B., Zheng, J., Ji, S., Yao, H., Luo, H. and Jin, P. (2015). Vanadium dioxide nanoparticle-based thermochromic smart coating: high luminous transmittance, excellent solar regulation efficiency, and near room temperature phase transition, *ACS Appl. Mater. Interfaces*, **7**, pp. 27796–27803.
30. Asayesh-Ardakani, H., Nie, A., Marley, P. M., Zhu, Y., Phillips, P. J., Singh, S., Mashayek, F., Sambandamurthy, G., Low, K.-b., Klie, R. F., Banerjee, S., Odegard, G. M. and Shahbazian-Yassar, R. (2015). Atomic origins of monoclinic-tetragonal (rutile) phase transition in doped VO_2 nanowires, *Nano Lett.*, **15**, pp. 7179–7188.
31. Huang, Z., Chen, C., Lv, C. and Chen, S. (2013). Tungsten-doped vanadium dioxide thin films on borosilicate glass for smart window application, *J. Alloys Compd.*, **564**, pp. 158–161.
32. Tan, X., Yao, T., Long, R., Sun, Z., Feng, Y., Cheng, H., Yuan, X., Zhang, W., Liu, Q., Wu, C., Xie, Y. and Wei, S. (2012). Unraveling metal-insulator transition mechanism of VO_2 triggered by tungsten doping, *Sci. Rep.*, **2**, p. 466.
33. Cho, J. H., Byun, Y. J., Kim, J. H., Lee, Y. J., Jeong, Y. H., Chun, M. P., Paik, J. H. and Sung, T. H. (2012). Thermochromic characteristics of WO_3-doped vanadium dioxide thin films prepared by sol-gel method, *Ceram. Int.*, **38**, pp. S589–S593.
34. Luo, Z., Wu, Z., Xu, X., Du, M., Wang, T. and Jiang, Y. (2011). Microstructures and thermochromic properties of tungsten doped vanadium oxide film prepared by using VOX–W–VOX sandwich structure, *Mater. Sci. Eng., B*, **176**, pp. 762–766.
35. Kong, F. Y., Li, M., Pan, S. S., Zhang, Y. X. and Li, G. H. (2011). Synthesis and thermal stability of W-doped VO_2 nanocrystals, *Mater. Res. Bull.*, **46**, pp. 2100–2104.
36. Batista, C., Ribeiro, R. M. and Teixeira, V. (2011). Synthesis and characterization of VO_2-based thermochromic thin films for energy-efficient windows, *Nanoscale Res. Lett.*, **6**, p. 301.
37. Ye, J., Zhou, L., Liu, F., Qi, J., Gong, W., Lin, Y. and Ning, G. (2010). Preparation, characterization and properties of thermochromic tungsten-doped vanadium dioxide by thermal reduction and annealing, *J. Alloys Compd.*, **504**, pp. 503–507.

38. Gao, Y. F., Cao, C. X., Dai, L., Luo, H. J., Kanehira, M., Ding, Y. and Wang, Z. L. (2012). Phase and shape controlled VO_2 nanostructures by antimony doping, *Energy Environ. Sci.*, **5**, pp. 8708–8715.
39. Dai, L., Chen, S., Liu, J., Gao, Y., Zhou, J., Chen, Z., Cao, C., Luo, H. and Kanehira, M. (2013). F-doped VO_2 nanoparticles for thermochromic energy-saving foils with modified color and enhanced solar-heat shielding ability, *Phys. Chem. Chem. Phys.*, **15**, pp. 11723–11729.
40. Kiri, P., Warwick, M. E. A., Ridley, I. and Binions, R. (2011). Fluorine doped vanadium dioxide thin films for smart windows, *Thin Solid Films*, **520**, pp. 1363–1366.
41. Burkhardt, W., Christmann, T., Franke, S., Kriegseis, W., Meister, D., Meyer, B. K., Niessner, W., Schalch, D. and Scharmann, A. (2002). Tungsten and fluorine co-doping of VO_2 films, *Thin Solid Films*, **402**, pp. 226–231.
42. Burkhardt, W., Christmann, T., Meyer, B. K., Niessner, W., Schalch, D. and Scharmann, A. (1999). W- and F-doped VO_2 films studied by photoelectron spectrometry, *Thin Solid Films*, **345**, pp. 229–235.
43. Chen, J.-M., Huang, Z.-L. and Liu, Y. (2009). Influence of annealing conditions on properties of $VO_{2-x}N_y$ thin films prepared by microwave plasma, *Semicond. Optoelectron.*, **30**, pp. 888–892.
44. Wei, J., Ji, H., Guo, W. H., Nevidomskyy, A. H. and Natelson, D. (2012). Hydrogen stabilization of metallic vanadium dioxide in single-crystal nanobeams, *Nat. Nanotechnol.*, **7**, pp. 357–362.
45. Chen, Y., Wang, Z., Chen, S., Ren, H., Wang, L., Zhang, G., Lu, Y., Jiang, J., Zou, C. and Luo, Y. (2018). Non-catalytic hydrogenation of VO_2 in acid solution, *Nat. Commun.*, **9**, p. 818.
46. Hong, W.-K., Park, J. B., Yoon, J., Kim, B.-J., Sohn, J. I., Lee, Y. B., Bae, T.-S., Chang, S.-J., Huh, Y. S., Son, B., Stach, E. A., Lee, T. and Welland, M. E. (2013). Hydrogen-induced morphotropic phase transformation of single-crystalline vanadium dioxide nanobeams, *Nano Lett.*, **13**, pp. 1822–1828.
47. Ji, H., Wei, J. and Natelson, D. (2012). Modulation of the electrical properties of VO_2 nanobeams using an ionic liquid as a gating medium, *Nano Lett.*, **12**, pp. 2988–2992.
48. Wu, C., Feng, F., Feng, J., Dai, J., Peng, L., Zhao, J., Yang, J., Si, C., Wu, Z. and Xie, Y. (2011). Hydrogen-incorporation stabilization of metallic VO_2(R) phase to room temperature, displaying promising low-temperature thermoelectric effect, *J. Am. Chem. Soc.*, **133**, pp. 13798–13801.
49. Andreev, V. N., Kapralova, V. M. and Klimov, V. A. (2007). Effect of hydrogenation on the metal-semiconductor phase transition in vanadium dioxide thin films, *Phys. Solid State*, **49**, pp. 2318–2322.

50. Eyert, V. (2011). VO_2: a novel view from band theory, *Phys. Rev. Lett.*, **107**, p. 016401.
51. Petrov, G. I., Yakovlev, V. V. and Squier, J. (2002). Raman microscopy analysis of phase transformation mechanisms in vanadium dioxide, *Appl. Phys. Lett.*, **81**, pp. 1023–1025.
52. Yuan, X., Zhang, W. and Zhang, P. (2013). Hole-lattice coupling and photoinduced insulator-metal transition in VO_2, *Phys. Rev. B*, **88**, p. 035119.
53. Richard, L. (2013). *Introduction to Computational Materials Science: Fundamentals to Applications*. Cambridge University Press, Cambridge.
54. Zhu, Z. and Schwingenschlögl, U. (2012). Comprehensive picture of VO_2 from band theory, *Phys. Rev. B*, **86**, p. 075149.
55. Wentzcovitch, R. M., Schulz, W. W. and Allen, P. B. (1994). VO_2-Peierls or Mott-Hubbard: a view from band theory, *Phys. Rev. Lett.*, **72**, pp. 3389–3392.
56. Liebsch, A., Ishida, H. and Bihlmayer, G. (2005). Coulomb correlations and orbital polarization in the metal-insulator transition of VO_2, *Phys. Rev. B*, **71**, p. 085109.
57. Laad, M. S., Craco, L. and Müller-Hartmann, E. (2006). Metal-insulator transition in rutile-based VO_2, *Phys. Rev. B*, **73**, p. 195120.
58. Belozerov, A. S., Korotin, M. A., Anisimov, V. I. and Poteryaev, A. I. (2012). Monoclinic M-1 phase of VO_2: Mott-Hubbard versus band insulator, *Phys. Rev. B*, **85**, p. 045109.
59. Sakuma, R., Miyake, T. and Aryasetiawan, F. (2008). First-principles study of correlation effects in VO_2, *Phys. Rev. B*, **78**, p. 075106.
60. Continenza, A., Massidda, S. and Posternak, M. (1999). Self-energy corrections in VO_2 within a model GW scheme, *Phys. Rev. B*, **60**, pp. 15699–15704.
61. Yuan, X., Zhang, Y., Abtew, T. A., Zhang, P. and Zhang, W. (2012). VO_2: orbital competition, magnetism, and phase stability, *Phys. Rev. B*, **86**, p. 235103.
62. Biermann, S., Poteryaev, A., Lichtenstein, A. I. and Georges, A. (2005). Dynamical singlets and correlation-assisted peierls transition in VO_2, *Phys. Rev. Lett.*, **94**, p. 026404.
63. Gatti, M., Bruneval, F., Olevano, V. and Reining, L. (2007). Understanding correlations in vanadium dioxide from first principles, *Phys. Rev. Lett.*, **99**, p. 266402.
64. Cyrot, M. (1972). Theory of mott transition: applications to transition metal oxides, *J. Phys.*, **33**, pp. 125–134.

65. Cui, Y., Shi, S., Chen, L., Luo, H. and Gao, Y. (2015). Hydrogen-doping induced reduction in the phase transition temperature of VO_2: a first-principles study, *Phys. Chem. Chem. Phys.*, **17**, pp. 20998–21004.
66. Cui, Y., Wang, Y., Liu, B., Luo, H. and Gao, Y. (2016). First-principles study on the phase transition temperature of X-doped (X = Li, Na or K) VO_2, *RSC Adv.*, **6**, pp. 64394–64399.
67. Zhang, J. J., He, H. Y., Xie, Y. and Pan, B. C. (2014). Boron-tuning transition temperature of vanadium dioxide from rutile to monoclinic phase, *J. Chem. Phys.*, **141**, p. 194707.
68. Warnick, K. H., Wang, B. and Pantelides, S. T. (2014). Hydrogen dynamics and metallic phase stabilization in VO_2, *Appl. Phys. Lett.*, **104**, p. 101913.
69. Zhang, J., He, H., Xie, Y. and Pan, B. (2013). Theoretical study on the tungsten-induced reduction of transition temperature and the degradation of optical properties for VO_2, *J. Chem. Phys.*, **138**, pp. 114705–114706.
70. Miyazaki, K., Shibuya, K., Suzuki, M., Wado, H. and Sawa, A. (2014). Correlation between thermal hysteresis width and broadening of metal-insulator transition in Cr-and Nb-doped VO_2 films, *Jpn. J. Appl. Phys.*, **53**, p. 071102.
71. Piccirillo, C., Binions, R. and Parkin, I. P. (2007). Nb-doped VO_2 thin films prepared by aerosol-assisted chemical vapour deposition, *Eur. J. Inorg. Chem.*, **2007**, pp. 4050–4055.
72. Li, D., Li, M., Pan, J., Luo, Y., Wu, H., Zhang, Y. and Li, G. (2014). Hydrothermal synthesis of Mo-doped VO_2/TiO_2 composite nanocrystals with enhanced thermochromic performance, *ACS Appl. Mater. Interfaces*, **6**, pp. 6555–6561.
73. Mai, L. Q., Hu, B., Hu, T., Chen, W. and Gu, E. D. (2006). Electrical property of Mo-doped VO_2 nanowire array film by melting-quenching sol-gel method, *J. Phys. Chem. B*, **110**, pp. 19083–19086.
74. He, X., Zeng, Y., Xu, X., Gu, C., Chen, F., Wu, B., Wang, C., Xing, H., Chen, X. and Chu, J. (2015). Orbital change manipulation metal-insulator transition temperature in W-doped VO_2, *Phys. Chem. Chem. Phys.*, **17**, pp. 11638–11646.
75. Andersson, G. (1956). Studies on vanadium oxides. II. The crystal structure of vanadium dioxide, *Acta. Chem. Scand.*, **10**, pp. 623–628.
76. Zhang, J., He, H., Xie, Y. and Pan, B. (2013). Giant reduction of the phase transition temperature for beryllium doped VO_2, *Phys. Chem. Chem. Phys.*, **15**, pp. 4687–4690.

77. Sun, C., Yan, L., Yue, B., Liu, H. and Gao, Y. (2014). The modulation of metal-insulator transition temperature of vanadium dioxide: a density functional theory study, *J. Mater. Chem. C*, **2**, pp. 9283–9293.

78. Cui, Y., Cao, C., Chen, Z., Luo, H. and Gao, Y. (2017). Atomic and electronic structures of thermochromic VO_2 with Sb-doping, *Comput. Mater. Sci.*, **130**, pp. 103–108.

79. Kresse, G. and Hafner, J. (1993). Ab initio molecular dynamics for liquid metals, *Phys. Rev. B*, **47**, pp. 558–561.

80. Kresse, G. and Furthmüller, J. (1996). Efficient iterative schemes for ab initio total-energy calculations using a plane-wave basis set, *Phys. Rev. B*, **54**, pp. 11169–11186.

81. Blöchl, P. E. (1994). Projector augmented-wave method, *Phys. Rev. B*, **50**, pp. 17953–17979.

82. Perdew, J. P., Burke, K. and Ernzerhof, M. (1996). Generalized gradient approximation made simple, *Phys. Rev. Lett.*, **77**, pp. 3865–3868.

83. Netsianda, M., Ngoepe, P. E., Catlow, C. R. A. and Woodley, S. M. (2008). The displacive phase transition of vanadium dioxide and the effect of doping with tungsten, *Chem. Mater.*, **20**, pp. 1764–1772.

84. Kittel, C. and McEuen, P. (1976). *Introduction to Solid State Physics* (Wiley, New York).

Chapter 3

VO_2 Nanocomposite Coatings for Smart Windows

Y. F. Gao and Z. Chen
School of Materials Science and Engineering, Shanghai University, Shanghai 200444, China
yfgao@shu.edu.cn

Nanocomposite (NC) is another kind of composite form of films/coatings besides multilayer structures. Different from multilayer structures, NC films (coatings) are obtained by distributing nanoparticles (NPs) into other materials (host matrix) to form a single layer/coating (Fig. 3.1). Interactions between NPs and a host matrix of this kind of composite films/coatings are more intensive, and both the physical and chemical properties of the NPs and the host matrix might be changed in this structure [1–4]. For an NC coating based on VO_2 (representing the M/R phase if there are no special instructions in this section), the biggest change is the greatly improved luminous performance (T_{lum}) and solar energy modulation ability (ΔT_{sol}), mainly contributed by the surface plasmon resonance (SPR) of the VO_2(R) NPs of around 1200 nm [5]. Besides, it is also convenient to composite VO_2 with other functional materials to enable, for example, color modification [6], self-cleaning [7, 8], low emission [9], visible-light modulation [10, 11], and

Vanadium Dioxide-Based Thermochromic Smart Windows
Edited by Yi Long and Yanfeng Gao

ISBN 978-981-4877-06-0 (Hardcover), 978-1-003-17731-9 (eBook)
www.jennystanford.com

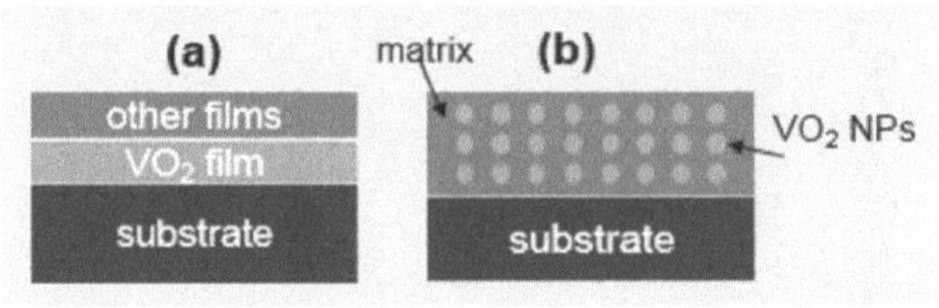

Figure 3.1 Schema of VO_2-based multilayered thin films (a) and VO_2 NC coating (b).

power generation [12, 13]. NC coatings also offer advantages in production—VO_2-based smart foils took seven years to develop after the group started related research work [14–16]—while the traditional pure VO_2 thin films, irrespective of whether physical vapor deposition [17–20] is used or solution-based methods [21, 22] are used, are still limited by their sizes [23, 24]. This section will introduce VO_2-based NC coatings for application in smart windows using their optical performance simulation, preparation and relatively properties, and multifunctional compounding with other materials.

3.1 Optical Simulation of VO_2-Based NC Coatings

3.1.1 Effective-Medium Theory

3.1.1.1 VO_2-based NC coatings

Li et al. constructed an optical model for this type of VO_2 films/coatings based on effective-medium theory (EMT, Fig. 3.2). Their simulation results illustrate the advantages of nanocomposite (NC) coatings in terms of luminous transmittance (T_{lum}) and solar energy modulation ability (ΔT_{sol}) [25]. This model is useful in predicting the optical performance of VO_2-based NC coatings with different matrixes, coating thicknesses, VO_2 contents, particle shapes, and distributions forms [25–27].

Here, we will give a brief introduction of Li et al.'s optical model. When the volume fraction (filling factor, f) of VO_2 nanoparticles (NPs) in an NC coating is small (e.g., 0.01), the dielectric constant

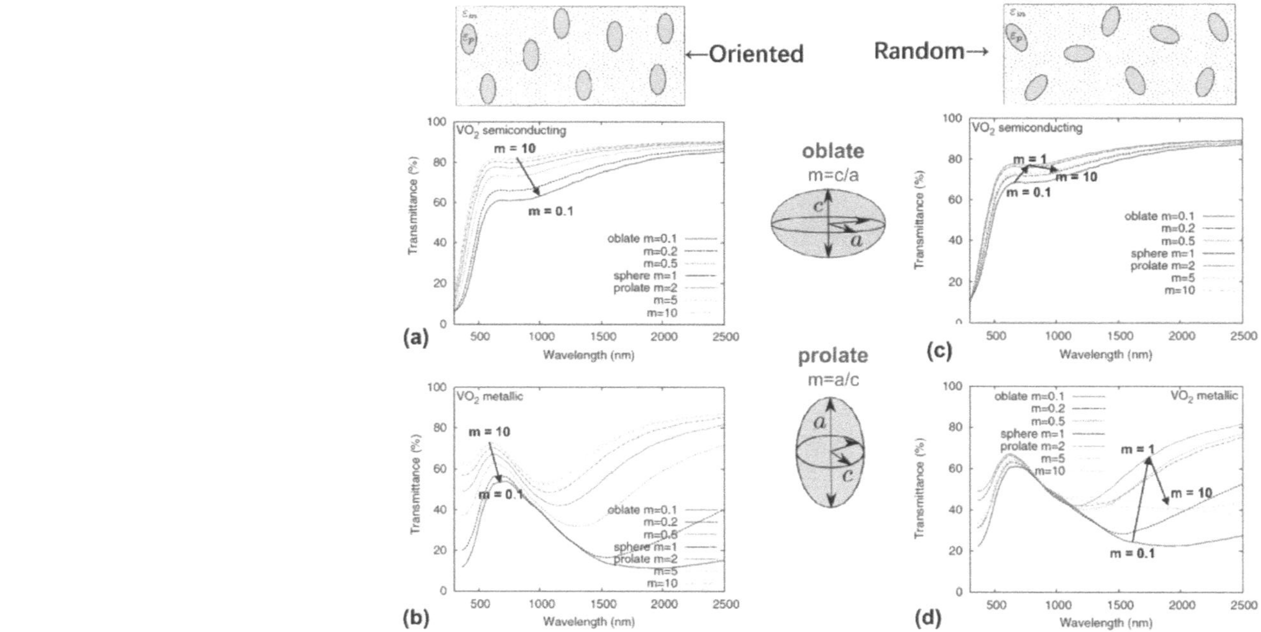

Figure 3.2 Spectra transmittance (250–2500 nm) of VO_2 NC coatings simulated by EMT [25]. (a, b) The transmittance of oriented VO_2 NPs with different m values (changing from oblate to prolate) before and after phase transition, respectively. (c, d) The results of random VO_2 NPs. For the oriented ones, the VO_2 content per unit area is different in different directions; that T_{lum} decreases as the NPs change from prolate to oblate can be attributed to the increased VO_2 content interacting with light. When the VO_2 NPs are randomly dispersed, the content is no longer nonuniform in different directions; thus the effect of NPs' shapes on optical performance is prominent. The result implies that spherical NPs perform better in terms of T_{lum} and ΔT_{sol}.

of the NC coating (ε^{NC}) could be governed as

$$\varepsilon^{NC} = \varepsilon_m \frac{1 + \frac{2}{3} f\alpha}{1 - \frac{1}{3} f\alpha} \tag{3.1}$$

and the polarizability α could be calculated by

$$\alpha = \frac{\varepsilon_p - \varepsilon_m}{\varepsilon_m + L\left(\varepsilon_p - \varepsilon_m\right)} \tag{3.2}$$

In the equations, ε_m and ε_p are the dielectric constants of the matrix and the VO_2 NPs, respectively, and L is the appropriate depolarization factor, related to the morphology of the NPs. For example, when the NPs are spheres, $L = 1/3$. Li et al. gave the calculation method for L for prolate spheroid and oblate spheroid NPs; details can be found in the paper. During simulation, VO_2 NPs can be treated approximately as these two topologies. Since the optical constants (refractive index n and extinction coefficient k) are related to the dielectric constant ε ($\varepsilon \equiv \varepsilon_1 + i\varepsilon_2, \varepsilon_1 = n^2 - k^2, \varepsilon_2 = 2nk$), we can calculate the n and k of the NC coating from those of the VO_2 and the matrix using this model. Then the transmittance or reflectance of NC coatings can be simulated by optical methods (e.g., the optical admittance recursive method [28]).

Li's results showed that the optical performance (T_{lum} and ΔT_{sol}) of VO_2-based NC coatings is morphology and orientation related and compared to pure VO_2 thin films, they show obvious advantages in T_{lum} and ΔT_{sol}. Zhou et al. analyzed this phenomenon and concluded that the main reason is the SPR of the metal-phase VO_2 NPs, which is located around 1200 nm and enlarges the transmittance difference at low and high temperatures [5].

To further understand the optical properties of VO_2-based NC coatings for practical applications, the author studied the relationship of T_{lum} and ΔT_{sol} with coating thickness, volume contents of VO_2 NPs, and the refractive index of the matrix, by the EMT method [16]. The main results are shown in Fig. 3.3. It is clear that the coating thickness and the VO_2 volume content have a similar effect on the optical performance of the VO_2 NC coating. A preferred result appears at a lower VO_2 volume content and when the coatings are thinner (about 1.2% and 2000 nm, respectively, in this simulation). And it was also found that the lower the matrix refractive index, the better is the optical performance of the NC coating.

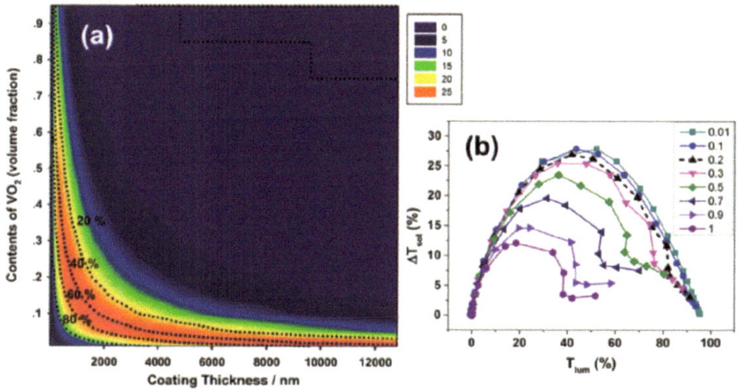

Figure 3.3 Optical performance ($T_{\rm lum}$ and $\Delta T_{\rm sol}$) of the NC coating calculated by EMT [16]. (a) The effect of coating thickness and VO_2 volume content on the optical performance of an NC coating. The dotted lines indicate the values of $T_{\rm lum}$, and the colors represent the values of $\Delta T_{\rm sol}$. The figure shows that the effect of volume content and coating thickness on optical performance is similar. (b) The effect of the refractive index of different matrixes on the optical performance of NC coatings; curves in the figure represent the different refractive indices of different matrixes. The results show that the lower the matrix refractive index, the better is the optical performance of the NC coating.

3.1.1.2 VO_2-based core/shell structures

Li et al. extended their model to VO_2-based core/shell structures to find if there were possibilities to further improve the $T_{\rm lum}$ and $\Delta T_{\rm sol}$ by microstructure designing (Fig. 3.4). Compared to pure VO_2 NPs, the optical model of the core/shell structure is different in terms of polarizability (α),

$$\alpha = 3\frac{(\varepsilon_{\rm s} - \varepsilon_{\rm m})(\varepsilon_{\rm c} + 2\varepsilon_{\rm s}) + \Omega(2\varepsilon_{\rm s} + \varepsilon_{\rm m})(\varepsilon_{\rm c} - \varepsilon_{\rm s})}{(\varepsilon_{\rm s} + 2\varepsilon_{\rm m})(\varepsilon_{\rm c} + 2\varepsilon_{\rm s}) + \Omega(2\varepsilon_{\rm s} - \varepsilon_{\rm m})(\varepsilon_{\rm c} - \varepsilon_{\rm s})} \tag{3.3}$$

Here, $\Omega = x^3/(2t + x)^3$ is the volume ratio of the inner and outer spheres, where x and t are the diameter of the core and the thickness of the shell, respectively, as shown in the figure. $\varepsilon_{\rm m}$, $\varepsilon_{\rm s}$, and $\varepsilon_{\rm c}$ are the dielectric constants of the matrix, the shell, and the core, respectively. The filling factor f in this model is set as $f = 0.01/(1 - \Omega)$. Since there are few advantages if the concentric spheroids with specific orientations compare to spheres, the simulation of the core/shell structure is only done using the spherical structure.

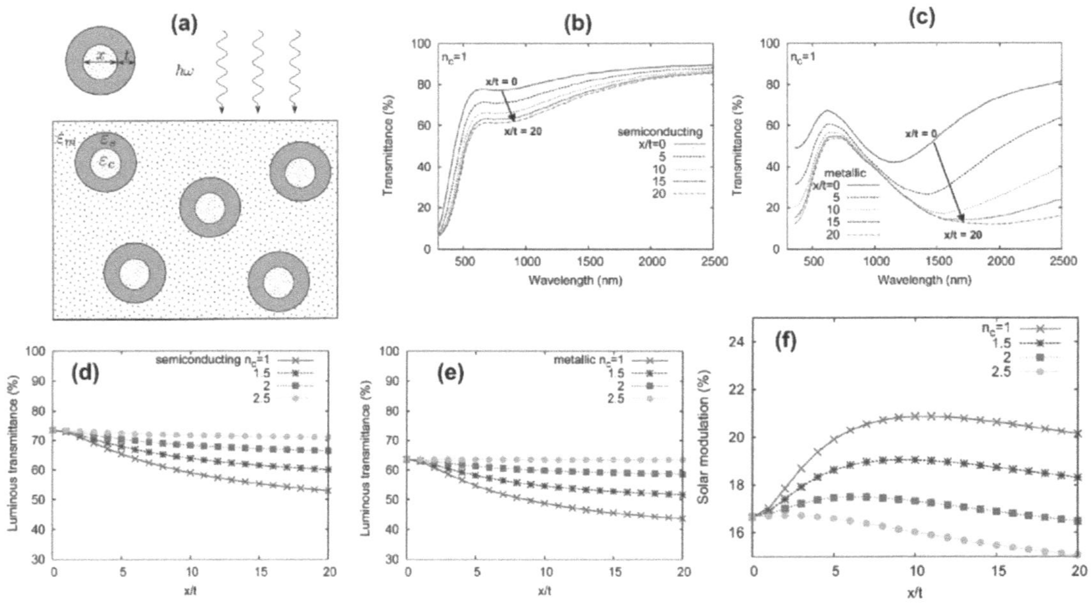

Figure 3.4 Simulated optical performance of an NC coating with core/shell structured VO_2 NPs [29]. (a) Schema of an NC coating with core/shell structured NPs; x is the diameter of the core, and t is the thickness of the shell; (d, e) the changing T_{lum} with different x/t values; (f) the Δhanging ΔT_{sol} with different x/t values (0–20); the higher the value, the thinner are the VO_2 shells. From (d, e, f), we found that optical performance increased with a decreased core refractive index (n_c) and an increased x/t value; a preferred performance was obtained when $n_c = 1$ and x/t is close to 10, which represents a VO_2 hollow ball with a very thin shell; (b, c) the transmittance spectra (250–2500 nm) of the NC coatings with a VO_2 hollow ball at 20°C and 90°C; as the x/t value increased, the transmittance decreased visibly both at 20°C and 90°C.

Li et al. simulated the optical performance of the core/shell structure with VO_2 as the core and then as the shell. Their results showed that moderately thinly walled ($x/t = 10$), hollow, spherical VO_2 nanoshells had higher values of ΔT_{sol} than pure VO_2 NPs but the T_{lum} becomes lower [29]. On the other hand, the structure with a VO_2 core didn't show advantages in terms of T_{lum} and ΔT_{sol}, although the spherical shells would protect VO_2 NPs from degradation, which is meaningful in practical usage.

3.1.2 Four-Flux Method

EMTs are suitable for simulating the optical performance of NP dispersions when the particle size $r \ll \lambda$; thus no scattering is considered in these works. Actually, particles scatter light no matter how small they are. Taking this into consideration, Laaksonen et al. adopted a four-flux method (FFM) to simulate the optical performance of VO_2 NC coatings (Fig. 3.5) [30]. The FFM considers diffuse and collimated radiation fluxes going through the coatings and obtains direct and diffuse transmittance (reflectance) from the fluxes. The fluxes depend on the scattering and absorption of NPs determined from the Mie theory. Laaksonen et al. found that scattering leads to higher absorption for VO_2 NC coatings. When the particle sizes are very small, the results obtained from effective-medium theory (EET) and the FFM are similar, but the differences become clear for particles of radii larger than 20 nm. The FFM detects light scattering effects on decreasing the transmittance and increasing the reflectance at short wavelengths. Hence, they suggested that to avoid light scattering effects in VO_2 NC coatings, the VO_2 particles should be less than 20 nm in radii.

3.1.3 Accuracy of a Theoretical Simulation

Compared to experimental results, T_{lum} and ΔT_{sol} from simulations are too high. Even Chen et al. greatly improved T_{lum} and ΔT_{sol} of NC coatings by using VO_2 NPs with crystallinity close to that of a single crystal. There are still gaps between theoretical and experimental results (Fig. 3.5c and e, $\Delta T_{sol} = 17.7\%$ and 13.2%, respectively, when T_{lum} is about 70%).

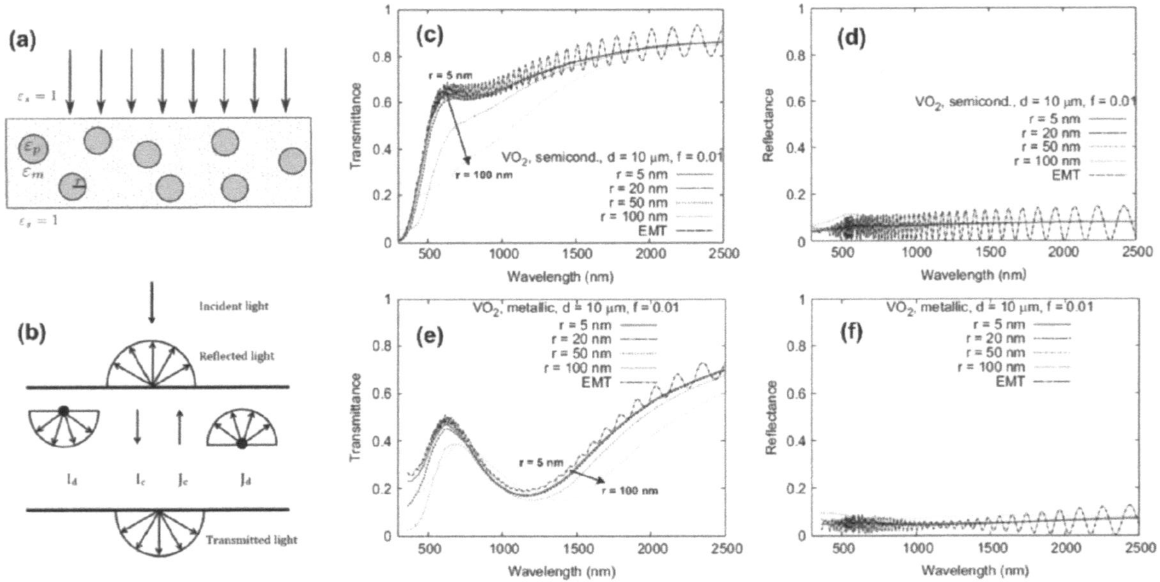

Figure 3.5 Comparison of the optical simulation of VO_2 NC coatings by the ETM and the FFM [30]. Schema of the ETM (a) and the FFM (b). Spectra transmittance and reflectance (250–2500 nm) of VO_2 NC coatings at 20°C, with radii in the 5–100 nm range, calculated by the FFM (c) and the ETM (d). (e, f) Spectra of VO_2 NC coatings at 90°C. The curves simulated by the ETM show obvious interferences. Besides, the average transmittance/reflectance is similar to that simulated by the FFM when the NPs' radii are small, which indicates the accuracy of the two methods. For NPs with larger radii, the transmittance decreases rapidly and the reflectance improves slightly because the scattered light propagates backward, and we observed this in our experiments when the VO_2 NPs were poorly dispersed.

Samples prepared in experiments did not reach the levels required in theoretical simulations, for example, in terms of crystallinity, dimensional uniformity, and dispersion uniformity. More importantly, it was thought that the optical constant used in simulations was not the actual optical constant of the NPs in the NC coatings. Chen et al. used the optical constants obtained from the VO_2 single crystal and the VO_2 thin film to simulate the transmittance of VO_2-based NC coatings by EMT. It can be seen that there is a huge gap between them. By comparing the $n(\lambda)$ and $k(\lambda)$ used, they inferred that, besides the error caused by data measurement and the used fitting model, the difference in optical constants caused by the sample itself is more pronounced. Compared to single crystal VO_2, thin film VO_2 had more defects due to the large number of interfaces. These defects will not only decrease the density of VO_2 (reduce n) but also increase absorption when VO_2 is in the insulation phase (higher k) and decrease the carrier migration rate in the metal phase (lower k), thereby lowering the transmittance difference between the low and high temperatures of NC coatings.

It is hard to measure the optical constants of VO_2 NPs. The current simulations of VO_2 NC coatings use the optical constants obtained from VO_2 thin films prepared by magnetron sputtering. Compared to the relatively independent particles in NC coatings, the VO_2 particles in the films prepared by magnetron sputtering are closely packed with each other and have fewer defects. This could explain why the results of current theoretical simulations are higher than those obtained by experiments.

It can also be seen from the simulation results (Fig. 3.6) that small differences in optical constants would lead to large changes in the optical performance of NC coatings. Since the optical constant of VO_2 NPs could be easily influenced by doping, nonstoichiometry, surface oxidation, crystallinity changes, and so on, it would be difficult to accurately simulate the optical performance of the NPs. Thus, the simulation was only used to qualitatively analyze the trend of NC coatings' optical performance, for example, that the optical properties vary with film thickness, solid content, and refractive index of the dispersed matrix. Later, simulation might be used as a tool to study the optical constants of VO_2 NPs based on precise control synthesis technology of VO_2 NPs.

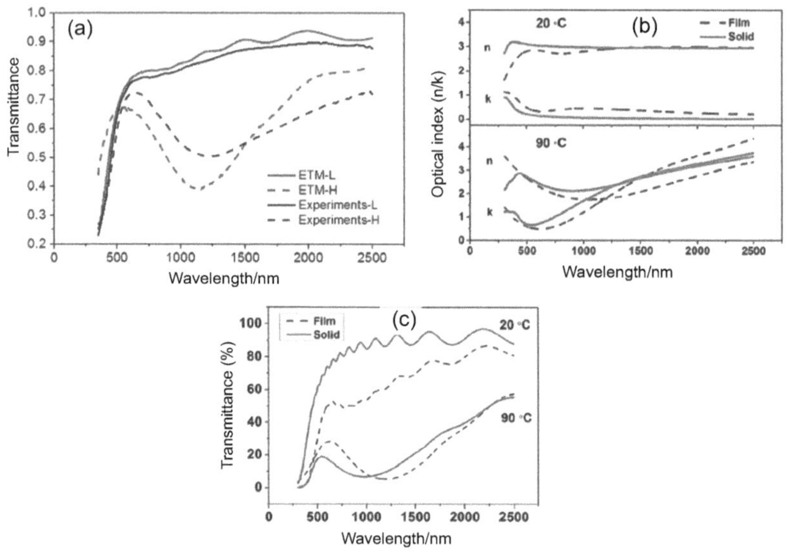

Figure 3.6 (a) Comparison of simulated spectra transmittance (250–2500 nm) by ETM and experiment (using high-crystallinity VO_2 NPs prepared by hydrothermal reaction with a rapid heating rate [27]). (b) Optical constants of VO_2 film and crystal (solid) at different wavelengths (250–2500 nm) before and after phase transition; (c) simulated spectra transmittance with two kinds of optical constants (the coating thickness is 2000 nm and the volume content of VO_2 NPs is 1%). It was found that solid VO_2 has a higher refractive index and a lower extinction coefficient at 20°C and a relatively higher refractive index but a higher extinction coefficient at 90°C, the small difference in the optical constant leads to a huge difference in the optical performance of the NC coatings.

3.2 Preparation of VO_2-Based NC Coatings for Luminous Transmittance and Solar Energy Modulation Ability Improvement

Luminous transmittance and solar energy modulation ability are two basic properties required for VO_2-based smart windows application. As predicted by theory, VO_2 NC coatings would show an obvious advantage in these two properties compared to pure VO_2 thin films. Various kinds of VO_2 NC coatings have been prepared successfully. Here, we would introduce these works by dividing them into three categories: nanoporous VO_2 thin films, VO_2-based inorganic NC coatings, and VO_2-based organic NC coatings.

3.2.1 Nanoporous VO_2 Thin Films

Nanoporous VO_2 thin films are composite films of VO_2 and air. Strictly speaking, most of the VO_2 thin films prepared by solution methods are VO_2 and air NC films. Because these films are not dense enough compared to pure VO_2 thin films (e.g., VO_2 prepared by magnetron sputtering [31]), these films perform better in terms of luminous transmittance (T_{lum}) and solar energy modulation ability (ΔT_{sol}), which could be well explained by the simulation results of NC films/coatings.

Kang et al. noticed that the nanopores in VO_2 thin films improve the optical performance of VO_2 thin films (Fig. 3.7a–c) [32]. They prepared VO_2 thin films with high porosity and studied them by morphology analysis and EMT simulation. Their work showed that VO_2 with a large number of nanopores would visibly improve the optical performance of VO_2 thin films and predicted that the optical performance of VO_2 thin films would be further improved by increasing the number of nanopores (porosity) in VO_2 thin films.

Cao et al. [33] further improved the number of nanopores (porosity improved to 45%) with the help of a freeze-drying process (Figs. 3.7d and 3.7e). By varying the prefreezing temperature and tailoring the coating thickness and the VO^{2+} concentration, T_{lum} and ΔT_{sol} of the obtained VO_2 thin films could be adjusted from 50% to 71.6% and from 5.6% to 14.7%, respectively. Their films showed up to 50% improvement in luminous transmittance (~41% in Kang's work) when keeping ΔT_{sol} larger than 14%.

It should be noticed that increasing the porosity will greatly weaken the strength of the film, which would limit its applications.

3.2.2 VO_2-Based Inorganic NC Coatings

Typically, VO_2-based inorganic NC coatings are VO_2 NPs dispersed in inorganic matrixes. The reported matrixes are SiO_2 [34–36], TiO_2 [7, 8, 26], ZrV_2O_7 [37], CeO_x [38], SnO_2 [39], and Si-Al oxide [40]. These materials would have different effects on VO_2 besides providing

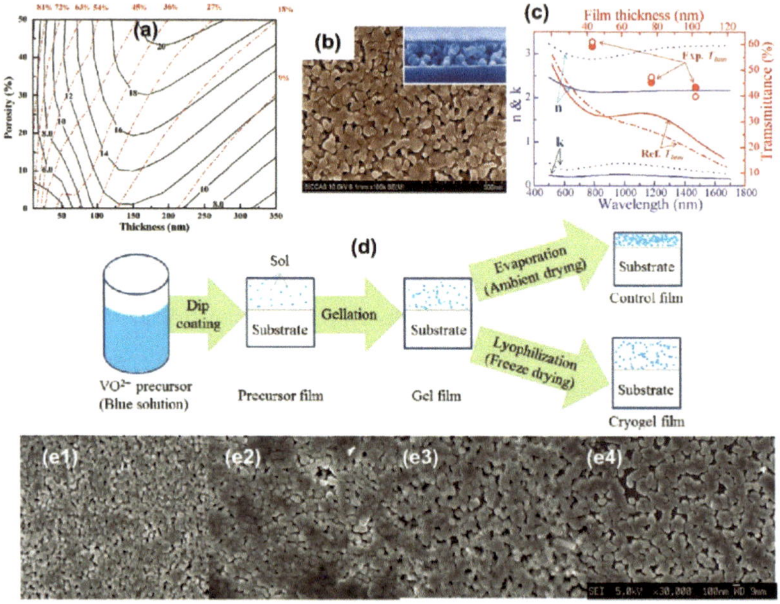

Sample	Prefreezing temperature (°C)	Freeze-drying	T_{lum}(20/90°C)	ΔT_{lum}	T_{sol}(20/90°C)	ΔT_{sol}	porosity
e	none	none	55.2 / 58.2 %	-3.0 %	61.8 / 58.9 %	2.9 %	2 %
f	none	yes	63.0 / 62.6 %	0.40 %	68.5 / 64.3 %	4.2 %	9 %
g	−20	yes	67.2 / 66.2 %	1.00 %	70.8 / 65.8 %	5.0 %	20 %
h	−30	yes	71.1 / 70.2 %	0.90 %	72.9 / 65.0 %	7.9 %	33 %

Figure 3.7 Morphology and optical properties of VO_2 thin films with nanopores. (a–c) The work of Kang et al.; nanopores were formed by a polymer-assisted deposition method [21]. (a) The simulated results (by ETM) of the relationship of films' porosity and thickness with their optical performance (T_{lum} and ΔT_{sol}), (b) the surface and cross-sectional SEM image, where nanopores can be seen clearly, and (c) the measured optical constant by ellipsometer (blue) and T_{lum} compared with references; compared with films by magnetron sputtering [25], the optical constants of the nanopores are much smaller because VO_2 is composited with air (pore) and this is the reason these films have much higher T_{lum} than the films of the same thickness made by magnetron sputtering. (d–h) The work of Cao et al. [33], who prepared VO_2 nanoporous films using a freeze-drying assistant solution deposition method. (d) The deposition processes (e1–e4) are the surface SEM images of samples with/without prefreezing and freeze-drying processes; the process parameters and optical performance can be seen in the table. Similar results with Kang's work are obtained, but they could get better optical performance due to the higher porosity obtained with the freeze-drying process.

protection. The SiO_2 in the SiO_2 and VO_2 composite coatings can adjust the VO_2-phase transition temperature and limit its grain growth [34], and the TiO_2 can impart self-cleaning properties to composite membranes [7, 8, 26].

According to the results by theoretical simulation, VO_2-based NC coatings would have better optical performance than pure VO_2 films. However, in the work [14], the NC coatings didn't show enhanced optical performance, mainly because the VO_2 in the NC coatings had poor crystallinity or poor dispersion. Thus, controlling the conditions during the preparation process is important to obtain good crystallinity and dispersion of VO_2 NPs.

VO_2-based NC coatings could be produced by chemical vapor deposition (CVD) or solution methods (Fig. 3.8). For the CVD method, precursors of VO_2 and matrix materials were brought onto the heated substrates, where they reacted and formed NC coatings [8, 39]. The solution methods can be divided into two types: (i) the precursor is premixed, then annealed (as shown in Fig. 3.8b) [7, 8, 34] and (ii) the crystallized VO_2 NPs are dispersed into the precursor of the composite materials to form a coating (as shown in Fig. 3.8c) [27, 40, 41]. For CVD and the first solution method, controlling the

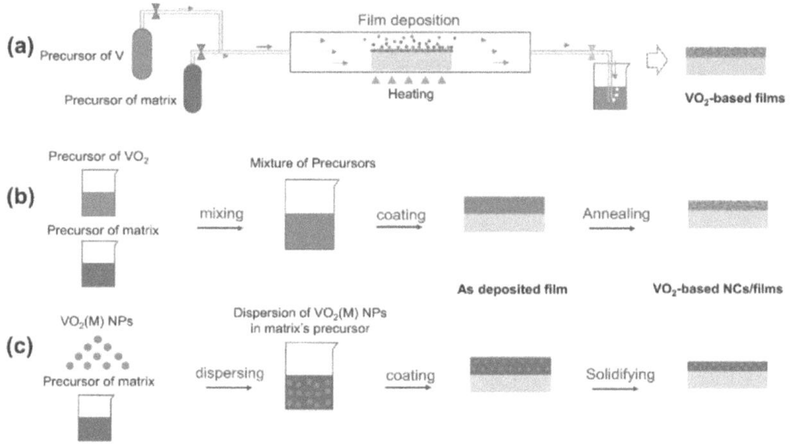

Figure 3.8 Schema of the typical methods for preparing NC coatings. (a) Chemical vapor deposition. (b, c) Solution deposition methods; (b) precursor coannealing method and (c) VO_2 NP predispersing method.

process is more difficult because it is necessary to comprehensively consider the effect of the annealing atmosphere of the matrix and the crystallization performance of VO_2, which is very sensitive to the annealing atmosphere and temperature. Good results are only found in matrix materials with little influence on the crystallinity of VO_2 during annealing [34, 36, 37].

The difficulties of the second solution method are prepreparing well-crystallized VO_2 NPs of a suitable size (usually less than 100 nm) and dispersing them well in the precursor of the matrix materials. If these two problems can be solved, NC coatings with an excellent performance can be obtained. This method is more suitable for large-scale preparation since it has lower requirements for annealing equipment. In the reports that used this method, the VO_2-Si-Al oxide NC coating has an average T_{lum} of $\sim$59% and ΔT_{sol} of 12% (Fig. 3.9d–g) [40] due to the relatively low refractive index of the matrix. The VO_2 and TiO_2 composite coatings should introduce mesopores to reduce the matrix's refractive index to achieve better optical performance (T_{lum} $\sim$60% and $\Delta T_{\mathrm{sol}} = 14.6\%$) (Fig. 3.9a–c) [26].

3.2.3 VO_2-Based Organic Nanocomposite Coating

Compared with inorganic systems, the most obvious advantages of organic NC coatings are the variety, low refractive index, and flexibility of the matrix materials. From a market perspective, an organic NC coating has a simpler preparation process and lower operation requirements and is very convenient for application in existing building windows. Since the organic matrix cannot be treated at a high temperature, an organic VO_2 NC coating is usually obtained using the second method discussed in the inorganic NC coatings section, that is dispersing the previously synthesized VO_2 NPs into an organic matrix, then coating it, and then allowing it to solidify. Thus the creation of high-quality VO_2 NPs and their surface modification and dispersion into matrix materials are the key problems in organic NC coatings [41].

High-quality VO_2 NPs require good crystallinity, subspheroidal morphology, and small sizes (usually less than 100 nm). Zhou et al. used a microemulsion method to synthesize small-sized precursors and then coated a SiO_2 layer on them, which would limit the growth of VO_2 during the followed annealing process. The obtained $V_{1-x}W_xO_2$@SiO_2 showed an average size of several tens of nanometers. By coating the NPs on a transparent Teflon tape, samples with $T_{lum} = 53.2\%$ and $\Delta T_{sol} = 12.6\%$ were obtained [42]. The transmission spectrum (250–2600 nm) of the samples at high temperatures showed obvious plasmon resonance absorption in the near-infrared (NIR) region, indicating that the VO_2 NPs had high dispersion [5]. Other researchers, including Xiao et al., have used thermal decomposition to prepare VO_2 NPs [43]. Lu et al. introduced a bead-milled process to obtain nanosized VO_2(M) NPs since the VO_2(M) NPs they produced by the thermal decomposition method were large [44], Li et al. adopted the hydrothermal method at a temperature as low as 220°C with Mo doping or rutile TiO_2 nanocrystals as seed [45], and Zhang et al. prepared VO_2 NPs by a solvothermal-annealing two-step method [46]. The matrixes in these works were Teflon tape [42], acrylic resin [43], poly(vinylpyrrolidone) [45, 47], and polysiloxane oligomers [44]. These matrixes were adopted because they had good compatibility with VO_2 and good film formation ability.

Gao's group carried out a lot of work in NC coating applications, making outstanding achievements. In early 2008, they successfully prepared single-crystal VO_2(M) by a one-step hydrothermal method [48]. Subsequently, they developed a doping method to control the particle sizes and morphology [27]. The VO_2 NPs were surface-modified and dispersed into polyurethane, and a flexible VO_2 NC coating was successfully prepared (Fig. 3.10a–d) [14]. These coatings had excellent optical performance ($T_{lum} = 55\%$ and $\Delta T_{sol} = 7.5\%$), and their optical properties can be further improved by improving the crystallinity of VO_2 NPs (Fig. 3.10e, $T_{lum} = 53\%$ and $\Delta T_{sol} = 21\%$) [27]. W-Zr dual element doping treatment of VO_2 NPs can improve the optical performance of the coating and simultaneously reduce their phase transition temperature [49]. The method developed is suitable for large-scale production. Smart NC coatings with a width of 1 m have been successfully prepared (the length prepared by the method is not limited) (Fig. 3.10f) [16]. This method also facilitates the synthesis of other functionalized VO_2-based NC coatings, such as the VO_2-Sb:SnO_2 composite coating exhibiting enhanced solar heat shielding properties [50].

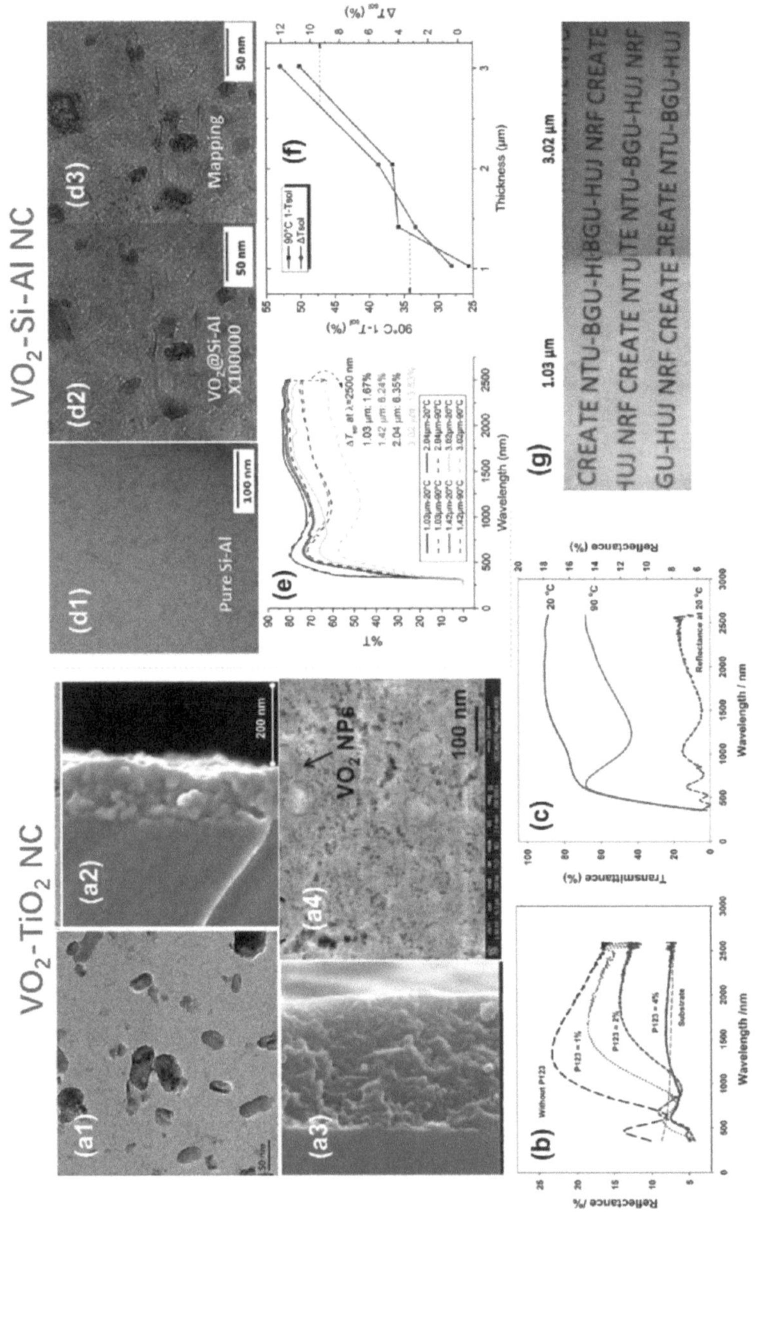
VO2-TiO2 NC
VO2-Si-Al NC
(a1)
(a2)
(a3)
(a4)
VO2 NPs
100 nm
200 nm
(b)
(c)
(d1)
Pure Si-Al
100 nm
(d2)
VO2@Si-Al
X100000
50 nm
(d3)
Mapping
50 nm
(e)
(f)
(g)
1.03 μm
3.02 μm
CREATE NTU-BGU-HUJ NRF CREATE
Wavelength (nm)
Thickness (μm)
Wavelength / nm
Transmittance (%)
Reflectance (%)
20 °C
90 °C

Figure 3.9 VO_2-based inorganic NC coatings prepared by the NP predispersing method. (a–c) Crystallized VO_2-TiO_2 NC coating [26]; (a1) TEM of TiO_2 sol with dispersed VO_2 NPs, (a2) cross-sectional SEM image of VO_2-TiO_2 sol NC coating, and (a3, a4) cross-sectional and surface SEM images of crystallized VO_2-TiO_2 NC coating with mesopores created by polyethylene oxide–polypropylene oxide–polyethylene (P123). (b) The reflectance curves (250–2500 nm) of a crystallized VO_2-TiO_2 NC coating with different solid contents of P123 in the TiO_2 sols; after adding 4 wt.% P123, the reflectance of the NC coating got close to that of the substrate, which means the refractive index of the NC coating is close to 1.5; ~40% porosity was estimated in this situation; (c) the transmittance curves after adding 4 wt.% P123 (a T_{lum} of about 60% and a ΔT_{sol} of 14.6% was achieved). (d–g) VO_2-Si-Al NC coating reported by Liu et al. [40]. VO_2 NPs were dispersed in the Si-Al sol and coated on the substrate without further annealing; (d1) the surface SME image of the Si-Al gel coating, (d2) the surface SME image of the VO_2-Si-Al NC coating, and (d3) the TEM ESI mapping image of (d2), where VO_2 NPs could be seen clearly. (e) The transmittance curves (250–2500 nm) of NC coatings of different thicknesses, (g) the optical images of 1.03 um and 3.02 um thick coatings, and (f) the ΔT_{sol} and the solar-heat shielding ability ($1-T_{sol}$, which will be discussed in the following section), which changes with coating thickness; the higher the coating thickness, the better are these two properties. However, the ΔT_{sol} is only 12% when T_{lum} is about 60%

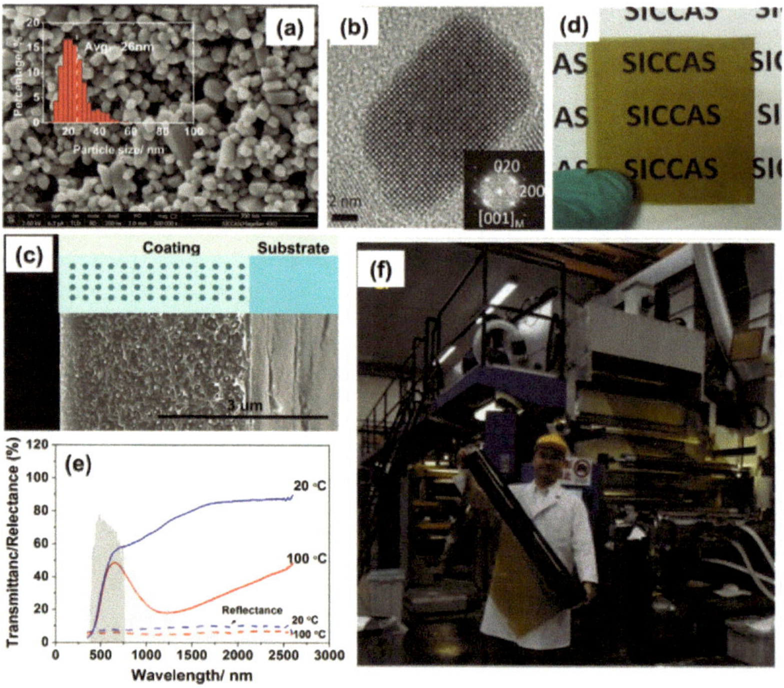

Figure 3.10 (a) SEM and (b) HRTEM of VO_2 NPs; the average particle size of VO_2 NPs is 26 nm. The VO_2-based nanocomposite foils prepared from these NPs are uniform, with an intrinsic yellowish-green color (d), and have a huge transmittance difference before and after phase transition (e). An SEM cross-section of this foil shows that the NPs were well-dispersed in the PU matrix (c). The shadow in (e) is the spectral distribution of solar energy. (f) The equipment and foils of the roll-to-roll method; the produced foil is 1 m wide and 2 km long [16].

Under the efforts of the researchers, the optical performance of VO_2-based NC coatings was greatly improved. Figure 3.11 summarizes the optical properties of the different VO_2-based thermochromic (TC) layers reported. The ordinate represents the solar energy modulation ability (ΔT_{sol}), and the abscissa represents the luminous transmittance (T_{lum}). The closer the value to the upper right of the figure, the better is the optical performance. Single layer pure VO_2 films have the lowest performance, and the ΔT_{sol} cannot exceed 10%. By introducing nanoporosity, the ΔT_{sol} of the

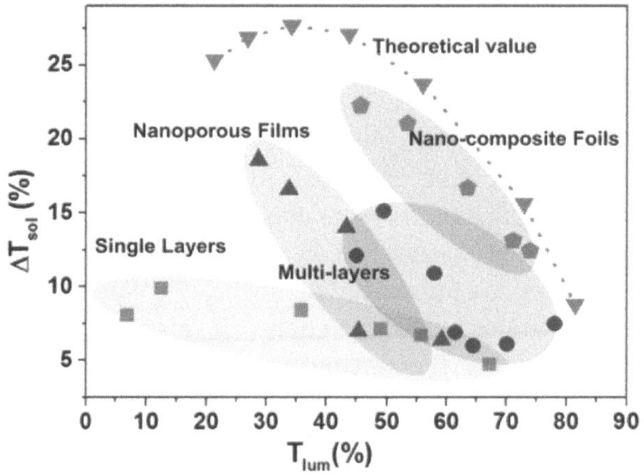

Figure 3.11 Solar energy modulation ability (ΔT_{sol}) and luminous transmittance (T_{lum}) of the different VO_2 films (coatings): single layer [21, 51, 52], multilayer by magnetron sputtering [53, 54] and polymer-assisted deposition [28], nanoporous films [32, 33], nanocomposite foils [14, 27], and theoretical values calculated from the optical index obtained from thin films by magnetron sputtering [25]. This image was edited on the basis of a graph in a previous work Ref. [24]. Copyright (2012) Elsevier.

films can be greatly improved, but the ΔT_{sol} decreases rapidly with improved T_{lum}, indicating that the structure is thickness sensitive. The performances of the multilayer films are reversed. Because the multilayer films are designed with the goal of antireflection, the reported data all have a relatively high T_{lum}. Considering only ΔT_{sol} and T_{lum}, the NC coating shows absolute advantage, high in both ΔT_{sol} and T_{lum}. The dotted line in the upper right of the figure is the theoretical calculation value of the NC coating (refractive index 1.5) using Li's method, representing the best possible performance of the NC coating that can be achieved by the current VO_2-based TC coatings. As shown in the Fig. 3.5, the NC coatings prepared are very close to the theoretical value.

Benefitting from their simple preparation process and good optical performance, VO_2-based organic NC coatings have good application prospects. However, from the point of view of practical application, attention must be paid to weather and scratch and water resistance.

3.3 VO_2-Based NC Coatings Modified with Other Light Functional Materials

Besides T_{lum} and ΔT_{sol}, there are other optical performance parameters that should be emphasized in applications of VO_2-based smart windows, such as color, solar shading ability, and emissivity. Though T_{lum} and ΔT_{sol} have been greatly improved in VO_2-based NC coatings, they are still not ideal for practical applications. Since visible light occupies half of the energy of solar energy, researchers have started turning their attention to the visible waveband of solar energy. These works include composite VO_2 with visible-light-regulated materials or solar cells. All these works will be described in detail next.

3.3.1 Adjustment of Other Optical Performance of VO_2-Based NC Coatings

3.3.1.1 VO_2-based NC coatings with color modulation

VO_2 films/coatings are yellowish green, which is not a popular color for windows. Two methods are adopted to modulate the color of VO_2 films/coatings. The first is absorption edge regulation by doping [49, 55–57]. This method lightens/darkens the color of VO_2 films/coatings, improving/reducing their T_{lum}, but the adjustment range is limited and heavy doping would severely damage the lattice of the VO_2, reducing the ΔT_{sol}. The other method is composite VO_2 with materials of other colors. Although this will greatly reduce the T_{lum}, changing the proportions and colors of the added materials would allow a wide range of color adjustment of the composite films/coatings.

Au is one kind of color material used to create a nanocomposite VO_2 coatings, where VO_2 is the matrix [6, 58, 59]. Au-VO_2 NC coatings were prepared by codepositing Au and VO_2 using magnetron sputtering [58, 59], sol-gel [60], pulsed laser deposition [61], or CVD [6]. By adjusting the concentration of the Au particles, the SPR of Au NPs would change and the composite coating can gradually change from brownish yellow to green and finally to blue (Fig. 3.12) [6]. As the concentration of Au particles increases, the reflection of

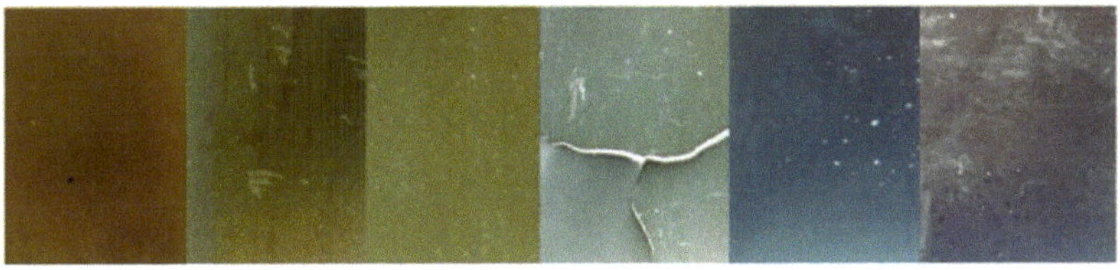

Figure 3.12 Examples of glass (3×5 cm^2 approximate size) with gold and VO_2 NC films. The films had a Au/V ratio determined by EDAX of (A) 0 (W-doped VO_2), (B) 0.09, (C) 0.15, (D) 0.30, (E) 0.36, and (F) (gold nanoparticle film) [6].

the coating in the infrared band is also enhanced, which helps to improve the thermal insulation performance of VO_2.

However, Au-VO_2 NC coatings are not economical enough, and the process is not suitable for large-scale preparation. Most of the Au-VO_2 NC coatings were used for studying the SPR of Au NPs on VO_2's optical performance and the influence of carrier concentration on the phase transition temperature of VO_2. The simplest way to adjust color is by adding pigments into the VO_2 NPs dispersion used for organic NC coating preparation. According to CIELAB color system regulations, the color can be modulated using VO_2 NPs and two other different pigments.

3.3.1.2 Improvement of solar-heat shielding ability (T_s) for VO_2 NPs

NIR light spans wavelengths of 760 to 2500 nm and contains about 47% of solar energy. T_s represents the ability to block this part of energy ($1 - T_{infr}$, where T_{infr} refers to infrared energy transmittance). It is a pity that the T_s of VO_2 at high temperatures is not high enough, which leads to its low energy saving efficiency in a hot area.

To solve this problem, Gao et al. studied VO_2-based organic NC coatings with improved T_s by adding Sb:SnO_2 (ATO) NPs (Fig. 3.13). Their results showed that both T_{lum} and the visual appearance of the coatings remain almost unchanged while their T_{infr} and ΔT_{sol} are monotonical as the ATO mass in the NC coatings increases, though the T_s of the NC coatings had visibly improved—about 20% at an ATO content of 9%. To test the applied properties, the NC

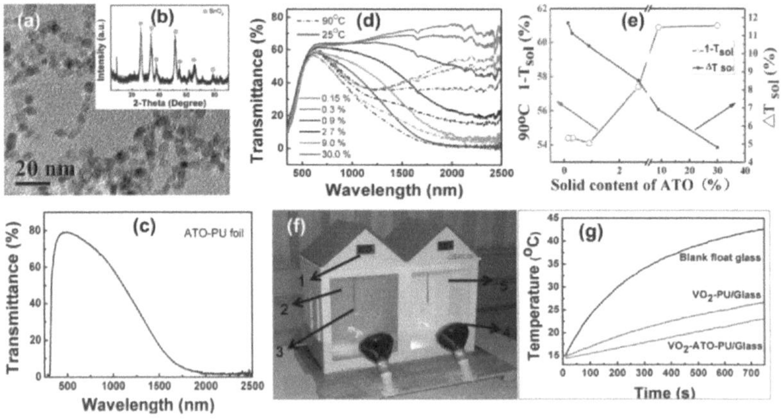

Figure 3.13 VO_2-based NC coatings with improved solar-heat shielding ability using ATO [50]. (a) TEM image and (b) XRD of ATO NPs; (c, d) the transmittance curve (350–2500 nm) of an ATO-PU foil and VO_2-ATO-PU NC coating (with different solid contents of ATO), respectively. (e) The ΔT_{sol} and the solar-heat shielding ability changing with the solid content of ATO; on increasing ATO, the T_s improved obviously. (f) Illustration of a testing system used to measure the solar thermal barrier effect of foils; 1 is a temperature monitor, 2 and 5 are test windows, 3 is a temperature probe, and 4 indicates infrared lamps, which are used to imitate solar light. (g) Temperature dependence on the irradiation time for blank float glass, VO_2-PU foils, and VO_2-ATO-PU foils (solid content of ATO is 30 wt.%). It was found that a VO_2-PU foil reduces the temperature by 17°C, compared with the blank float glass, while the VO_2-ATO-PU foil further reduces the temperature by 3.5°C.

coatings were applied to the blank float glass (4 mm in thickness) of a model house. The results showed that VO_2 NC coatings reduced the temperature by 17°C compared with the blank float glass, while adding ATO (30% by mass) further reduced the temperature by 3.5°C due to the enhancement of the infrared shielding ability.

3.3.1.3 VO_2 NC coatings with low emission

Emission represents the ability to transfer heat. High values of emissivity imply strong energy exchange between the NC coatings and their surroundings, which leads to poor thermal insulation ability and high energy consumption requirement for buildings. Pure, dense VO_2 films can reach an emissivity value as low as

0.11–0.26 (metal-phase) and 0.63–0.67(insulator phase) [62], but the value would increase as the density of VO_2 films decreases. For VO_2-based NC coatings with dielectric materials (SiO_2, TiO_2, PU, etc.) as the matrix, the emissivity is very high (usually >0.8) due to the low content of VO_2 NPs.

There are reports in which emissivity of VO_2 films could be reduced by coating low-emissivity materials on the VO_2 films, such as the noble metal Pt [62] or a transparent conductive oxide (AZO [63], FTO [64], etc.). These would reduce the emissivity of solution prepared VO_2 films from $\sim$0.70 to $\sim$0.2. Li et al. studied the VO_2-ATO composite coating. Their work showed that pure ATO had a emissivity of about 0.2 and the addition of VO_2 contents would increase the emissivity a little, but the composite films can serve as low-emittance coatings, especially for the lower VO_2 fractions. However, the preparation process in these methods is a little complicated and not suited for large-scale production.

Ag nanowires (Ag NWs) have been successfully prepared in recent years, which makes it possible to prepare low-emission coatings by the solution method. Luo et al. adopted this method [9], coated a Ag NW layer on top of a VO_2 NC coating, and obtained a multilayered VO_2 NC coating with low emissivity (Fig. 3.14). Compared to the results reported for VO_2/AZO, their VO_2/Pt structure samples showed obvious advantages in reducing the emissivity while maintaining the ΔT_{sol}. In the case of a single-layer VO_2 coating on quartz glass, the emissivity can be reduced to 0.21 by optimizing the coating thickness of Ag NWs, while the ΔT_{sol} remains 5.8%. This method has low operational difficulty and is suitable for large-scale production of low-emission VO_2 based NC coatings.

3.3.2 VO_2-Based NC Coatings with Infrared and Visible-Light Utilization

3.3.2.1 VO_2/hydrogel hybrid nanothermochromic material with ultrahigh solar modulation and luminous transmission

As discussed above, although VO_2 showed huge potential in TC smart windows, intrinsic restrictions made it hard to further improve

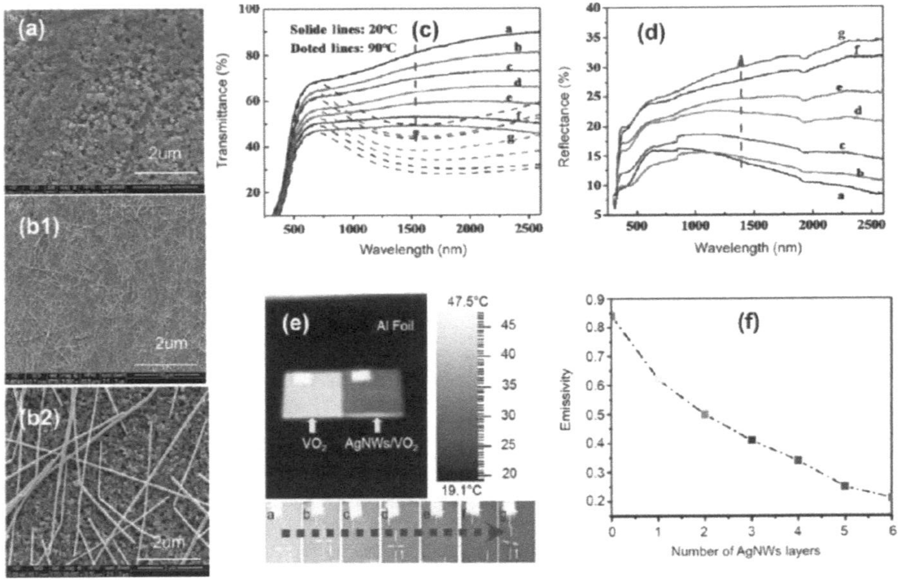

Figure 3.14 Low-emission VO_2 NC coatings with the help of Ag NWs. (a) Surface SEM image of VO_2 NC coatings prepared by dispersing VO_2 NPs into isopropanol and then coating; (b1, b2) different magnification surface SEM images of VO_2 NC coatings coated with Ag NWs; (c, d) transmittance and reflectance curves (250–2500 nm) of VO_2-Ag NW NC coatings with different layers of Ag NWs, respectively. As the Ag NWs layers increased, the transmittance of the NC coatings decreased and the reflectance increased (especially in the infrared waveband). (e) Infrared image of the VO_2-Ag NW NC coatings with different layers in (c); with the help of an infrared imager, the emissivity of the samples could also be measured. (f) The emissivity of the samples with different Ag NW layers. The emissivity decreased as the Ag NW layers increased. An emissivity of about 0.2 could be obtained with 6 layers of Ag NWs [9].

the ΔT_{sol} and T_{lum}. Since visible light occupies half of the solar energy, some of the researchers have turned to visible-light-utilizing materials to composite with VO_2 to further increase the utilization of solar energy.

Zhou et al. composited VO_2 with poly(N-isopropylacrylamide) (PNIPAm) (Figs. 3.15a and 3.15b) [11, 65]. The PNIPAm had a hydrophilic- to hydrophobic-phase transition (a lower critical solution temperature [LCST] of 32°C). Below the LCST, the coating of the PNIPAm is transparent, but when the temperature is above the LCST, the coating turns milky, blocking a large amount of light because of scattering. With this characteristic, the PNIPAm coating could reach TC properties with $\Delta T_{sol} = 25.5\%$ and $T_{lum} = 70.7\%$. The scattering strength depends on the wavelength of the incident light—the shorter the wavelength, the stronger is the scattering. Thus the transmittance changes in the PNIPAm coating through the LCST were mainly in the visible-light waveband, while the transmittance changes in the VO_2 coating were mainly in short NIR waveband. Combining these two TC materials would greatly improve the ΔT_{sol}. VO_2-PNIPAm NC coatings reported by Zhou et al. reached a dramatically high ΔT_{sol} (34.7%) while maintaining a moderate average T_{lum} (62.6%), much higher than when NC coatings have only VO_2 NPs. This novel concept of a TC NC coating opens up a new phase for TC smart windows applications.

Zhu et al. reported NC coatings of VO_2 and ionic liquid-nickel-chlorine (IL-Ni-Cl) (Figs. 3.15c and 3.15d) [10]. The IL-Ni-Cl used in their work was composed of di-(1-butyl-3-methylimidazolium)]$_2$$NiCl_4$ dissolved in an ionic liquid (1-(3-hydroxyethyl)-3-methylimidazolium tetrafluoroboride ($C_2OHmimBF_4$)). On being heated, the absorption (at 656–705 nm) of the IL-Ni-Cl complexes increased and they changed color from colorless to blue gradually. The color change mechanism is

$$\underbrace{\left[\mathrm{NiCl}_{6-x}(\mathrm{C_2OH\,min\,BF_4})_x\right]^{x-4}}_{\text{cold state, octahedral, colorless}} + (x-2)\mathrm{Cl}^- \leftrightarrow \underbrace{[\mathrm{NiCl_4}]^{2-} + x\mathrm{C_2OH\,min\,BF_4}}_{\text{hot state, terahedral, blue}}, \quad (3.4)$$

where $0 < x \leq 2$. The VO_2/IL-Ni-Cl NC coating showed a large ΔT_{sol} (26.45%) and high luminous transmittance ($T_{lum,low} = 66.4\%$, $T_{lum,high} = 43.9\%$). Since the IL-Ni-Cl changes transmittance at

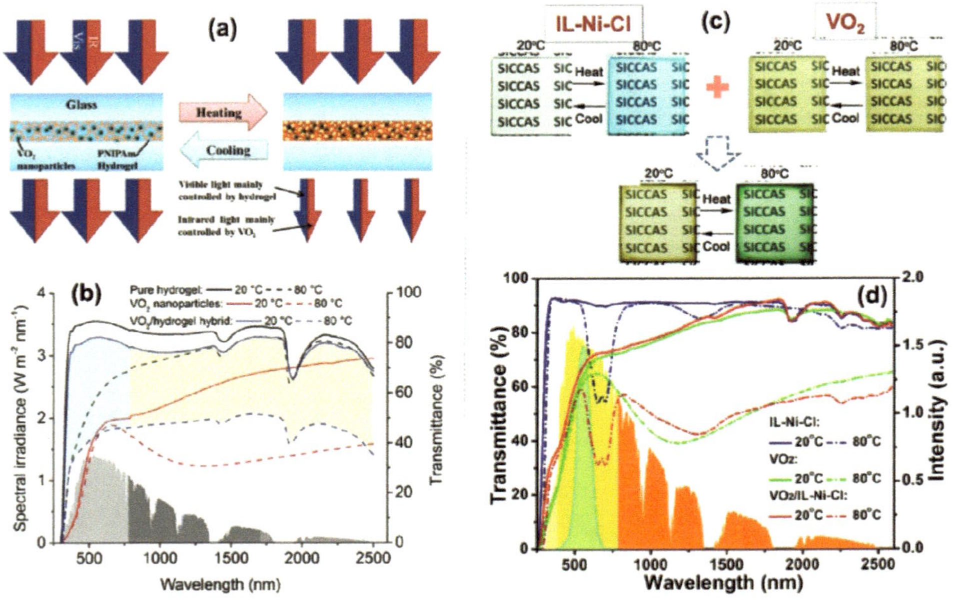

Figure 3.15 A VO_2 composite with other thermochromic materials for visible-light utilization. (a, b) A VO_2 composite with PNIPAm [65]; (a) the schema of the regulation mechanism of NC coating, and (b) the transmittance curves of pure PNIPAm, VO_2 NC, and VO_2-PNIPAm coatings at 20°C and 80°C; it is clear that the PNIPAm changes its transmittance mainly in the visible-light range, while the VO_2 does so in the infrared light; thus a combination of them would regulate the whole spectra from 250 to 2500 nm. (c, d) A VO_2 composite with IL-Ni-Cl [10]; (c) a schema of the regulation mechanism of NC coating, and (d) the transmittance curves of pure IL-Ni-Cl, VO_2 NC, and VO_2-IL-Ni-Cl NC coatings; the IL-Ni-Cl changes its transmittance round 656 and 705 nm and thus the coating has a color change; however, this would also improve the regulation range of the NC coatings.

656–705 nm, the NC coating changes its color from light brown to dark green as the temperature increases from room temperature.

Composite TC materials with visible-light adjustment ability could greatly improve the NC coatings' ability to modulate solar energy but also bring a visual change in the coating. As shown above, the VO_2/PNIPAm NC coatings would change from a transparent state to an opaque state, while the VO_2/IL-Ni-Cl NC coatings would change from light brown to dark green. This characteristic would limit their applications.

3.3.2.2 VO_2 NC coating with electrochromism-thermochromism dual-response properties

Electrochromic (EC) and TC smart windows are the mostly widely studied smart windows. But they work in different wavebands. EC smart windows mainly adjust in visible light, while TC smart windows adjust in NIR light. This offers the possibility to combine these two kinds of smart windows to build a new type of smart window with a higher ΔT_{sol}.

Yang et al. successfully prepared an EC-TC dual-response smart device using a VO_2-TiO_2 NC coating (Fig. 3.16). The NC coating was prepared by dispersing VO_2 NPs in a TiO_2 sol, which was then coated and annealed. The optimized NC coating showed a T_{lum} of 66.1% at 20°C and 59.3% at 90°C and a ΔT_{sol} of 11.2%. And by applying a voltage (5 V), it showed an EC transmittance modulation (ΔT) of 19.1% at 630 nm and the total ΔT_{sol} reached 17.1% (an increase of 52% when compared to the total ΔT_{sol} when only heat was applied). The color of the NC coating also changed from a yellowish brown to a light blue. Different from the color change of a VO_2-(IL-Ni-Cl) NC coating, the color change in this NC coating could be controlled by changing the applied voltage, offering the users more choices.

Compared to the VO_2-PNIPAm and VO_2-(IL-Ni-Cl) NC coatings, the EC-TC dual-response NC coating didn't show advantages in ΔT_{sol}. However, the EC material Yang used is TiO_2, which is not the best EC material. On using other EC materials with high EC properties (for WO_3, the ΔT at 630 nm could reach as high as 80%, while the ΔT of TiO_2 in this work is only 19.1%), the total ΔT_{sol} could be greatly improved. This implies the huge potential of the EC-TC dual-

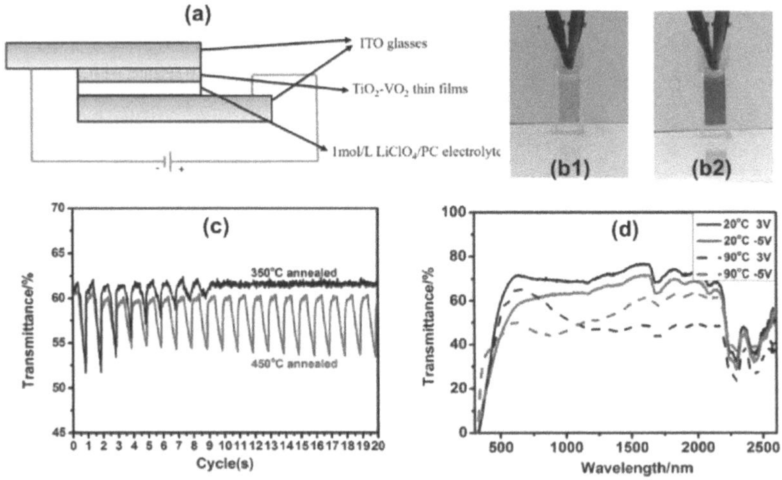

Figure 3.16 A VO_2 NC coating with EC-TC dual-response properties [66]. (a) A schema of the structure of the dual-response device. (b1, b2) Photographs of bleached and colored VO_2-TiO_2 dual-response films, (c) the cycling stability of the VO_2-TiO_2 dual-response films annealed at 350°C and 450°C, and (d) transmittance of the device at different temperatures and voltages.

response smart device in solar energy modulation, which may be an important breakthrough in energy saving smart windows.

3.3.2.3 VO_2 NC coatings for energy saving and generation

For all the smart windows mentioned above, solar energy was treat as a light or heat source and this energy was blocked by the smart window if the indoor environment didn't need it. To a certain extent, this is a waste of solar energy. NC coatings for energy saving and generation aim to improve solar utilization by using the extra solar energy to generate electricity.

Zhou et al. reported a kind of VO_2-based smart window for this purpose (Fig. 3.17a–c). The window could scatter light to the solar cell installed around the window for electricity generation. The size of the VO_2 NPs in the NC coatings was optimized so that they could scatter a part of the light and still retain their TC performance. The NC layer was sandwiched between two pieces of transparent board with a higher refractive index than the matrix of the NCs, which

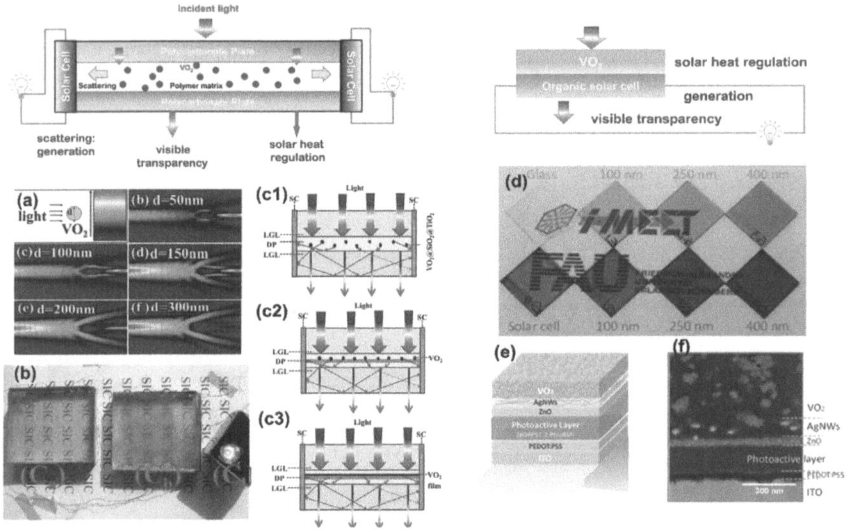

Figure 3.17 VO_2 NC coatings for energy saving and generation. (Top left) Device based on scattering [12], (top right) device that simply assembles VO_2 NC coatings with an organic film solar cell [13]. (a) The FDTD simulated scatter electric field distribution interacted with light; as the particle size increased, the scattering was enhanced. (b) Optical photo of the device. (c1–c3) Different structures of the scatter device; (c1) VO_2 pure film on quartz serving as a scattering medium, (c2) an NC coating as the scattering medium, created by dispersing $VO_2@SiO_2@TiO_2$ core/shell-shell NPs in PU and then cast in PDMS, and (c3) a smooth VO_2 thin film as the scattering medium. SC, DP, and LGL in the schema refer to solar cell, low-reflective-index medium, and light guider layer, respectively. (d) Different layers of VO_2 NC coatings and VO_2 NC coatings on solar cells, (e) Schematic architecture of the smart photovoltaic window. (f) A cross-sectional SEM image of a fabricated smart window.

was used to conduct the scattering light to the surrounding of the window, where solar cells were installed (the higher the refractive index, the better is the conduction effect of the scattered light). The ΔT_{sol} of the device could reach 7.5% and it had a T_{lum} of about 43% and showed an energy generation efficiency of 0.52%. Although the efficiency is small, the successful design and preparation of the window combining electricity generation and energy saving achieved comprehensive utilization of solar energy, which provides a new insight into resolving the energy consumption issues.

Adding VO_2 NC coatings on thin film solar cells is also an idea to obtain smart windows with electricity generation and energy saving ability. Strictly speaking, it is a kind of device that simply assembles VO_2 NC coatings with solar cells, but it is worth an introduction because of its potential application value. This method requires a high-transmittance thin film solar cell and should comprehensively consider its visible-light transmittance and energy generation efficiency. Guo et al. and Gao et al. worked together to develop the device using organic thin film solar cells (Fig. 3.17d–f) [13, 67]. This device utilized part of the visible light to generate electrical energy while regulating the infrared portion of the solar energy by VO_2. The energy generation efficiency could reach 3.07%, the ΔT_{sol} was 3.9%, and the T_{lum} was 40% [67], which is comparable to that of the smart window using the scattering mechanism above. The most important point is that both organic solar cells and VO_2 NC coatings could be prepared by printing, so it is easy to produce on a large scale.

3.4 Conclusion

VO_2-based NC coating is a new type of VO_2-based smart coating, rapidly developed in recently years, and product (VO_2-based smart foils) had been sonly developed, when product pure VO_2 thin films are still under development. The main reasons are their excellent performance in luminous transmittance (T_{lum}) and solar energy modulation ability (ΔT_{sol}) and the simple preparation process, making it suitable for large-scale production. Besides, they show advantages when composited with other functional materials (color modulation, infrared shielding, low emission, self-cleaning, wear resistant, etc.). However, since their optical performance (T_{lum} and ΔT_{sol}) has been successfully enhanced to reach a value close to the theoretical value by improving the crystallinity and dispersion in the matrix [25], their further improvement had become a huge challenge. Modifying the intrinsic properties of VO_2 (e.g., doping [55, 56] and size effect [68, 69]) has offered the possibility to overcome the theoretical limits in its optical performance, but current works have not shown the expected results; even the TC properties of

VO_2 coatings have been weakened due to the destroyed lattice integrity of VO_2 by introducing defects. Composition with materials to utilize the visible part of solar energy, which occupies half of the total energy, is currently the most effective way to increase the utilization of solar energy and improve the application value of VO_2-based smart windows. With the help of composite materials, the visible light could be adjusted or used to generate electricity so the utilization of solar energy can be greatly increased.

References

1. Liu, Z. X., Wang, H. H., Li, H. and Wang, X. M. (1998). Red shift of plasmon resonance frequency due to the interacting Ag nanoparticles embedded in single crystal SiO_2 by implantation, *Appl. Phys. Lett.*, **72**, pp. 1823–1825, doi:10.1063/1.121196.
2. He, T., Ma, Y., Cao, Y., Yang, W. S. and Yao, J. N. (2001). Enhanced electrochromism of WO3 thin film by gold nanoparticles, *J. Electroanal. Chem.*, **514**, pp. 129–132, doi:10.1016/s0022-0728(01)00616-7.
3. Naoi, K., Ohko, Y. and Tatsuma, T. (2004). TiO_2 films loaded with silver nanoparticles: control of multicolor photochromic behavior, *J. Am. Chem. Soc.*, **126**, pp. 3664–3668, doi:10.1021/ja039474z.
4. Subramanian, V., Wolf, E. E. and Kamat, P. V. (2004). Catalysis with TiO_2/gold nanocomposites. Effect of metal particle size on the Fermi level equilibration, *J. Am. Chem. Soc.*, **126**, pp. 4943–4950, doi:10.1021/ja0315199.
5. Zhou, Y., et al. (2013). Surface plasmon resonance induced excellent solar control for VO_2@SiO_2 nanorods-based thermochromic foils, *Nanoscale*, **5**, pp. 9208–9213, doi:10.1039/C3NR02221H.
6. Saeli, M., Piccirillo, C., Parkin, I. P., Ridley, I. and Binions, R. (2010). Nano-composite thermochromic thin films and their application in energy-efficient glazing, *Sol. Energy Mater. Sol. Cells*, **94**, pp. 141–151, doi:10.1016/j.solmat.2009.08.010.
7. Qureshi, U., Manning, T. D. and Parkin, I. P. (2004). Atmospheric pressure chemical vapour deposition of VO_2 and VO_2/TiO_2 films from the reaction of $VOCl_3$, $TiCl_4$ and water, *J. Mater. Chem.*, **14**, pp. 1190–1194, doi:10.1039/b316531k.
8. Qureshi, U., Manning, T. D., Blackman, C. and Parkin, I. P. (2006). Composite thermochromic thin films: (TiO_2)-(VO_2) prepared from

titanium isopropoxide, $VOCl_3$ and water, *Polyhedron*, **25**, pp. 334–338, doi:10.1016/j.poly.2005.08.018.

9. LinLin, L., et al. (2017). Investigation on preparation and properties of quasicontinous AgNWs micromesh regulated thermochromic and low emissivity VO_2 energy-efficient windows (in Chinese), *Scientia Sinica (Technologica)*, **47**, p. 6, doi:10.1360/N092016-00141.
10. Zhu, J., et al. (2016). composite film of vanadium dioxide nanoparticles and ionic liquid-nickel-chlorine complexes with excellent visible thermochromic performance, *ACS Appl. Mater. Interfaces*, **8**, pp. 29742–29748, doi:10.1021/acsami.6b11202.
11. Zhou, Y., Cai, Y. F., Hu, X. and Long, Y. (2014). Temperature-responsive hydrogel with ultra-large solar modulation and high luminous transmission for "smart window" applications, *J. Mater. Chem. A*, **2**, 13550–13555, doi:10.1039/c4ta02287d.
12. Zhou, J. D., et al. (2013). VO_2 thermochromic smart window for energy savings and generation, *Sci. Rep.*, **3**, p. 3029, doi:10.1038/Srep03029.
13. Guo, F., et al. (2015). Printed smart photovoltaic window integrated with an energy-saving thermochromic layer, *Adv. Opt. Mater.*, **3**, pp. 1524–1529, doi:10.1002/adom.201500314.
14. Gao, Y. F., et al. (2012). Enhanced chemical stability of VO_2 nanoparticles by the formation of SiO_2/VO_2 core/shell structures and the application to transparent and flexible VO_2-based composite foils with excellent thermochromic properties for solar heat control, *Energy Environ. Sci.*, **5**, pp. 6104–6110, doi:10.1039/c2ee02803d.
15. Zong Tao, Z., Hong Jie, L. and Yan Feng, G. (2015). VO_2 thermochromic smart windows: from nanoparticles to flexible foils (in Chinese), *Scientia Sinica (Technologica)*, **60**, p. 13, doi:10.1360/N972015-00580.
16. Zhang, C., Chuan Xiang, C., Hong Jie, L. and Yan Feng, G. (2016). VO_2-based thermochromic smart window: from energy savings to generation (in Chinese), *Chin. Sci. Bull.*, **61**, pp. 1661–1678, doi:10.10.1360/N972015-01403.
17. Mlyuka, N. R., Niklasson, G. A. and Granqvist, C. G. (2009). Thermochromic multilayer films of VO_2 and TiO_2 with enhanced transmittance, *Sol. Energy Mater. Sol. Cells*, **93**, pp. 1685–1687.
18. Jin, P., Nakao, S. and Tanemura, S. (1998). Tungsten doping into vanadium dioxide thermochromic films by high-energy ion implantation and thermal annealing, *Thin Solid Films*, **324**, pp. 151–158.
19. Binions, R., Hyett, G., Piccirillo, C. and Parkin, I. P. (2007). Doped and undoped vanadium dioxide thin films prepared by atmospheric pressure

chemical vapour deposition from vanadyl acetylacetonate and tungsten hexachloride: the effects of thickness and crystallographic orientation on thermochromic properties, *J. Mater. Chem.*, **17**, pp. 4652–4660, doi:10.1039/b708856f.

20. Kana, J. B. K., et al. (2008). Thermochromic VO_2 thin films synthesized by RF-inverted cylindrical magnetron sputtering, *Appl. Surf. Sci.*, **254**, pp. 3959–3963, doi:10.1016/j.apsusc.2007.12.021.
21. Kang, L. T., Gao, Y. F. and Luo, H. J. (2009). A novel solution process for the synthesis of VO_2 thin films with excellent thermochromic properties, *ACS Appl. Mater. Interfaces*, **1**, pp. 2211–2218, doi:10.1021/am900375k.
22. Liu, C., Wang, N. and Long, Y. (2013). Multifunctional overcoats on vanadium dioxide thermochromic thin films with enhanced luminous transmission and solar modulation, hydrophobicity and anti-oxidation, *Appl. Surf. Sci.*, **283**, pp. 222–226, doi:10.1016/j.apsusc.2013.06.085.
23. Zheng, J. Y., Bao, S. H. and Jin, P. (2015). TiO_2(R)/VO_2(M)/TiO_2(A) multilayer film as smart window: combination of energy-saving, antifogging and self-cleaning functions, *Nano Energy*, **11**, pp. 136–145, doi:10.1016/j.nanoen.2014.09.023.
24. Gao, Y., et al. (2012). Nanoceramic VO_2 thermochromic smart glass: a review on progress in solution processing, *Nano Energy*, **1**, pp. 221–246, doi:10.1016/j.nanoen.2011.12.002.
25. Li, S. Y., Niklasson, G. A. and Granqvist, C. G. (2010). Nanothermochromics: calculations for VO_2 nanoparticles in dielectric hosts show much improved luminous transmittance and solar energy transmittance modulation, *J. Appl. Phys.*, **108**, p. 063525, doi:06352510.1063/1.3487980.
26. Chen, Z., Cao, C., Chen, S., Luo, H. and Gao, Y. (2014). Crystallised mesoporous TiO_2 (A)–VO_2 (M/R) nanocomposite films with self-cleaning and excellent thermochromic properties, *J. Mater. Chem. A*, **2**, pp. 11874–11884.
27. Chen, Z., et al. (2014). Fine crystalline VO_2 nanoparticles: synthesis, abnormal phase transition temperatures and excellent optical properties of a derived VO_2 nanocomposite foil, *J. Mater. Chem. A*, **2**, pp. 2718–2727, doi:10.1039/c3ta14612j.
28. Chen, Z., et al. (2011). VO_2-based double-layered films for smart windows: optical design, all-solution preparation and improved properties, *Sol. Energy Mater. Sol. Cells*, **95**, pp. 2677–2684, doi:10.1016/j.solmat.2011.05.041.

29. Li, S. Y., Niklasson, G. A. and Granqvist, C. G. (2011). Nanothermochromics with VO_2-based core-shell structures: calculated luminous and solar optical properties, *J. Appl. Phys.*, **109**, p. 113515, doi:11351510.1063/1.3592350.
30. Laaksonen, K., et al. (2014). Nanoparticles of TiO_2 and VO_2 in dielectric media: conditions for low optical scattering, and comparison between effective medium and four-flux theories, *Sol. Energy Mater. Sol. Cells*, **130**, pp. 132–137, doi:10.1016/j.solmat.2014.06.036.
31. Jin, P., Xu, G., Tazawa, M. and Yoshimura, K. (2002). A VO_2-based multifunctional window with highly improved luminous transmittance, *Jpn. J. Appl. Phys., Part 2*, **41**, pp. L278–L280.
32. Kang, L., et al. (2011). Nanoporous thermochromic VO_2 films with low optical constants, enhanced luminous transmittance and thermochromic properties, *ACS Appl. Mater. Interfaces*, **3**, pp. 135–138, doi:10.1021/am1011172.
33. Cao, X., et al. (2014). Nanoporous thermochromic VO_2 (M) thin films: controlled porosity, largely enhanced luminous transmittance and solar modulating ability, *Langmuir*, **30**, pp. 1710–1715, doi:10.1021/la404666n.
34. Chen, H.-K., Hung, H.-C., Yang, T. C. K. and Wang, S.-F. (2004). The preparation and characterization of transparent nano-sized thermochromic VO_2-SiO_2 films from the sol-gel process, *J. Non-Cryst. Solids*, **347**, pp. 138–143, doi:10.1016/j.jnoncrysol.2004.07.065.
35. Lopez, R., Boatner, L. A., Haynes, T. E., Haglund, R. F. and Feldman, L. C. (2004). Switchable reflectivity on silicon from a composite VO_2-SiO_2 protecting layer, *Appl. Phys. Lett.*, **85**, pp. 1410–1412, doi:10.1063/1.1784546.
36. Zhao, L., et al. (2014). Solution-processed VO_2-SiO_2 composite films with simultaneously enhanced luminous transmittance, solar modulation ability and anti-oxidation property, *Sci. Rep.*, **4**, p. 7000, doi:10.1038/srep07000.
37. Du, J., et al. (2011). Formation and metal-to-insulator transition properties of VO_2-ZrV_2O_7 composite films by polymer-assisted deposition, *Sol. Energy Mater. Sol. Cells*, **95**, pp. 1604–1609, doi:10.1016/j.solmat.2011.01.009.
38. Song, L., et al. (2013). Characteristics of CeOx-VO_2 composite thin films synthesized by sol-gel process, *J. Mater. Sci. Mater. Electron.* **24**, pp. 3496–3503, doi:10.1007/s10854-013-1275-2.

39. Manning, T. D., Parkin, I. P., Blackman, C. and Qureshi, U. (2005). APCVD of thermochromic vanadium dioxide thin films - solid solutions $V_{2-x}M_xO_2$ (M = Mo, Nb) or composites VO_2: SnO_2, *J. Mater. Chem.*, **15**, pp. 4560–4566.
40. Liu, C., et al. (2014). VO_2/Si-Al gel nanocomposite thermochromic smart foils: largely enhanced luminous transmittance and solar modulation, *J. Colloid Interface Sci.*, **427**, pp. 49–53, doi:10.1016/j.jcis.2013.11.028.
41. Valmalette, J. C. and Gavarri, J. R. (1994). Vanadium dioxide polymer composites - thermochromic behavior and modeling of optical transmittance, *Sol. Energy Mater. Sol. Cells*, **33**, pp. 135–144.
42. Zhou, Y., et al. (2014). Microemulsion-based synthesis of $V_{1-x}W_xO_2$@SiO_2 core-shell structures for smart window applications, *J. Mater. Chem. C*, **2**, pp. 3812–3819, doi:10.1039/c3tc32282c.
43. Xiao, X., et al. (2014). A cost-effective process to prepare VO_2 (M) powder and films with superior thermochromic properties, *Mater. Res. Bull.*, **51**, pp. 6–12, doi:10.1016/j.materresbull.2013.11.051.
44. Lu, Y. F., Zhou, S. X., Gu, G. X. and Wu, L. M. (2013). Preparation of transparent, hard thermochromic polysiloxane/tungsten-doped vanadium dioxide nanocomposite coatings at ambient temperature, *Thin Solid Films*, **534**, pp. 231–237, doi:10.1016/j.tsf.2013.02.130.
45. Li, D. B., et al. (2014). Hydrothermal synthesis of Mo-doped VO_2/TiO_2 composite nanocrystals with enhanced thermochromic performance, *ACS Appl. Mater. Interfaces*, **6**, pp. 6555–6561, doi:10.1021/am500135d.
46. Zhang, H., et al. (2015). A cost-effective method to fabricate VO_2 (M) nanoparticles and films with excellent thermochromic properties, *J. Alloys Compd.*, **636**, pp. 106–112, doi:10.1016/j.jallcom.2015.01.277.
47. Madida, I. G., et al. (2014). Submicronic VO_2-PVP composites coatings for smart windows applications and solar heat management, *Sol. Energy*, **107**, pp. 758–769, doi:10.1016/j.solener.2014.06.025.
48. Cao, C. X., Gao, Y. F. and Luo, H. J. (2008). Pure single-crystal rutile vanadium dioxide powders: synthesis, mechanism and phase-transformation property, *J. Phys. Chem. C*, **112**, pp. 18810–18814, doi:10.1021/jp8073688.
49. Shen, N., et al. (2014). The synthesis and performance of Zr-doped and W-Zr-codoped VO_2 nanoparticles and derived flexible foils, *J. Mater. Chem. A*, **2**, pp. 15087–15093, doi:10.1039/C4TA02880E.
50. Gao, Y. F., et al. (2012). VO_2-Sb:SnO_2 composite thermochromic smart glass foil, *Energy Environ. Sci.*, **5**, pp. 8234–8237, doi:10.1039/c2ee21119j.

51. Tazawa, M., Jin, P. and Tanemura, S. (1998). Optical constants of $V_{1-x}W_xO_2$ films, *Appl. Opt.*, **37**, pp. 1858–1861.
52. Zhang, Z. T., et al. (2010). Thermochromic VO_2 thin films: solution-based processing, improved optical properties, and lowered phase transformation temperature, *Langmuir*, **26**, pp. 10738–10744, doi:10.1021/la100515k.
53. Jin, P., Xu, G., Tazawa, M. and Yoshimura, K. (2003). Design, formation and characterization of a novel multifunctional window with VO_2 and TiO_2 coatings, *Appl. Phys. A*, **77**, pp. 455–459.
54. Granqvist, C. G., Lansåker, P. C., Mlyuka, N. R., Niklasson, G. A. and Avendaño, E. (2009). Progress in chromogenics: new results for electrochromic and thermochromic materials and devices, *Sol. Energy Mater. Sol. Cells*, **93**, pp. 2032–2039, doi:10.1016/j.solmat.2009.02.026.
55. Zhou, J. D., et al. (2013). Mg-doped VO_2 nanoparticles: hydrothermal synthesis, enhanced visible transmittance and decreased metal-insulator transition temperature, *Phys. Chem. Chem. Phys.*, **15**, pp. 7505–7511, doi:10.1039/c3cp50638j.
56. Dai, L., et al. (2013). F-doped VO_2 nanoparticles for thermochromic energy-saving foils with modified color and enhanced solar-heat shielding ability, *Phys. Chem. Chem. Phys.*, **15**, pp. 11723–11729, doi:10.1039/c3cp51359a.
57. Mlyuka, N. R., Niklasson, G. A. and Granqvist, C. G. (2009). Mg doping of thermochromic VO_2 films enhances the optical transmittance and decreases the metal-insulator transition temperature, *Appl. Phys. Lett.*, **95**, p. 171909, doi:17190910.1063/1.3229949.
58. Kana, J. B. K., et al. (2010). Thermochromic nanocrystalline Au-VO_2 composite thin films prepared by radiofrequency inverted cylindrical magnetron sputtering, *Thin Solid Films*, **518**, pp. 1641–1647, doi:10.1016/j.tsf.2009.11.074.
59. Kana, J. B. K., et al. (2008). Combined thermochromic and plasmonic: optical responses in novel nanocomposite Au-VO_2 films prepared by RF inverted cylindrical magnetron sputtering, in *Laser and Plasma Applications in Materials Science*, Amara, E. H., Boudjemai, S. and Doumaz, D. eds., Vol. 1047, *AIP Conference Proceedings* (American Institute of Physics), pp. 119–122.
60. Cavanna, E., Segaud, J. P. and Livage, J. (1999). Optical switching of Au-doped VO_2 sol-gel films, *Mater. Res. Bull.*, **34**, pp. 167–177, doi:10.1016/s0025-5408(99)00017-3.

61. Maaza, M., Nemraoui, O., Sella, C., Beye, A. C. and Baruch-Barak, B. (2005). Thermal induced tunability of surface plasmon resonance in Au-VO_2 nano-photonics, *Opt. Commun.*, **254**, pp. 188–195, doi:10.1016/j.optcom.2004.08.056.
62. Kang, L., et al. (2010). Pt/VO_2 double-layered films combining thermochromic properties with low emissivity, *Sol. Energy Mater. Sol. Cells*, **94**, pp. 2078–2084, doi:10.1016/j.solmat.2010.06.023.
63. Kang, L. T., et al. (2011). Thermochromic properties and low emissivity of ZnO:Al/ VO_2 double-layered films with a lowered phase transition temperature, *Sol. Energy Mater. Sol. Cells*, **95**, pp. 3189–3194, doi:10.1016/j.solmat.2011.06.047.
64. Zhang, Z. T., et al. (2011). Solution-based fabrication of vanadium dioxide on F:SnO_2 substrates with largely enhanced thermochromism and low-emissivity for energy-saving applications, *Energy Environ. Sci.*, **4**, pp. 4290–4297, doi:10.1039/c1ee02092g.
65. Zhou, Y., Cai, Y., Hu, X. and Long, Y. (2015). VO_2/hydrogel hybrid nanothermochromic material with ultra-high solar modulation and luminous transmission, *J. Mater. Chem. A*, **3**, pp. 1121–1126, doi:10.1039/c4ta05035e.
66. Yang, F., et al. (2016). Crystallized TiO_2 (A)-VO_2 (M/R) nanocomposite films with electrochromism-thermochromism dual-response properties, *RSC Adv.*, **6**, pp. 32176–32182, doi:10.1039/c6ra05700d).
67. Guo, F., et al. (2015). Printed smart photovoltaic window integrated with an energy-saving thermochromic layer, *Adv. Opt. Mater.*, **3**, pp. 1524–1529, doi:10.1002/adom.201500314.
68. Dai, L., Cao, C. X., Gao, Y. F. and Luo, H. J. (2011). Synthesis and phase transition behavior of undoped VO_2 with a strong nano-size effect, *Sol. Energy Mater. Sol. Cells*, **95**, pp. 712–715, doi:10.1016/j.solmat.2010.10.008.
69. Lopez, R., Haynes, T. E., Boatner, L. A., Feldman, L. C. and Haglund, R. F. (2002). Size effects in the structural phase transition of VO_2 nanoparticles, *Phys. Rev. B*, **65**, p. 224113.

Chapter 4

Antireflection for the Performance of VO_2 Thermochromic Thin Films

Xiudi Xiao
Guangzhou Institute of Energy Conversion, Chinese Academy of Sciences, Guangzhou 510640, China
xiaoxd@ms.giec.ac.cn

4.1 The Principle of Antireflection

The optical properties of thin films result from the transportation of light in the multilayer medium. According to the Fresnel equation, antireflection (AR) coatings are utilized to suppress Fresnel reflection between different media and help photons to enter the thin films. In the 19th century, the concept of AR coatings was proposed by Lord Rayleigh when he found that a piece of glass with a tarnished surface can increase the transmittance of incident light [1]. Joseph Fraunhofer developed the first antireflective coating on glass by acid etching [2]. There are at least two aims of AR coatings on VO_2 thin films: suppressing reflectance and shifting the transmittance peak toward the vicinity of the wavelength with the maximum luminous efficiency for photopic vision (about 555 nm)

Vanadium Dioxide-Based Thermochromic Smart Windows
Edited by Yi Long and Yanfeng Gao

ISBN 978-981-4877-06-0 (Hardcover), 978-1-003-17731-9 (eBook)
www.jennystanford.com

without weakening the thermochromic property. Factually, there are several methods for suppressing the reflection: single-layer AR coatings, double-layer AR coatings, multilayer AR coatings, gradient AR coatings, and nanostructure AR coatings. Generally, AR is based on the substrate and the number of AR layers is the thin films on the substrate. As for AR coatings on VO_2 thin films, the layer number of the AR coating is the film upper or under the VO_2 thin film.

4.2 Antireflection on a VO_2 Thin Film

4.2.1 Single-Layer and Double-Layer AR Coatings

The AR effect can be achieved by light destructive interference. When the effective optical thickness is an integral multiple of quarter wavelength, destructive interference can be observed. When a single-layer AR coating is designed, the antireflective layer is sandwiched between air and substrate. For this layer structure, the refractive index of the intermediate film should be equal to $\sqrt{n_0 n_\mathrm{s}}$ in order to attain minimized reflectance, as dictated by the following equation:

$$R = \left(\frac{n_0 n_\mathrm{s} - n_\mathrm{i}^2}{n_0 n_\mathrm{s} + n_\mathrm{i}^2}\right)^2 \tag{4.1}$$

where R is reflectance and n_0, n_s, and n_i are the refractive indexes of air, substrate, and intermediate films, respectively. To fulfill Eq. 4.1, the materials selected for single-layer AR coatings are usually dielectrics with proper n_i.

In a single-layer AR coating, light destructive interference appears at the location of the integral multiple of quarter wavelength, while the reflection is still high in the other wavelengths. To further improve the transmission in a wide-wavelength region, double-layer or multilayer AR coatings are used. The quarter-quarter-waved structure and the half-quarter-waved structure are the most typical and simple models for double-layer AR coatings. According to the double-layer model based on the optical admittance recursive method [3], the refractive index based on the quarter-quarter-waved structure should fulfill the following equation:

$$n_2 = n_1\sqrt{\frac{n_s}{n_0}} \tag{4.2}$$

Here, n_0, n_s, n_1, and n_2 are the refractive indexes of air, substrate, the first antireflective layer, and the second antireflective layer, respectively. For VO_2 thin films on a glass substrate, the AR layer is meant to reduce the reflection of the VO_2 surface and not that of the substrate surface, so the AR coatings of VO_2 thin films begin with a double-layer model. According to Eq. 4.2, the refractive indexes of a glass substrate (K9) and VO_2 are about 1.52 and 3.14 [4] at a wavelength of 550 nm, respectively. Thus, the refractive index of an AR coating is around 2.5. By adjusting the thickness and the refractive index, the reflectance minima can be achieved.

The first calculation of transmission enhancement as the effect of a quarter wavelength AR coating on a VO_2 thin film of a specific refractive index (n_i) and thickness was done by Babulanam [5]. According to the calculation, the experimental enhancement of transmission in a VO_2 thin film with a SiO_2 AR coating was confirmed but the thermochromic property of VO_2 was locked by the SiO_2 AR coating. Though Lee and Cho [6] fabricated SiO_2-coated VO_2 thin films with increased luminous transmittance without deterioration of the thermochromism, SiO_2 is not the best AR material for VO_2 AR. On the basis of the transfer-matrix method [3], the refractive index (shown in Table 4.1) and the thickness of the AR coating were optimized by Xu et al. [7]. The calculation demonstrated that the optimal n_i value changes with thickness and the highest transmission, about 55%, was obtained at $n_i = 2.2$ with $d = 55$ nm for 50 nm VO_2. The optimal values of refractive index and thickness of AR coatings are shown in Fig. 4.1. According to the contour map, when $2 \leq n_i \leq 2.4$ and 50 nm $\leq d \leq$ 60 nm, the maximum luminous transmission is about 55%, which is much higher than that of VO_2 without an AR coating (32%). The experimental result of the structure of 50 nm VO_2 on quartz glass using ZrO_2 with $n = 2.2$ as an AR coating was also investigated. The structure, ZrO_2 (56 nm)/VO_2 (50 nm)/quartz, was optimized by sputtering and the luminous transmission was up to 50.5%. The infrared (IR) switching properties did not deteriorate much.

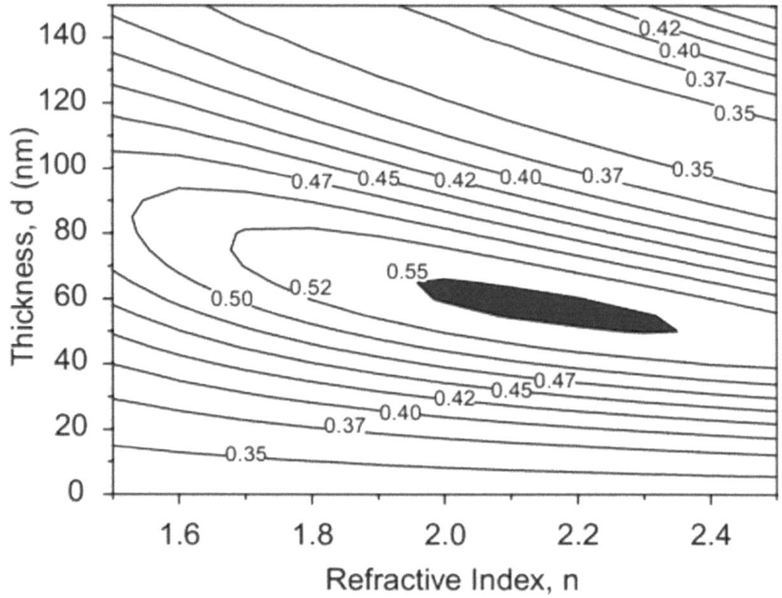

Figure 4.1 Contour map of calculated integrated luminous transmittance as a function of refractive index *n* and thickness *d* of an AR coating for 50 nm VO_2 on quartz glass. The shadowed area indicates the optimal values for *n* and *d* [7].

On the basis of the typical double-layer model, Chen et al. [8] designed and prepared the quarter-quarter-waved structure, the half-quarter-waved structure, and the quarter-3-quarter-waved structure by the solution method. The half-quarter-waved structure of the TiO_2/VO_2 double-layered thin film showed the largest improvement in luminous transmission (from 40.3% to 61.5%). By adjusting the position of AR peaks, solar modulation of up to 15.1% can be achieved.

4.2.2 Multilayer Antireflection Coating

Though a single-layer AR coating is confirmed to improve the luminous transmission and thermochromic properties, the luminous transmission is still low and the thermochromic properties are less than what is desired for practical application. Introducing

Table 4.1 Material candidates as single-layer AR coatings for a 50 nm thick VO_2 film

Film	n (at $\lambda = 0.55$ μm)
ZnS, Nb_2O_5	2.35
TiO_2	2.2–2.5
ZrO_2, HfO_2, Ta_2O_5, CeO_2	2.2
Si_3N_4, Sc_2O_3, ZnO	2
SnO_2, In_2O_3, ITO, Y_2O_3	1.95

Source: [7]

oxidation resistance and multifunctionality will close this gap, for which multilayer AR coatings with structures based on specific requirements are being investigated.

On the basis of VO_2 on glass, Jin et al. [9] proposed the sandwiched structure $TiO_2/VO_2/TiO_2$, taking into account AR and photocatalytic properties. According to calculations, the maximum luminous transmission can be obtained for TiO_2 (25 nm)/VO_2(50 nm)/TiO_2(25 nm), and sandwiched thin films with a luminous transmission of about 57.6% was confirmed by sputtering. The phase transition temperature was also reduced by about 1°C. The sandwiched thin film structure a-SiO_x/VO_2/a-SiO_x proposed by Kakiuchida et al. [10] enhanced the transmission sharply through the broad wavelength region. However, the luminous transmission was depressed infinitely, as shown in Fig. 4.2.

Beyond the sandwiched structure $TiO_2/VO_2/TiO_2$, Mlyuka et al. [11] proposed the five-layer structure $TiO_2/VO_2/TiO_2/VO_2/TiO_2$. The transmission around the solar spectral region increased sharply, accompanied by an enhancement in thermochromic properties, as shown in Fig. 4.3. For enhancing the thermal stability and production efficiency, Zhan et al. [12] proposed the seven-layer structure $SiN_x/NiCrO_x/SiN_x/VO_x/SiN_x/NiCrO_x/SiN_x$. The luminous transmission was improved by 30% (from 24.11% to 32.7%) and the solar modulation by 50% (from 12.75% to 18.02%) compared with those of a VO_2 single-layer thin film.

Due to the dense thin film prepared by sputtering, the refractive index of VO_2 is nearly constant. Though a variety of sandwiched structures have been prepared, such as TiO_2(R)/VO_2/TiO_2(A)

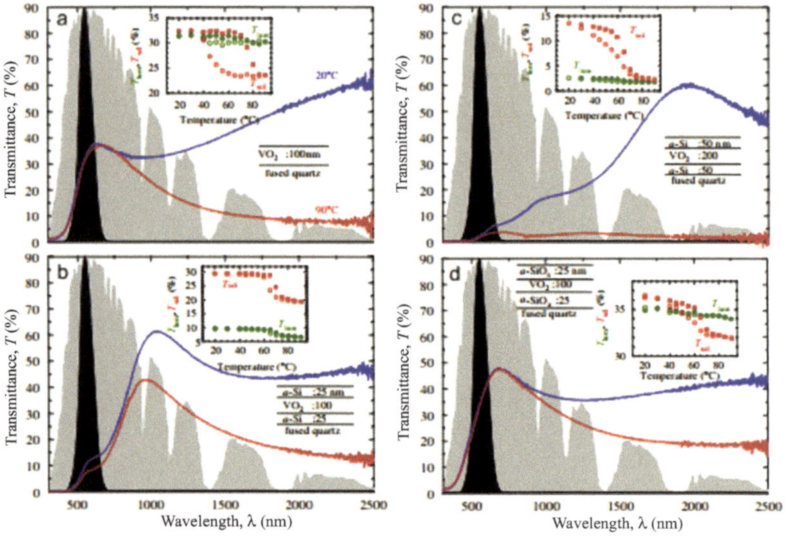

Figure 4.2 Optical transmittance spectra of VO_2 single and SVS multilayers deposited on a fused-quartz substrate at temperatures of 20°C and 90°C, as shown by blue and red curves, respectively. The black and gray areas indicate the normalized values of the luminous efficiency function and the spectral solar irradiation, respectively. The green and red circles show the temperature dependence of the luminous and solar transmittances, which were estimated using Eqs. 4.1 and 4.2, respectively, where the closed and open circles are the measurements during the rise and drop in temperature. (a) Single-layer VO_2, (b) VO_2 sandwiched between a-Si layers, (c) VO_2 sandwiched between a-Si layers where the thicknesses of the respective layers were designed to be two times that of the layers schematically shown in (b), and (d) VO sandwiched between a-SiO_x layers with $x = 0.3$ [10].

[13], $WO_3/VO_2/WO_3$ [14], $ZrO_2/V_{1-x}W_xO_2/ZrO_2$ [15] (shown in Fig. 4.4), $SiO_2/VO_2/TiO_2$ [16], and $SiN_x/VO_2/SiN_x$ [17], the luminous transmission was not improved beyond 60%. The different materials were chosen mainly for multifunctionality [13, 14, 17] and stability [13, 17].

4.2.3 Gradient AR Coatings

Reflection results from the different refractive indexes at the interface between different dielectric mediums. If the difference

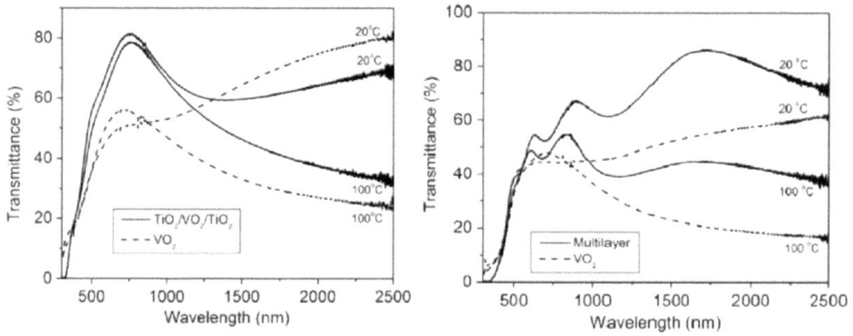

Figure 4.3 Spectral normal transmittance for three-layer $TiO_2/VO_2/TiO_2$ films (left panel) and five-layer $TiO_2/VO_2/TiO_2/VO_2/TiO_2$ films (right panel), as specified in the main text, at two temperatures, one lying below the metal–insulator transition temperature and the other above this temperature. The data are compared with those for corresponding VO_2 films [11].

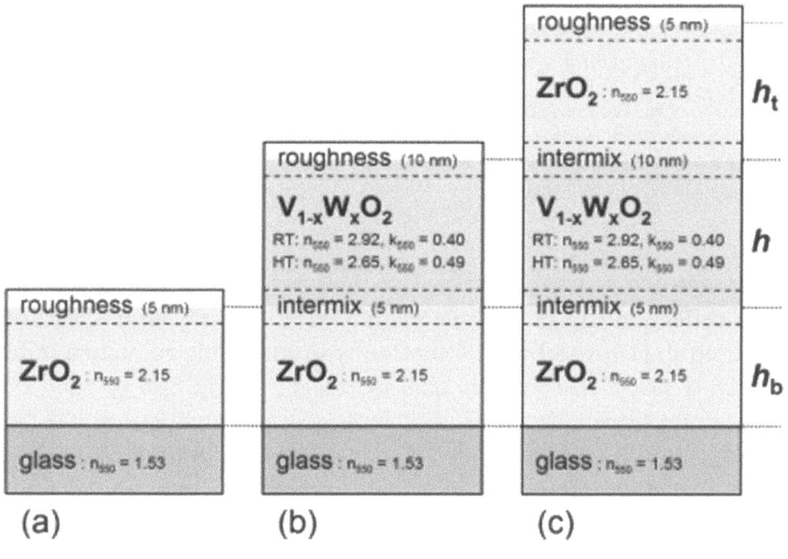

Figure 4.4 An optical model representing the individual stages of the coating preparation and used to predict the coating transmittance. Half of each surface roughness layer (top layer + void) contributes to the thickness of the corresponding top layer. The thickness of each expected intermix layer is given by a typical roughness of the layer below it and contributes to the thicknesses of both corresponding layers. Selected optical constants, in the case of $V_{1-x}W_xO_2$ both below and above T_{tr} (RT and HT, respectively), are shown as well [14].

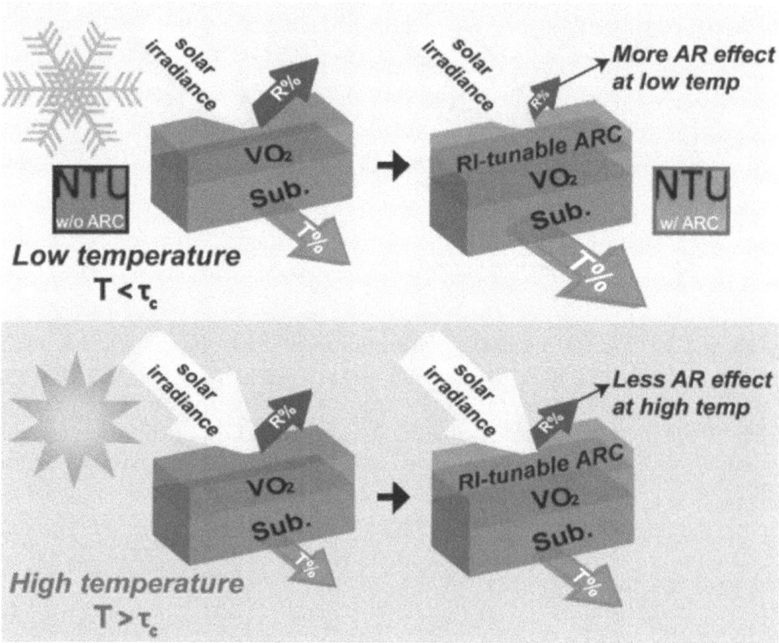

Figure 4.5 Scheme of the RI-tunable ARC applied on the VO_2 substrate [19]. Copyright © 2018 Elsevier.

between the refractive indexes of the interfaces can be weakened, the reflection can be sharply decreased. Using the sol-gel method, Zhu et al. [18] used an AR coating with a tunable refractive index (1.16–1.42 at 700 nm) on the surface of a VO_2 film to tune the luminous transmission to 50% with a solar modulation of 8.5%. Due to the VO_2 film deposited by sputtering, the difference between the refractive indexes of VO_2 and AR coatings is large. To make the interface smooth, an AR coating with a tunable refractive index (1.47–1.92 at 550 nm) on a VO_2 composite thin film is prepared (as shown in Fig. 4.5). The best-performing coatings could maximize the solar modulation (from 15.7% to 18.9%) and increase the luminous transmission from 39% to 44% simultaneously [19]. Apart from adjusting the concentration of the AR coating, a porous AR coating is another way to acquire a tunable refractive index, as shown in Fig. 4.6. A mesoporous SiO_2 layer with a refractive index ranging

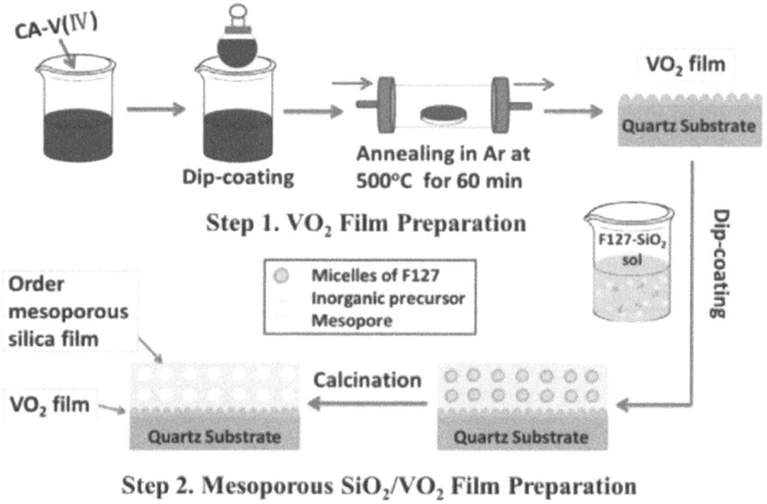

Figure 4.6 The preparation procedure of a mesoporous SiO_2/VO_2 film [20].

from 1.243 to 1.354 was used to reduce the surface reflection of a porous VO_2 thin film. A luminous transmission of up to 80% can be achieved, with a solar modulation of 10.2% as well as a reduced phase transition temperature at 53.8°C [20].

4.2.4 Nanostructure AR Coating

Apart from the AR layer on the VO_2 thin film, decreasing the refractive index of the VO_2 thin film itself is another way to reduce the surface reflection of a VO_2 thin film. By adjusting the porosity in VO_2 thin films, Sun et al. [21] prepared sculptured VO_2 thin films with slanted columnar structure and pores, as shown in Fig. 4.7. The luminous transmission improved by 26% compared with that of normal VO_2 thin films. Compared with physical vapor deposition, the solution method is more flexible when adjusting the refractive index of VO_2 thin films. Using polyvinylpyrrolidone as the pore forming material, Kang et al. [22] prepared VO_2 porous thin films with low optical constants and improved the luminous transmission with superior solar modulation Fig. 4.8). On the basis of effective medium theory, Li et al. [23] calculated that dilute

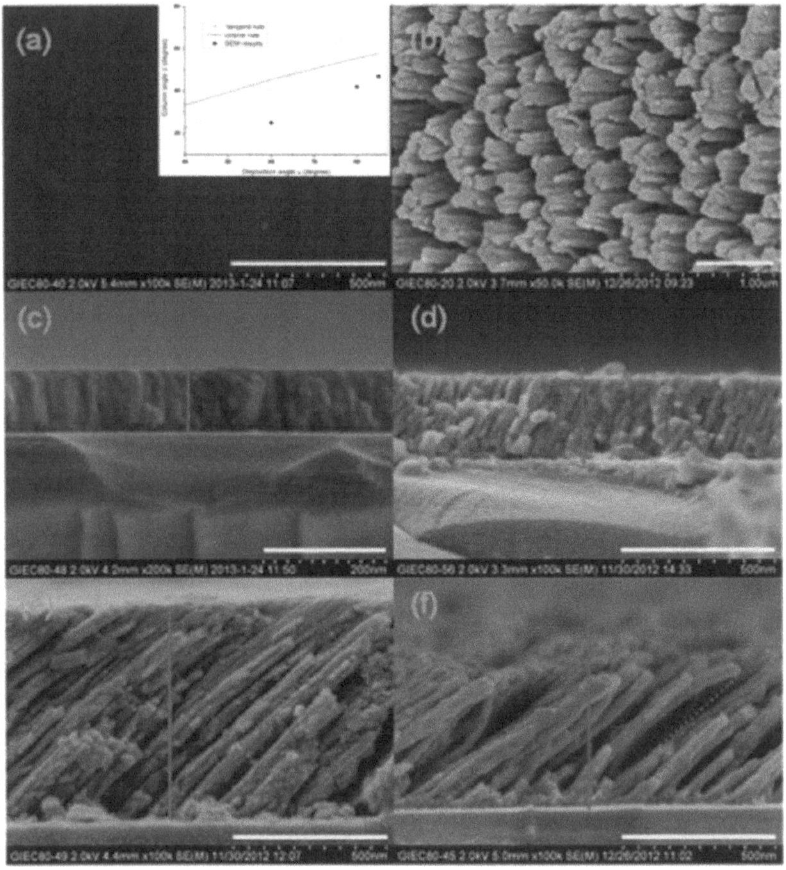

Figure 4.7 SEM surface and cross-sectional images of as-deposited V_2O_5 sculptured thin films deposited at different oblique angles. (a, c) 0°; (b, f) 85°; (d) 60°; (e) 80°. The scale bar in Fig. 4.7c is 200 nm. The scale bar in others is 500 nm [21].

composites with VO_2 nanoparticles embedded in transparent hosts can improve the luminous transmission and solar modulation. Accordingly, composite VO_2 thin films with polymer matrixes were produced. Gao et al. [24] prepared a VO_2@SiO_2/polyurethane coating on a polyethylene terephthalate substrate (Fig. 4.9). Due to the homogeneous dispersion and protection by a SiO_2 shell, the foil showed superior luminous transmittance and solar modulation

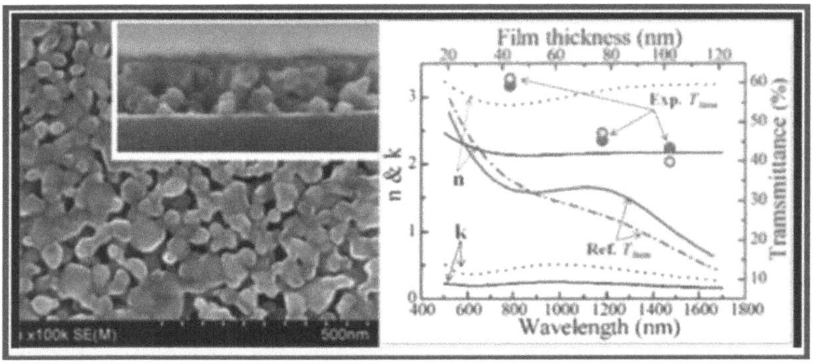

Figure 4.8 Nanoporous thermochromic VO_2 films with low optical constants and tunable thicknesses, which exhibit high integrated luminous transmittance ($T_{lum,l} = 43.3\%$, $T_{lum,h} = 39.9\%$) and solar modulation ($\Delta T_{sol} = 14.1\%$, from $T_{sol,l} = 42.9\%$ to $T_{sol,h} = 28.8\%$), have been prepared by polymer-assisted deposition [22]. Copyright © 2011 American Chemical Society.

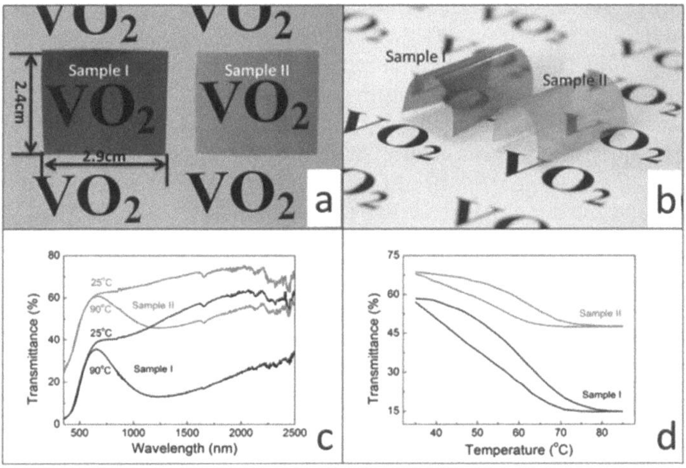

Figure 4.9 (a, b) Photographs of sample films at room temperature, (c) optical transmittance spectra of sample films in UV-visible-NIR range, and (d) thermal hysteresis of sample films at a wavelength of 1500 nm [24].

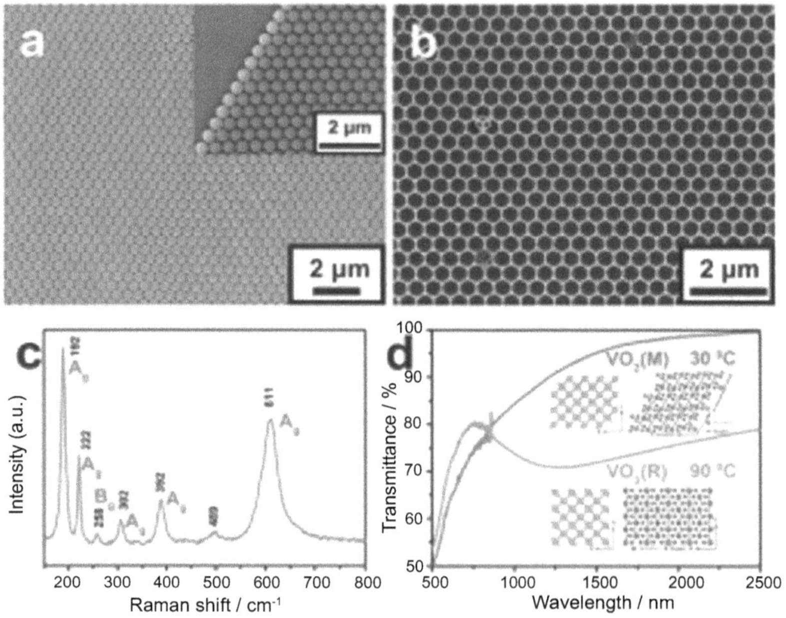

Figure 4.10 SEM images of (a) an MCC template with an edge view inset, and (b) periodic ordered porous VO_2(M) film. (c) Raman spectra of a periodic ordered porous VO_2(M) film. (d) Optical transmittance spectra of the final film. (Inset) The crystal structure of monoclinic VO_2(M) projected along [100], the supercell structure projected along [100], and the supercell structure projected along [010]; vanadium atoms form the zigzag-type chains along the c_R axis of the structure; the crystal structure of rutile VO_2(R), the crystal structure of tetragonal rutile VO_2(R) projected along [001], and the supercell structure projected along [010]; vanadium atoms form chains parallel to the c axis of the structure [25].

with special antioxidation. This is the first experimental result on a flexible VO_2 thin film. Apart from the irregular porous and composite VO_2 thin films, the periodic structure is another way to reduce the refractive index of VO_2 thin films. With monolayer colloidal polystyrene spheres as the templates, the hexagonal array of VO_2 thin films was first investigated by Zhou et al. [25] (Fig. 4.10). The maximum visible transmission is up to 80%. However, due to the crossover of the transmission curve, the solar modulation

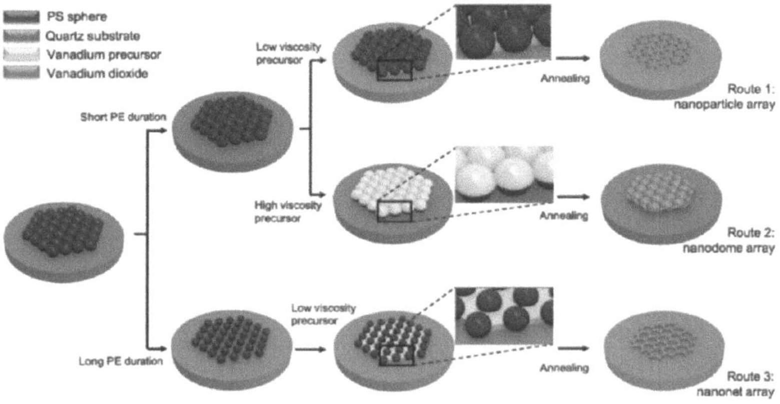

Figure 4.11 Effect of synthesis conditions on the morphology evolution [27]. Route 1: Nanoparticle arrays are prepared via short PE duration and low viscosity precursor. Route 2: Nanodome arrays are produced, using high viscosity precursor that can stick on the tops of PS spheres. Route 3: Nanonet arrays are fabricated by controlling the interval space between adjacent spheres via prolonging PE duration.

is in a moderate range (~7.5%). Though both Cao et al. [26] and Ke et al. [27] used polystyrene spheres as the template (shown in Fig. 4.11), the transmittance was not up to the level mentioned above.

Different from the sacrificial template method, 2D photonic crystal structures of VO_2@SiO_2 thin films using hexagonal-close-packed SiO_2 spheres as the template were prepared by Ke et al. [28] (Fig. 4.12). The transmission peaks could be tuned by the period of SiO_2 spheres, and an undesirable yellow-brown color was obtained. By controlling the reaction process, ordered VO_2 honeycomb structures without a template have been fabricated by a dual-phase transformation technique, as shown in Fig. 4.13. The visible light transmittance at 700 nm is up to 95.4%, with a solar modulation ability of 5.5% [29].

Tables 4.2–4.5 provide a summary of the optical properties when different methods are used to put AR coatings on VO_2 thin films.

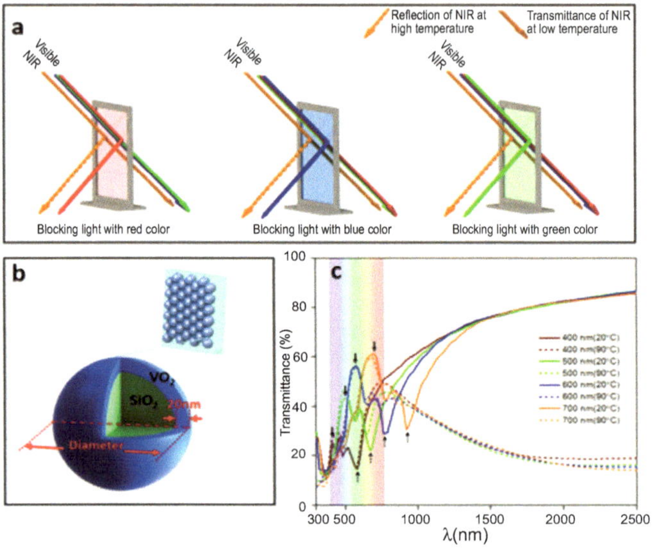

Figure 4.12 (a) Illustration of how a photonic VO_2 smart window works, (b) illustration of designed structures for simulation, and (c) simulated transmittance spectrum, where the transmittance peaks and troughs are indicated by the solid and dashed arrows, respectively. The colorful background in (c) denotes the visible spectrum from 370 to 770 nm [28].

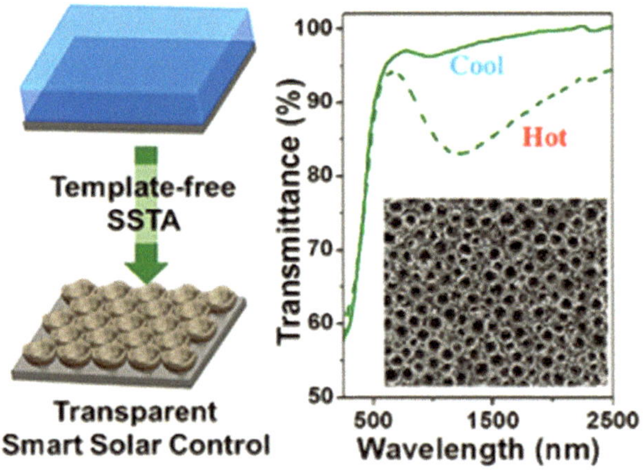

Figure 4.13 The $VO_2(M)$ honeycomb-structured coatings, which have excellent visible light transmittance at 700 nm (T_{vis}) up to 95.4% with decent solar modulating ability (ΔT_{sol}) of 5.5%, has been developed [29]. Copyright © 2017 American Chemical Society.

Table 4.2 AR coatings on VO_2 continuous thin films

Double-layer films	Ref.	Fabrication techniques		$T_{lum(avg)}$%		ΔT_{sol}		ΔT%@2000 nm	
		VO_2 film	ARC	w/o ARC	w/ ARC	w/o ARC	w/ ARC	w/o ARC	w/ ARC
TiO_2/VO_2 (2002)	[30]	RF sputtering	RF sputtering	32	47↑	4.4	7	53↑	39↓
ZrO_2/VO_2 (2004)	[7]	RF sputtering	RF sputtering	32	51↑	—	—	45	39↓
SiO_2/VO_2 (2011)	[8]	Sol-gel and thermal annealing	Sol-gel	40	51↑	5.9	7.2↑	54	48↓
TiO_2/VO_2 (2011)	[8]	Sol-gel and thermal annealing	Sol-gel	42	47↑	13.2	15.1↑	56	46↓
Si-Al/VO_2 (2013)	[31]	Sol-gel and thermal annealing	Sol-gel	44	48↑	6	7.7↑	14	18↑
CeO_2/VO_2 (2013)	[32]	Pulse laser deposition	Pulse laser deposition	30	31↑	11.6	15.3↑	48	46↓

Table 4.3 AR coatings on VO_2 continuous thin films

		Fabrication techniques		$T_{lum(avg)}$%		ΔT_{sol}		ΔT%@2000 nm	
Multilayer films	**Ref.**	**VO_2 film**	**ARC**	**w/o ARC**	**w/ ARC**	**w/o ARC**	**w/ ARC**	**w/o ARC**	**w/ ARC**
$TiO_2/VO_2/TiO_2$ (2003)	[9]	RF sputtering	RF sputtering	31	56↑	3.9	2.9↓	51	36↓
a-SiO_x/VO_2/a-SiO_x $x = 0.3$ (2008)	[10]	RF sputtering	RF sputtering	31	35↑	6.5	4.5↓	43	20↓
$TiO_2/VO_2/TiO_2$ $/VO_2/TiO_2$ (2009)	[11]	Reactive DC magnetron sputtering	Reactive DC magnetron sputtering	41	44↑	6.7	12.1↑	39	38↓
TiO_2(R)/VO_2(M)/ TiO_2 (A) (2015)	[13]	RF sputtering	RF sputtering	32.4		8.6	10.2	24.6	49.5
$WO_3/VO_2/WO_3$ (2016)	[14]	RF sputtering	RF sputtering	37.2	55.4↑	6.1	4.3↓	44.4	34.7↓
$VO_2/SiO_2/TiO_2$ (2016)	[16b]	Chemical vapor deposition	Chemical vapor deposition	14.15	44.24↑	3.14	7.63↑	—	—
$SiN_x/VO_x/SiN_x$ (2018)	[12]	Reactive magnetron sputtering	Reactive magnetron sputtering	24.11	29.21↑	12.75	15.65↑	—	—
$SiN_x/NiCrO_x/SiN_x/$ $VO_x/SiN_x/$ $NiCrO_x/SiN_x$ (2018)	[12]	Reactive magnetron sputtering	Reactive magnetron sputtering	24.11	32.7↑	12.75	18.02↑	—	—

Table 4.4 AR coatings on VO_2 continuous/porous thin films

Gradient AR coating	Ref.	Fabrication techniques: VO_2 film	Fabrication techniques: ARC	$T_{lum(avg)}$% w/o ARC	$T_{lum(avg)}$% w/ ARC	ΔT_{sol} w/o ARC	ΔT_{sol} w/ ARC	ΔT%@2000 nm w/o ARC	ΔT%@2000 nm w/ ARC
Mesoporous SiO_2/porous VO_2 (2017)	[20]	Sol-gel method and thermal annealing	Sol-gel method and thermal annealing	69.6	80↑	12.6	10.2↑	32.1	29↑
Si-Al/VO_2 (2018)	[19]	Sol-gel method	Sol-gel method and thermal annealing	39.9	46.3↑	15.7	18.9↑	30.8	36↑
SiO_2/VO_2 (2018)	[18]	Sputtering	Sol-gel method	37.06	50.11↑	6.25	8.51↑	—	—

Table 4.5 Nanostructure AR coating on VO_2 thin films

Nanostructure AR coating	Ref.	Fabrication techniques: VO_2 film	Fabrication techniques: ARC	$T_{lum(avg)}$% w/o ARC	$T_{lum(avg)}$% w/ ARC	ΔT_{sol} w/o ARC	ΔT_{sol} w/ ARC	ΔT%@2000 nm w/o ARC	ΔT%@2000 nm w/ ARC
VO_2 sculptured thin films (2013)	[21]	Electron beam evaporation	Porous	25.1	40↑	10	4.6↓	44.7	35.6↓
Porous VO_2 thin films (2011)	[22]	Sol-gel method and thermal annealing	Porous	—	43.3	—	14.1	—	—
VO_2 composite foil (2012)	[24]	Sol-gel method	Composite	—	55.3	—	7.5	—	—
Periodic porous VO_2 (2013)	[25]	Colloidal lithography	PS template	—	81%	—	—	—	23%
Patterned VO_2 (2017)	[27]	Sol-gel method and thermal annealing	PS template	—	61.5	—	7.2	—	32.6
VO_2-SiO_2 patterned thin film (2017)	[26]	Sol-gel method and thermal annealing	PS template	46.1	55.6↑	3.03	8.42↑	—	—
SiO_2-VO_2 photonic crystals (2016)	[28]	Sol-gel method and thermal annealing	SiO_2 template	53.1	48.5	2.8	1.5	37.3@ 2500 nm	24.4@ 2500 nm
SiO_2-VO_2Patterned thin films (2014)	[33]		SiO_2 template	39.2	43.6↑	5.7	7.1↑	17.9	19.4↑
Micropatterned VO_2 (2016)	[34]	Mesh printing	—	60	67↑	6.9	8.8↑	—	—
VO_2 honeycomb (2018)	[29]	Dual-phase transformation			95.4@ 700 nm		5.5		

4.3 Conclusions

To sum up, AR coating is a useful method to improve luminous transmission. Though a nanostructure AR design can vastly improve luminous transmission, solar modulation is generally sacrificed due to a reduction in the effective thickness of VO_2 thin films. With special optical effects, such as surface plasmon resonance of nanostructures, solar modulation can be remedied to a moderate level. Due to the limitation of the VO_2 refractive index, the luminous transmission of multilayer AR coatings on VO_2 continuous thin films is confined. Jointing different methods like gradient AR coating on porous thin films is a superior way to solve the contradiction of luminous transmittance and solar modulation.

References

1. Rayleigh, L. (1880). On reflection of vibrations at the confines of two media between which the transition is gradual, *Proc. London Math. Soc.*, **S1–11**, pp. 51–56.
2. Fraunhofer, J. V. (1888). *Joseph von Fraunhofer Gesannekte Schriften*, Vol. 1 (Munich, Germany).
3. Heavens, O. S. (1991). *Optical Properties of Thin Solid Films*, 2nd ed. (Dover Publications, New York).
4. Tazawa, M., Asada, H., Xu, G., Jin, P. and Yoshimura, K. (2003). Optical constants of vanadium dioxide films and design of a solar energy control window, *MRS Proc.*, **785**, pp. 345–350.
5. Babulanam, S. M., Eriksson, T. S., Niklasson, G. A. and Granqvist, C. G. (1987). Themrmochromic VO_2 films for energy-efficient windows, *Sol. Energy Mater.*, **16**, pp. 347–363.
6. Lee, M. H. and Cho, J. S. (2000). Better thermochromic glazing of windows with anti-reflection coating, *Thin Solid Films*, **365**, pp. 5–6.
7. Xu, G., Jing, P., Tazawa, M. and Yoshimura, K. (2004). Optimization of antireflection coating for VO_2-based energy efficient window, *Sol. Energy Mater. Sol. Cells*, **83**, pp. 29–37.
8. Chen, Z., Gao, Y. F., Kang, L. T., Du, J., Zhang, Z. T., Luo, H. J., Miao, H. Y. and Tan, G. Q. (2011). VO_2-based double-layered films for smart windows: optical design, all-solution preparation and improved properties, *Sol. Energy Mater. Sol. Cells*, **95**, pp. 2677–2684.

9. Xu, G., Jin, P., Tazawa, M. and Yoshimura, K. (2003). Design, formation and characterization of a novel multifunctional window with VO_2 and TiO_2 coatings, *Appl. Phys. A*, **77**, pp. 455–459.
10. Kakiuchida, H., Jing, P. and Tazawa, M. (2004). Control of thermochromic spectrum in vanadium dioxide by amorphous silicon suboxide layer, *Sol. Energy Mater. Sol. Cells*, **92**, pp. 1279–1284.
11. Mlyuka, N. R., Niklasson, G. A. and Granqvist, C. G. (2009). Thermochromic multilayer films of VO_2 and TiO_2 with enhanced transmittance, *Sol. Energy Mater. Sol. Cells*, **93**, pp. 1685–1687.
12. Zhan, Y. J., Xiao, X. D., Lu, Y., Cao, Z. Y., Qi, S., Huan, C. M., Cheng, H. L., Shi, J. F. and Xu, G. (2018). Enhanced thermal stability and thermochromic properties of VO_x-based thin films by room-temperature magnetron sputtering, *Sol. Energy Mater. Sol. Cells*, **174**, pp. 102–111.
13. Zheng, J. Y., Bao, S. H. and Jing, P. (2015). TiO_2(R)/VO_2(M)/TiO_2(A) multilayer film as smart window: combination of energy-saving, antifogging and self-cleaning functions, *Nano Energy*, **11**, pp. 136–145.
14. Long, S. W., Zhou, H. J., Bao, S. H., Xin, Y. C., Cao, X. and Jing, P. (2016). Thermochromic multilayer films of $WO_3/VO_2/WO_3$ sandwich structure with enhanced luminous transmittance and durability, *RSC Adv.*, **6**, pp. 106435–106442.
15. Houska, J., Kolenaty, D., Vlcek, J., Barta, T., Rezek, J. and Cerstvy, R. (2019). Significant improvement of the performance of $ZrO_2/V_{1-x}WO_2/ZrO_2$ thermochromic coatings by utilizing a second-order interference, *Sol. Energy Mater. Sol. Cells*, **191**, pp. 365–371.
16. (a) Wang, H. Y., He, W. J., Yuan, G. L., Wang, X. Y. and Chen, Q. (2013). Large change of visible transmittance with VO_2 phase transition in VO_2/TiO_2 polycrystalline films, *Thin Solid Films*, **540**, pp. 168–172; (b) Powell, M. J., Quesada-Cabrera, R., Taylor, A., Teixeira, D., Papakonstantinou, I., Palgrave, R. G., Sankar, G. and Parkin, I. P. (2016). Intelligent multifunctional $VO_2/SiO_2/TiO_2$ coatings for self-cleaning, energy-saving window panels, *Chem. Mater.*, **28**, pp. 1369–1376.
17. Long, S. W., Cao, X., Li, N., Xin, Y. C., Sun, G. Y., Chang, T. C, Bao, S. H. and Jing, P. (2019). Application-oriented VO_2 thermochromic coatings with composite structures: optimized optical performance and robust fatigue properties, *Sol. Energy Mater. Sol. Cells*, **189**, pp. 138–148.
18. Zhu, M. D., Qi, H. J., Wang, B., Wang, H., Zhang, D. P. and Lv, W. Z. (2019). Enhanced visible transmittance and reduced transition temperature for VO_2 thin films modulated by index-tunable SiO_2 anti-reflection coatings, *RSC Adv.*, **8**, pp. 28953–28959.

19. Liu, C., Wang, S. C., Zhou, Y., Yang, H. B., Lu, Q., Mandler, D., Magdassi, S., Tay, C. Y. and Long, Y. (2018). Index-tunable anti-reflection coatings: maximizing solar modulation ability for vanadium dioxide-based smart thermochromic glazing, *J. Alloys Compd.*, **731**, pp. 1197–1207.
20. Zhang, J., Wang, J., Yang, C. M., Jia, H. B., Cui, X. M., Zhao, S. C. and Xu, Y. (2017). Mesoporous SiO_2/VO_2 double-layer thermochromic coating with improved visible transmittance for smart window, *Sol. Energy Mater. Sol. Cells*, **162**, pp. 134–141.
21. Sun, Y. M., Xiao, X. D., Xu, G., Dong, G. P., Chai, G. Q., Zhang, H., Liu, P. Y., Zhu, H. M. and Zhan, Y. J. (2013). Anisotropic vanadium dioxide sculptured thin films with superior thermochromic properties, *Sci. Rep.*, **3**, pp. 2756–2766.
22. Kang, L. T, Gao, Y. F, Luo, H. J, Chen, Z., Du, J. and Zhang, Z. T. (2011). Nanoporous thermochromic VO_2 films with low optical constants, enhanced luminous transmittance and thermochromic properties, *ACS Appl. Mater. Interfaces*, **3**, pp. 135–138.
23. Li, S.Y., Niklasson, G. A. and Granqvist, C. G. (2010). Nanothermochromics: calulations for VO_2 nanoparticle in dielectric hosts show much improved luminous transmittance and solar energy transmittance modulation, *J. Appl. Phys.*, **108**, p. 063525.
24. Gao, Y. F., Wang, S. B., Luo, H. J., Dai, L., Cao, C. X., Liu, Y. L., Chen, Z. and Kanehira, M. (2012). Enhanced chemical stability of VO_2 nanoparticles by the formation of SiO_2/VO_2 core/shell structures and the application to transparent and flexible VO_2-based composite foils with excellent thermochromic properties for solar heat control, *Energy Environ. Sci.*, **5**, pp. 6104–6110.
25. Zhou, M. Bao, Tao, J. M., Zhu, R., Lin, Y. T., Zhang, X. D. and Xie, Y. (2013). Periodic porous thermochromic VO_2(M) films with enhanced visible transmittance, *Chem. Commun.*, **49**, pp. 6021–6023.
26. Cao, Z. Y., Lu, Y., Xiao, X. D., Zhan, Y. J., Cheng, H. L. and Xu, G. (2017). Tunable simultaneously visible-light and near-infrared transmittance for VO_2/SiO_2 composite films to enhance thermochromic properties, *Mater. Lett.*, **209**, pp. 609–612.
27. Ke, Y. J., Wen, X. L., Zhao, D. Y., Che, R. C., Xiong, Q. H. and Long, Y. (2017). Controllable fabrication of two-dimensional patterned VO_2 nanoparticle, nanodome and nanonet arrays with tunable temperature-dependent localized surface plasmon resonance, *ACS Nano*, **11**, pp. 7542–7551.

28. Ke, Y. J., Balin, I., Wang, N., Lu, Q., Tok, A. L. Y., White, T. J., Magdassi, S., Abdulhalim, I. and Long, Y. (2016). Two-dimensional SiO_2/VO_2 photonic crystals with statically visible and dynamically infrared modulated for smart window deployment, *ACS Appl. Mater. Interfaces*, **8**, pp. 33112–33120.

29. Liu, M. S., Su, B., Kaneti, Y. V., Chen, Z., Tang, Y., Yuan, Y., Gao, Y. F., Jiang, L., Jiang, X. C. and Yu, A. B. (2017). Dual-phase transformation: spontaneous self-template surface-patterning strategy for ultra-transparent VO_2 solar modulating coatings, *ACS Nano*, **11**, pp. 407–415.

30. Jing. P, Xu, G., Masato, T. and Kazuki, Y. (2002). A VO_2-based multifunctional window with highly improved luminous transmittance, *Jpn. J. Appl. Phys.*, **41**(Part 2, No. 3A), pp. L278–L280.

31. Liu, C., Wang, N. and Long, Y. (2013). Multifunctional overcoats on vanadium dioxide thermochromic thin films with enhanced luminous transmission and solar modulation, hydrophobicity and anti-oxidation, *Appl. Surf. Sci.*, **283**, pp. 222–226.

32. Koo, H., Shin, D. M., Bae, S. H., Ko, K. E., Chang, S. H. and Park, C. (2013). The effect of CeO_2 antireflection layer on the optical properties of thermochromic VO_2 film for smart window system, *J. Mater. Eng. Perform.*, **23**, pp. 402–407.

33. Qian, X. K., Wang, N., Li, Y. F., Zhang, J. H., Xu, Z. C. and Long, Y. (2014). Bioinspired multifunctional vanadium dioxide: improved thermochromism and hydrophobicity, *Langmuir*, **30**, pp. 10766–10771.

34. Lu, Q., Liu, C., Wang, N., Magdassi, S., Mandler, D. and Long, Y. (2016). Periodic micro-patterned VO_2 thermochromic films by mesh printing, *J. Mater. Chem. C*, **4**, pp. 8385–8391.

Chapter 5

Controllable Synthesis of Porous Vanadium Dioxide Nanostructures

Ning Wang,[a] Litao Kang,[b] Yanfeng Gao,[c] and Yi Long[d]

[a] *Shenzhen Institutes of Advanced Technology, Chinese Academy of Sciences 1068 avenue, Shenzhen 518055, China*
[b] *School of Environment and Material Engineering, Yantai University, Yantai 264005, Shandong, China*
[c] *School of Materials Science and Engineering, Shanghai University, Shanghai 200444, China*
[d] *School of Materials Science and Engineering, Nanyang Technological University, 50 Nanyang Avenue, Singapore 639798, Singapore*
longyi@ntu.edu.cn, yfgao@shu.edu.cn

The porosity in vanadium dioxide (VO_2) nanostructures has a significant effect when their thermochromic performance is being tuned with respect to the solar modulating ability and luminous transmittance. There is an abundance of research on the construction and characterization of porous VO_2 nanostructures that aims to improve their thermochromic properties by controllable porous synthesis. In this chapter, we will elaborate on the porosity design and controllable synthesis of porous VO_2 nanostructures from the fundamental and experimental aspects, referring to the most recent literature.

Vanadium Dioxide-Based Thermochromic Smart Windows
Edited by Yi Long and Yanfeng Gao

ISBN 978-981-4877-06-0 (Hardcover), 978-1-003-17731-9 (eBook)
www.jennystanford.com

5.1 Introduction

As a promising thermochromic material, vanadium dioxide (VO_2) has outstanding infrared (IR) modulation abilities where the IR light can be transmitted or blocked by the VO_2 thin films below or above the phase transition temperature of $\sim$68°C [1–10]. Due to the large IR modulation ability, VO_2 has been regarded as an excellent candidate for smart windows for energy saving buildings [11–14]. However, the trade-off between visible transmission (T_{lum}) and the solar modulating ability (ΔT_{sol}) is a great challenge that needs to be addressed before VO_2 can be actually used in such applications [15–17]. Porous design in the fabrication of VO_2 smart windows has generated intense interest among materials scientists because it enables simultaneously improved T_{lum} and ΔT_{sol} [18–21]. In the fundamental research studies, a porous design has been found to be effective in reducing the refractive index of VO_2, which could help to narrow the refractive index gap between the VO_2 ($\sim$3) and the air ($\sim$1) on the thin film interfaces, thereby improving the T_{lum} [21–23]. In the meanwhile, the ΔT_{sol} is not compromised since the thin film's thickness is not reduced [23].

In research on porous VO_2 nanostructures, scientists have developed various approaches for the controlled fabrication of porous structures, for example, polymer-assisted deposition (PAD) [14, 21, 24, 25], freeze-drying [26], colloidal lithography [7, 27, 28], and dual-phase transformation [29]. The development of new techniques that allow controlling size as well as porosity is promising for porous VO_2 nanostructures.

In this chapter, we will summarize the recent developments in the controllable synthesis of porous VO_2 nanostructures with enhanced thermochromic properties from both fundamental and experimental aspects.

5.2 Porous Design for Property Enhancement

For VO_2 thin films, the thermochromic performance is determined by the optical parameters, that is, the refractive index n and the extinction coefficient k [16]. The optical performance of nanoporous

VO_2 thin films can be calculated with an optical-admittance recursive method, where it is assumed that the optical parameters should be linearly dependent on the volume fraction (in other words, the n and the k decrease linearly with the porosity) [21]. Figure 5.1 shows the relationship between the thermochromic performance and the optical parameters. In the randomly distributed nanoporous VO_2 thin films (Fig. 5.1a), there was an obvious decrease in the optical constants (n, k) in the nanoporous structure compared with the normal or densely packed thin films, and the calculation results for nanoporous VO_2 showed largely improved T_{lum} and ΔT_{sol} with the porosity [21].

In the porous design for VO_2 thin films, the degree of order/disorder and the periodicity have a great impact on the performance improvement. In the randomly porous design, the thermochromic performance could be improved to $T_{\text{lum}} \approx 40\%$ and $\Delta T_{\text{sol}} \approx 14\%$ [21, 30]. In contrast, the performance could be further enhanced to $T_{\text{lum}} \approx 46\%$ and $\Delta T_{\text{sol}} \approx 13\%$ with the periodic porous design [28, 31, 32]. The periodic design should be more applicable for practical applications since a controllable or programmable porosity design is critical for the efficient optimization of thermochromic properties.

5.3 Approaches for Porosity Construction

Scientists have developed four effective approaches for constructing porous VO_2 thin films: colloidal lithography assembly [7, 27, 28], PAD [14, 21, 24, 25], dual-phase transformation [26], and freeze-drying preparation [29].

5.3.1 Colloidal Lithography Assembly

In colloidal lithography assembly, a closely packed monolayer colloidal crystal (MCC) is first used as the sacrificial template for constructing periodic continuous thin films and then the periodic porous structure is attained after the subsequent removal of the MCC template via chemical etching or high-temperature sintering, an attractive approach for preparing periodic porous structures.

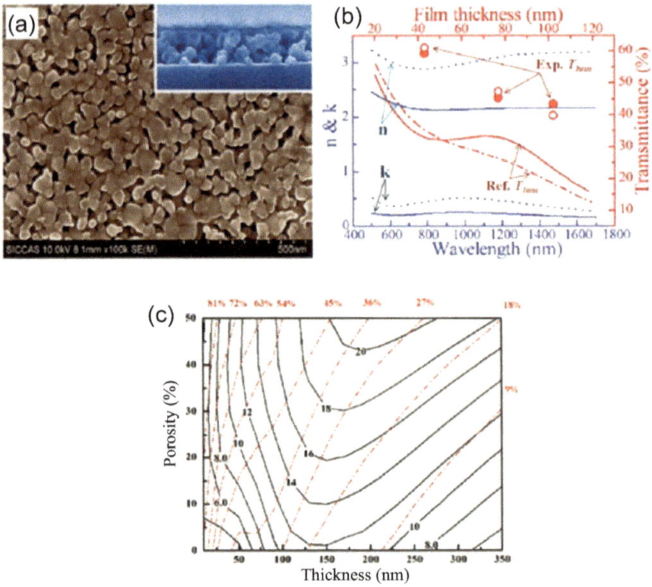

Figure 5.1 (a) Nanoporous VO_2 thin film. (b) Experimental (solid) and reference (dash) *n*, *k* (versus wavelength) and experimental/reference T_{lum} (versus film thickness). The reference data are from Ref. [25] (c) Optical calculations of nanoporous VO_2 thin films based on an optical-admittance recursive method, where the red dotted lines and the black solid lines represent the T_{lum} in the insulating state and the ΔT_{sol}, respectively [21].

Xie et al. fabricated periodic VO_2 thin films with a polystyrene (PS) MCC template, which was first infiltrated by a $VOSO_4$ (vanadium source)-NH_4HCO_3 (precipitator) solution and then removed by atmospheric annealing, leaving the periodic nanoporous VO_2 thin films on the substrate (Fig. 5.2) [31].

Recently, Long et al. proposed an alternative colloidal lithography approach for nanoporous VO_2 thin films assembly, where nanoparticle, nanonet, and nanodome arrays with tunable porosity could be achieved by sacrificing the MCC template [32]. This synthetic method is illustrated in Fig. 5.3. With short-duration plasma etching (PE), VO_2 nanoparticle and nanodome arrays are attained by tuning the precursor viscosity (see Route 1 and Route 2). The VO_2 nanonet arrays involved relatively long-duration PE (see Route 3). By tuning the periodicity, a decent thermochromic performance

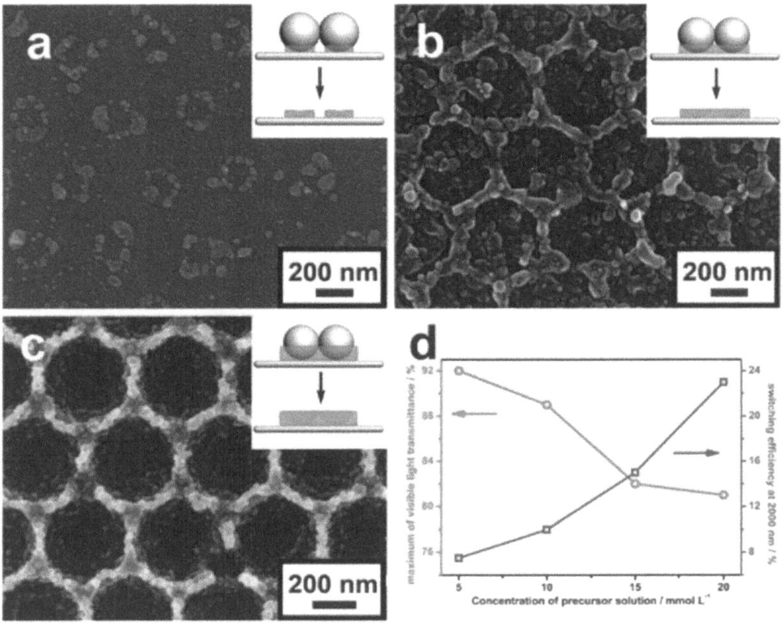

Figure 5.2 Typical SEM images of the morphology evolution of the VO_2(M) film at different precursor concentrations. (a) Nanoring array, 5 mmol L^{-1}, (b) nanonet array, 10 mmol L^{-1}, and (c) thicker nanonet array, 15 mmol L^{-1}. (d) Corresponding optical performances of the samples shown in (a–c) [31].

($\Delta T_{sol} = 13.2\%$ and $T_{lum} = 46\%$) is achieved. As depicted in this work, the colloidal lithography approach could result in a more uniform or less defective periodic porous structure than the template-free methods and the pitch resolution could be sub-100 nm, which is promising for the fabrication of porous VO_2 coatings with a large ΔT_{sol} and a high T_{lum}.

Besides, a 2D SiO_2-VO_2 core–shell monolayer could be used to construct a photonic structure with the colloidal lithography method (Figs. 5.4a and 5.4b) [28]. The color of the periodic structure could be tuned by the periodicity, with the dimension falling in the visible-light range, which could selectively reflect visible light of a certain color (Fig. 5.4c). Benefiting from this ability, the nanostructure displays controllable appearances and exhibits a good thermochromic performance with a combination of

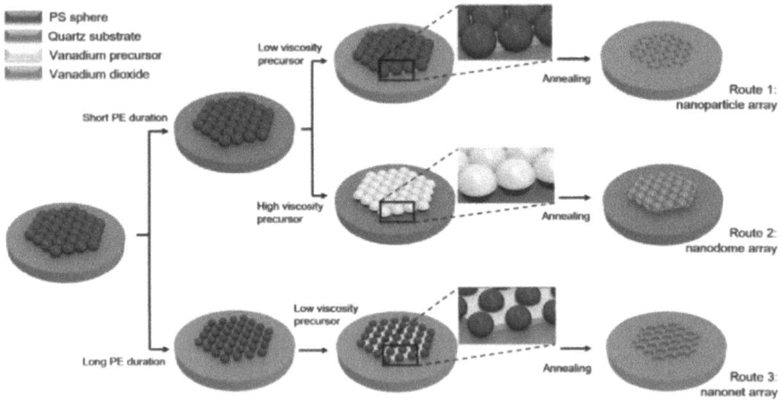

Figure 5.3 Effect of synthesis conditions on the morphology evolution. Route 1: Nanoparticle arrays are prepared by short-duration PE and low-viscosity precursor. Route 2: Nanodome arrays are produced using a high-viscosity precursor that can stick on the tops of PS spheres. Route 3: Nanonets are fabricated by controlling the interval space between adjacent spheres by long-duration PE [31].

$T_{\text{lum}} = 49.6\%$ and $\Delta T_{\text{sol}} = 11\%$ calculated by 3D finite-difference time domain.

5.3.2 Polymer-Assisted Deposition

In PAD of nanoporous VO_2 structures, several strategies have been developed for porous construction. For films, Kang et al. [33] developed a facile PAD method inspired by the works of Jia [34] and Kozuka and Kishimoto [35, 36]. In this process, cheap inorganic vanadium salts (e.g., $VOCl_2$) and functional polymer additives are dissolved in water to form a starting solution. The introduction of proper polymer additives is a crucial step for the coating process that follows. Studies show that polymers with amide groups can give aqueous solutions of inorganic salts (e.g., $Ti(SO_4)_2$ and $ZrCl_2O_3$) [35, 36] a strong film-forming ability, ensuring the formation of uniform hybrid precursor coatings of salts and polymers after coating. After annealing, polymers are burned while salts are decomposed and crystallized, resulting in crystalline surface coatings of oxides [35, 36].

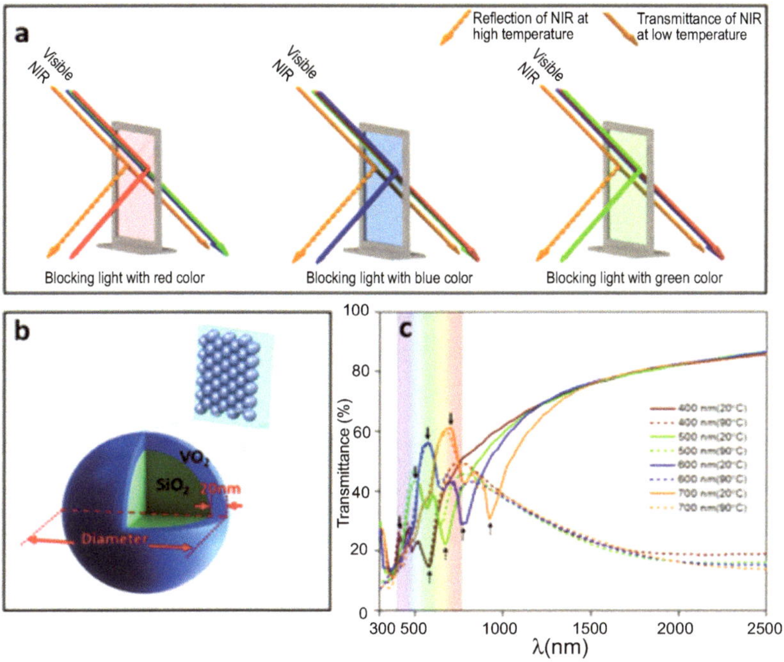

Figure 5.4 (a) Illustration of how color-changed thermochromic smart windows work. (b) Illustration of designed structures for simulation. (c) Calculated transmittance spectrum. The colorful background in (c) denotes the visible spectrum from 370 nm to 770 nm [28].

The film-forming ability of the precursor solution is believed to stem from the unique molecular structures of the polymers (Fig. 5.5). In the amide groups of the polymers, N atoms partially donate their lone pair of electrons to O atoms through orbit conjugation, resulting in a positively charged amine (–NH) group and a negatively charged carbonyl (–C=O) group in each amide group [37]. The electron transfer between the amine and carbonyl groups in the polymer helps the film deposition via the electrostatic interactions between the charged groups and the metal ions. On the other hand, the negatively charged –C=O can bind with positively charged metal cations, ensuring the formation of highly uniform salt/polymer hybrid films [33, 38].

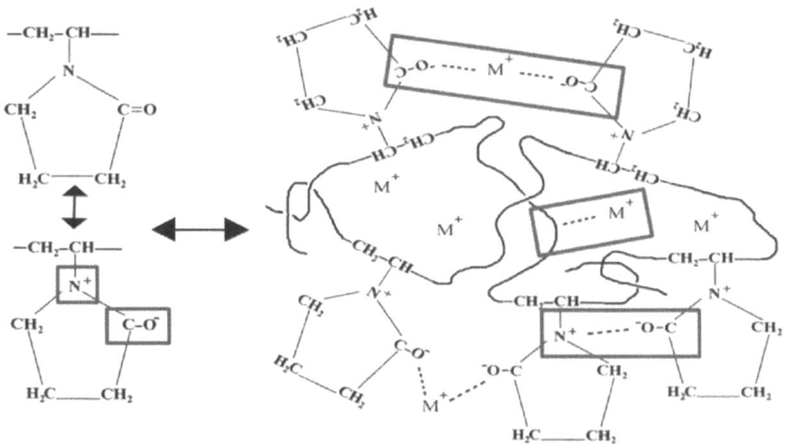

Figure 5.5 Schematic illustration of the film-forming mechanism in the PAD process of a PVP-$VOCl_2$ system [33]. Copyright 2009, Trans Tech Publications Ltd.

During annealing of PAD precursor films, there is remarkable volume shrinkage due to the decomposition of both polymers and vanadium salts [24]. If a low temperature is applied to avoid a strong recrystallization process, nanoporous structures can be produced without using any sacrificial templates (Fig. 5.6) [33, 39]. These porous structures have high specific surface areas, large surface roughness [40], and low optical constants [21], which are highly desirable for applications such as chromogenic [41]/antireflection

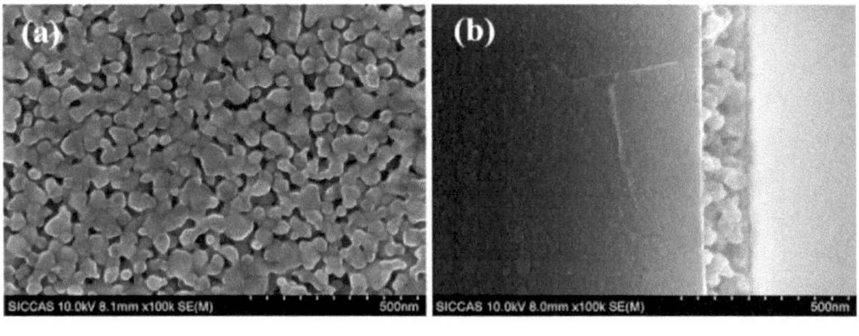

Figure 5.6 Top (a) and cross-sectional (b) SEM images of a 147 nm thick nanoporous VO_2 film on fused silica glass by the PAD method [21].

coatings [42], superhydrophobic/superhydrophilic coatings [43], and electrochemical energy storage [44]. Increasing the annealing temperature results in films with larger and more isolated VO_2 particles, which deteriorate the optical performance due to strong visible light scattering and strong light absorption of the particles [39]. With high enough annealing temperatures and single-crystal substrates, even crick-free dense (or even epitaxial) films can be obtained [34, 45], thanks to the remarkable mass-transfer and densification of oxide films, as in the case of ceramic sintering. In dense thermochromic VO_2 films, the tightly bonded particles minimize their interface resistance, highly desired for the achievement of large electrical resistance variation during metal–insulator transformation, which that is important for applications such as temperature-resistant switching elements [46, 47].

5.3.3 Dual-Phase Transformation

A newly developed template-free method to prepare nanoporous VO_2 structures is dual-phase transformation, which could give rise to ultrahigh visible transmittance. In this method, a porous structure is constructed by the transformation between the colloids and ionic states stimulated by the moisture (Fig. 5.7). First, the $VOCl_2$ precursor is spin-coated onto the fused silica substrates, and then hydrous colloids are formed through water evaporation. In the second step, quick annealing at 300°C is applied to solidify the thin film, which is followed by annealing at 500°C in N_2, resulting in a honeycomb-like nanoporous structure that shows high visible transmittance of above 90% at a wavelength of ~700 nm and a decent solar modulating ability ($\Delta T_{sol} \approx 5.5\%$). The ratio control between the HCl and the N_2H_4 in the precursor is believed to be critical for forming the initial hydrous spheres (colloids) and the final honeycomb-like nanoporous structures [29].

5.3.4 Freeze-Drying Preparation

Freeze-drying is also an efficient alternative way for preparing porous nanoporous VO_2 nanostructures. In a normal sol-gel process, the inherent porosity of the thin film products is always very low.

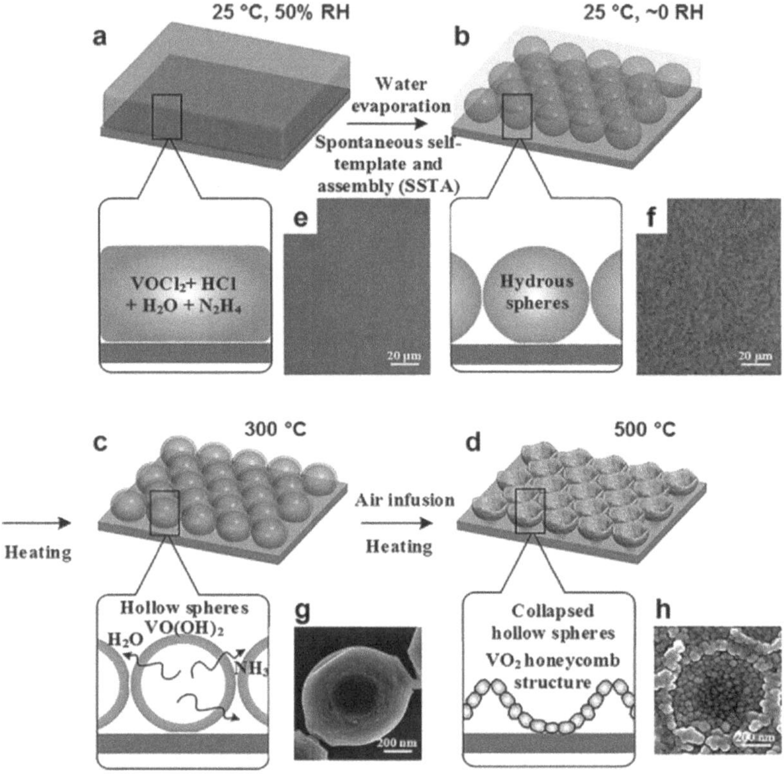

Figure 5.7 The formation mechanism of the highly ordered honeycomb-structured VO_2(M) film was inspired by the spontaneous self-template and assembly (SSTA) during the dual-phase transformation process. (a) A homogeneous, fully solution-based precursor film was deposited at room condition (25°C, 50% RH). (b) The precursor was spontaneously self-templated and assembled into hydrous sphere arrays after water evaporation in dry nitrogen (25°C, ~0 RH). (c) Hydrous spheres became hollow $VO(OH)_2$ spheres after instant heating to 300°C and (d) finally collapsed into honeycomb structures after being heated at a rate of 2°C and maintained at 500°C for 1 h. Microscopic photos of (e) the precursor film and (f) the film after the SSTA process. SEM images of (g) captured hollow $VO(OH)_2$ spheres and (h) final honeycomb structures [29].

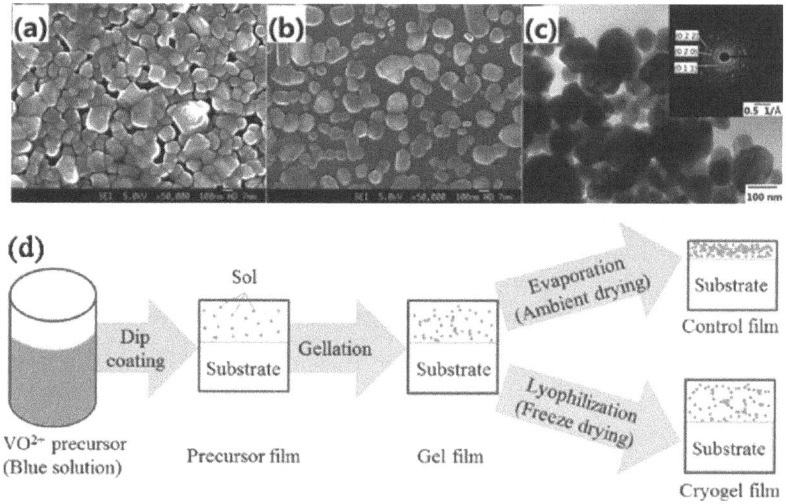

Figure 5.8 FESEM image for freeze-dried nanoporous VO_2 films with 7.5 mL of H_2O_2 (a) and 17.5 mL of H_2O_2 (b) in the precursor. (c) TEM image of (b) and the corresponding SAED (inset). (d) Schematic illustration of the freeze-drying process for the nanoporous design [30].

Comparatively, in the freeze-drying process, the precursor is quickly frozen and then dried in vacuum. In this process, the solvents in the precursor could be sublimated and removed quickly from the structure without the skeleton collapsing, normally seen in the routine drying process, which enormously enlarges the porosity in the final products (Fig. 5.8d). In the freeze-drying work of Long et al., the precursor V_2O_5-H_2O_2-ox (oxalic acid) is first dip-coated onto fused silica substrates and quickly frozen in liquid nitrogen and a nanoporous thin film is attained by freeze-drying at −80°C and 0.01 mbar [30]. After a postannealing in Ar flow at 550°C for 2 h, nanoporous VO_2 thin films are finally obtained (Fig. 5.8a–c).

5.4 Conclusion and Outlook

In summary, in this chapter, we have demonstrated and discussed the fabrication of nanoporous VO_2 nanostructures and the principle for enhancing their thermochromic properties by using the porosity design. After increasing the porosity in the VO_2 nanostructures, especially for the periodic porous design, the optical constants can be modified, which could largely increase the visible transmittance. Compared with other methods for enhancing the thermochromic performance of VO_2, the porous design is superior in terms of structural flexibility, easy to handle, and low in terms of consumption of VO_2 source materials, which could effectively reduce the cost in real applications. With respect to the fabrication methods of nanoporous VO_2 thin films, dual-phase transformation, PAD, and freeze-drying are effective approaches for random nanoporous construction with controllable porosity, while colloidal lithography with a sacrificial MCC template is a powerful method for periodic nanoporous construction. In the near future, it will be challenging but necessary to develop new cost-effective and mass-production methods for the construction of porous nanoporous VO_2 structures.

References

1. Cui, Y., Ke, Y., Liu, C., Chen, Z., Wang, N., Zhang, L., Zhou, Y., Wang, S., Gao, Y. and Long, Y. (2018). Thermochromic VO_2 for energy-efficient smart windows, *Joule*, **2**, pp. 1707–1746.
2. Wang, N., Duchamp, M., Xue, C., Dunin-Borkowski, R. E., Liu, G. and Long, Y. (2016). Single-crystalline W-doped VO_2 nanobeams with highly reversible electrical and plasmonic responses near room temperature, *Adv. Mater. Interfaces*, **3**, p. 1600164.
3. Wang, N., Duchamp, M., Dunin-Borkowski, R. E., Liu, S., Zeng, X., Cao, X. and Long, Y. (2016). Terbium-doped VO_2 thin films: reduced phase transition temperature and largely enhanced luminous transmittance, *Langmuir*, **32**, pp. 759–764.
4. Lu, Q., Liu, C., Wang, N., Magdassi, S., Mandler, D. and Long, Y. (2016). Periodical micro-patterned VO_2 thermochromic films by mesh printing, *J. Mater. Chem. C*, **4**, pp. 8385–8391.
5. Wang, N., Liu, S., Zeng, X. T., Magdassi, S. and Long, Y. (2015). Mg/W-codoped vanadium dioxide thin films with enhanced visible transmittance and low phase transition temperature, *J. Mater. Chem. C*, **3**, pp. 6771–6777.

6. Otto, M. R., René de Cotret, L. P., Valverde-Chavez, D. A., Tiwari, K. L., Émond, N., Chaker, M., Cooke, D. G. and Siwick, B. J. (2019). How optical excitation controls the structure and properties of vanadium dioxide, *PNAS*, **116**, pp. 450–455.
7. Li, J., Liu, W., Zhang, X., Chu, P. K., Cheung, K. M. C. and Yeung, K. W. K. (2019). Temperature-responsive tungsten doped vanadium dioxide thin film starves bacteria to death, *Mater. Today*, **22**, pp. 35–49.
8. Xu, F., Cao, X., Luo, H. and Jin, P. (2018). Recent advances in VO_2-based thermochromic composites for smart windows, *J. Mater. Chem. C*, **6**, pp. 1903–1919.
9. Goodenough, J. B. (1971). The two components of the crystallographic transition in VO_2, *J. Solid State Chem.*, **3**, pp. 490–500.
10. Morin, F. (1959). Oxides which show a metal-to-insulator transition at the Neel temperature, *Phys. Rev. Lett.*, **3**, pp. 34–36.
11. Gao, Y. F., Wang, S. B., Luo, H. J., Dai, L., Cao, C. X., Liu, Y. L., Chen, Z. and Kanehira, M. (2012). Enhanced chemical stability of VO_2 nanoparticles by the formation of SiO_2/VO_2 core/shell structures and the application to transparent and flexible VO_2-based composite foils with excellent thermochromic properties for solar heat control, *Energy Environ. Sci.*, **5**, pp. 6104–6110.
12. Gao, Y. F., Wang, S. B., Kang, L. T., Chen, Z., Du, J., Liu, X. L., Luo, H. J. and Kanehira, M. (2012). VO_2-Sb:SnO_2 composite thermochromic smart glass foil, *Energy Environ. Sci.*, **5**, pp. 8234–8237.
13. Gao, Y. F., Cao, C. X., Dai, L., Luo, H. J., Kanehira, M., Ding, Y. and Wang, Z. L. (2012). Phase and shape controlled VO_2 nanostructures by antimony doping, *Energy Environ. Sci.*, **5**, pp. 8708–8715.
14. Zhang, Z., Gao, Y., Luo, H., Kang, L., Chen, Z., Du, J., Kanehira, M., Zhang, Y. and Wang, Z. L. (2011). Solution-based fabrication of vanadium dioxide on F:SnO_2 substrates with largely enhanced thermochromism and low-emissivity for energy-saving applications, *Energy Environ. Sci.*, **4**, p. 4290.
15. Wang, N., Peh, Y. K., Magdassi, S. and Long, Y. (2018). Surface engineering on continuous VO_2 thin films to improve thermochromic properties: top-down acid etching and bottom-up self-patterning, *J. Colloid Interface Sci.*, **512**, pp. 529–535.
16. Wang, N., Goh, Q. S., Lee, P. L., Magdassi, S. and Long, Y. (2017). One-step hydrothermal synthesis of rare earth/W-codoped VO_2 nanoparticles: reduced phase transition temperature and improved thermochromic properties, *J. Alloys Compd.*, **711**, pp. 222–228.

17. Qian, X., Wang, N., Li, Y., Zhang, J., Xu, Z. and Long, Y. (2014). Bioinspired multifunctional vanadium dioxide: improved thermochromism and hydrophobicity, *Langmuir*, **30**, pp. 10766–10771.
18. Li, M., Magdassi, S., Gao, Y. F. and Long, Y. (2017). Hydrothermal synthesis of VO_2 polymorphs: advantages, challenges and prospects for the application of energy efficient smart windows, *Small*, **13**, p. 1701147.
19. Wang, N., Huang, Y., Magdassi, S., Mandler, D., Liu, H. and Long, Y. (2013). Formation of VO_2 zero-dimensional/nanoporous layers with large supercooling effects and enhanced thermochromic properties, *RSC Adv.*, **3**, pp. 7124–7128.
20. Ding, S., Liu, Z., Li, D., Zhao, W., Wang, Y., Wan, D. and Huang, F. (2013). Tunable assembly of vanadium dioxide nanoparticles to create porous film for energy-saving applications, *ACS Appl. Mater. Interfaces*, **5**, pp. 1630–1635.
21. Kang, L., Gao, Y., Luo, H., Chen, Z., Du, J., Zhang, Z. (2011). Nanoporous thermochromic VO_2 films with low optical constants, enhanced luminous transmittance and thermochromic properties, *ACS Appl. Mater. Interfaces*, **3**, pp. 135–138.
22. Liu, C., Balin, I., Magdassi, S., Abdulhalim, I. and Long, Y. (2015). Vanadium dioxide nanogrid films for high transparency smart architectural window applications, *Opt. Express*, **23**, pp. A124–A132.
23. Taylor, A., Parkin, I., Noor, N., Tummeltshammer, C., Brown, M. S. and Papakonstantinou, I. (2013). A bioinspired solution for spectrally selective thermochromic VO_2 coated intelligent glazing, *Opt. Express*, **21**, pp. A750–A764.
24. Kang, L. T., Gao, Y. F. and Luo, H. J. (2011). In *Functional and Electronic Materials*, Li, C. M., Jiang, C. B., Zhong, Z. Y. and Zhou, Y., eds. (Trans Tech Publications Ltd., Stafa-Zurich), pp. 791–797.
25. Xu, G., Jin, P., Tazawa, M. and Yoshimura, K. (2004). Tailoring of luminous transmittance upon switching for thermochromic VO_2 films by thickness control, *Jpn. J. Appl. Phys.*, **43**, pp. 186–187.
26. Cao, X., Thet, M. N., Zhang, Y., Joachim Loo, S. C., Magdassi, S., Yan, Q. and Long, Y. (2015). Solution-based fabrication of VO_2(M) nanoparticles via lyophilisation, *RSC Adv.*, **5**, pp. 25669–25675.
27. Ke, Y., Yin, Y., Zhang, Q., Tan, Y., Hu, P., Wang, S., Tang, Y., Zhou, Y., Wen, X., Wu, S., White, T. J., Yin, J., Peng, J., Xiong, Q., Zhao, D. and Long, Y. (2019). Adaptive thermochromic windows from active plasmonic elastomers, *Joule*, **3**(3), pp. 858–871.

28. Ke, Y. J., Balin, I., Wang, N., Lu, Q., Tok, A. T. Y., White, T. J., Magdassi, S., Abdulhalim, I. and Long, Y. (2016). Two-dimensional SiO_2/VO_2 photonic crystals with statically visible and dynamically infrared modulated for smart window deployment, *ACS Appl. Mater. Interfaces*, **8**, pp. 33112–33120.
29. Liu, M., Su, B., Kaneti, Y. V., Chen, Z., Tang, Y., Yuan, Y., Gao, Y., Jiang, L., Jiang, X. and Yu, A. (2017). Dual-phase transformation: spontaneous self-template surface-patterning strategy for ultra-transparent VO_2 solar modulating coatings, *ACS Nano*, **11**, pp. 407–415.
30. Cao, X., Wang, N., Law, J. Y., Loo, S. C. J., Magdassi, S. and Long, Y. (2014). Nanoporous thermochromic VO_2(M) thin films: controlled porosity, largely enhanced luminous transmittance and solar modulating ability, *Langmuir*, **30**, pp. 1710–1715.
31. Zhou, M., Bao, J., Tao, M., Zhu, R., Lin, Y., Zhang, X. and Xie, Y. (2013). Periodic porous thermochromic VO_2(M) films with enhanced visible transmittance, *Chem. Commun.*, **49**, pp. 6021–6023.
32. Ke, Y., Wen, X., Zhao, D., Che, R., Xiong, Q. and Long, Y. (2017). Controllable fabrication of two-dimensional patterned VO_2 nanoparticle, nanodome, and nanonet arrays with tunable temperature-dependent localized surface plasmon resonance, *ACS Nano*, **11**, pp. 7542–7551.
33. Kang, L., Gao, Y. and Luo, H. (2009). A novel solution process for the synthesis of VO_2 thin films with excellent thermochromic properties, *ACS Appl. Mater. Interfaces*, **1**, pp. 2211–2218.
34. Jia, Q. X., McCleskey, T. M., Burrell, A. K., Lin, Y., Collis, G. E., Wang, H., Li, A. D. Q. and Foltyn, S. R. (2004). Polymer-assisted deposition of metal-oxide films, *Nat. Mater.*, **3**, pp. 529–532.
35. Kozuka, H. and Kishimoto, T. (2001). Aqueous solution-based ceramic thin film deposition using organic polymers with amide groups, *Chem. Lett.*, **30**, pp. 1150–1151.
36. Kishimoto, T. and Kozuka, H. (2003). Sol-gel preparation of TiO_2 ceramic coating films from aqueous solutions of titanium sulfate (IV) containing polyvinylpyrrolidone, *J. Mater. Res.*, **18**, pp. 466–474.
37. Lu, G. Y. (1999). *Organic Chemistry* (Nanjing University Press).
38. Hao, C., Zhao, Y., Zhou, Y., Zhou, L., Xu, Y., Wang, D. and Xu, D. (2007). Interactions between metal chlorides and poly (vinyl pyrrolidone) in concentrated solutionsand solid-state films, *J. Polym. Sci., Part B: Polym. Phys.*, **45**, pp. 1589–1598.
39. Kang, L., Gao, Y., Zhang, Z., Du, J., Cao, C., Chen, Z. and Luo, H. (2010). Effects of annealing parameters on optical properties of thermochromic

VO_2 films prepared in aqueous solution, *J. Phys. Chem. C*, **114**, pp. 1901–1911.

40. Du, J., Gao, Y., Chen, Z., Kang, L., Zhang, Z. and Luo, H. (2013). Enhancing thermochromic performance of VO_2 films via increased microroughness by phase separation, *Sol. Energy Mater. Sol. Cells*, **110**, pp. 1–7.
41. Lee, S. H., Deshpande, R., Parilla, P. A., Jones, K. M., To, B., Mahan, A. H. and Dillon, A. C. (2006). Crystalline WO_3 nanoparticles for highly improved electrochromic applications, *Adv. Mater.*, **18**, pp. 763–766.
42. Walheim, S., Schäffer, E., Mlynek, J. and Steiner, U. (1999). Nanophase-separated polymer films as high-performance antireflection coatings, *Science*, **283**, pp. 520–522.
43. Wu, W., Wang, X., Liu, X. and Zhou, F. (2009). Spray-coated fluorine-free superhydrophobic coatings with easy repairability and applicability, *ACS Appl. Mater. Interfaces*, **1**, pp. 1656–1661.
44. Yu, G., Xie, X., Pan, L., Bao, Z. and Cui, Y. (2013). Hybrid nanostructured materials for high-performance electrochemical capacitors, *Nano Energy*, **2**, pp. 213–234.
45. Burrell, A. K., McCleskey, T. M. and Jia, Q. (2008). Polymer assisted deposition, *Chem. Commun.*, pp. 1271–1277.
46. Gurvitch, M., Luryi, S., Polyakov, A., Shabalov, A., Dudley, M., Wang, G., Ge, S. and Yakovlev, V. (2007). VO_2 films with strong semiconductor to metal phase transition prepared by the precursor oxidation process, *J. Appl. Phys.*, **102**, p. 033504.
47. Nagashima, K., Yanagida, T., Tanaka, H. and Kawai, T. (2006). Stress relaxation effect on transport properties of strained vanadium dioxide epitaxial thin films, *Phys. Rev. B*, **74**, p. 172106.

Chapter 6

Biomimetic, Gridded Structure, and Hybridation

Yujie Ke,[a] Yang Zhou,[a] and Yi Long[a,b,c]
[a]*School of Materials Science and Engineering, Nanyang Technological University, 50 Nanyang Avenue, Singapore 639798, Singapore*
[b]*Singapore-HUJ Alliance for Research and Enterprise (SHARE), Nanomaterials for Energy and Energy-Water Nexus (NEW), Campus for Research Excellence and Technological Enterprise (CREATE), Singapore 138602, Singapore*
[c]*Sino-Singapore International Joint Research Institute (SSIJRI), GuangZhou 510000, China*
yujie.ke@ntu.edu.sg, zhouyang@ntu.edu.sg, longyi@ntu.edu.sg

6.1 Biomimetic

There is a long-standing interest among researchers to learn about and understand the surprising properties of biomaterials from nature. In nature, the moth eye (Fig. 6.1a,b) and the cicada wing (Fig. 6.1c,d) exhibit intriguing physical properties, such as wide-angle antireflection, polarization insensitivity, and superhydrophobicity [1–3]. These amazing functions are proved to correlate with the densely packed cone morphology on the surface, which inspires

Vanadium Dioxide-Based Thermochromic Smart Windows
Edited by Yi Long and Yanfeng Gao

ISBN 978-981-4877-06-0 (Hardcover), 978-1-003-17731-9 (eBook)
www.jennystanford.com

Figure 6.1 (a) Photograph of the compound eyes of *Calliphora*. (b) The antireflection structures on its ommatidium surface. (c) Photograph of a butterfly and (d) the two distinct simultaneous structure-induced colors on its wings. Reprinted from Ref. [1].

the development of biomimetic-patterned surfaces for functional materials [2–4].

The moth-eye structure consists of hexagonally patterned bumps with a height of ~200 nm and a periodicity of ~300 nm (Fig. 6.1a,b) [1]. These subwavelength nanostructures provide a continuous refractive index gradient between the air and the medium, effectively decreasing the reflections by reducing the refractive index gap at the air-medium interface [3]. The refractive index gradient follows a specific fifth-order-polynomial dependence on depth:

$$n = n_s - (n_s - 1)(10u^3 - 15u^4 + 6u^5), \tag{6.1}$$

where u represents the normalized optical thickness (0 on the medium surface; 1 in air) and n_s is the refractive index of the medium (substrate). The grating criterion for efficient antireflection is $p < \lambda/n_s$ and $h > \lambda/2$, where p is the periodicity while h is the height.

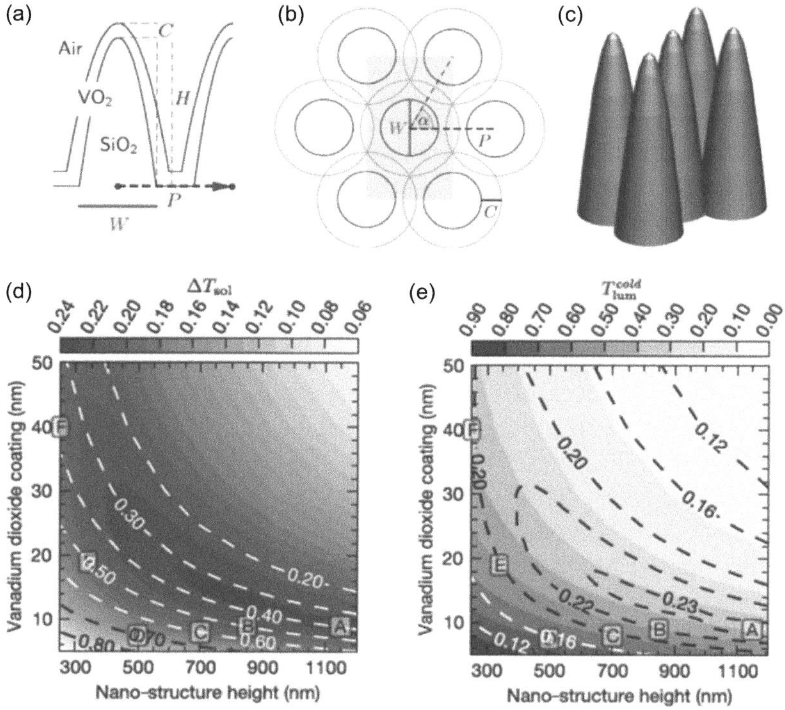

Figure 6.2 Side (a) and top (b) elevations of a nanotextured surface with hexagonally arranged circular paraboloid cones (c); (a, b) pitch P, base width W, and VO_2 coating thickness C. (d) FDTD parameter search for hexagonally arranged VO_2-coated paraboloid nanostructures. (e) Contour lines of T_{lum} are overlaid upon a heat-map of ΔT_{sol} and vice-versa. Reprinted from Ref. [5].

The incorporation of biomimetic structures in a VO_2 coating is believed to be a promising way to enhance its thermochromic performance because biomimetic structures enable tailoring of optical properties. In 2013, Taylor et al. calculated the intelligent glazing of VO_2-coated silica nipple arrays on a glass substrate (Fig. 6.2) [5].

This biomimetic design involved hexagonal-close-packed VO_2-coated tapered corn arrays on a glass substrate (Fig. 6.2a–c). The structure is assigned to achieve the highest area density that maintains the peak-to-trough height difference. The base width (W), pitch size (P), and coating thickness (C) of VO_2 are designed to correlate with each other by the following rule:

$$W = P\sqrt{\frac{4}{3}} - 2C, \quad (6.2)$$

where the W, P, and C are illustrated in Fig. 6.2a,b. To be unresolved by the incident light, the P value was designed to be $\sim$130 nm, since 99.9% and 99% of visible light up to angles of 25.9° and 42°, respectively, cannot resolve the structure under this pitch size. A height range of 250–1200 nm was identified for efficient antireflection. With this biomimetic model, the finite-difference time-domain (FDTD) simulation reveals that the solar modulating ability can be increased to 23.1% and the visible transmission can be enlarged to 70.3% (Fig. 6.2d,e), which is of great importance for practical applications.

In the year 2014, Qian et al. designed a real case of VO_2-coated silica tapered cone arrays on a glass substrate to successfully apply the biomimetic antireflective properties to VO_2-based thermochromic smart windows (Fig. 6.3a) [6]. The tapered cone arrays were first fabricated on a fused silica substrate by reactive ion etching using sacrificial templates of hexagonal-close-packed polystyrene (PS) spheres. The periodicity (P) could be adjusted by the size of the PS nanospheres. Prepared arrays are demonstrated to be nipple arrays, as shown in Fig. 6.3a. By adjusting the PS template, nipple arrays with different periodicities (210 nm, 440 nm, 580 nm, and 1000 nm) are achievable (Fig. 6.3b). Qian et al. further calculated the T_{lum} and ΔT_{sol} and found that both these parameters are enhanced on the 210 nm sample, compared with the 440, 580, and 1000 nm samples (Fig. 6.3c,d). In the research, patterns with small periodicities (210 nm and 440 nm) and $\sim$40 nm VO_2 coating thickness could increase visible/infrared transmittance (T_{lum}/T_{IR}) at 25°C and 90°C. This is because the P is smaller than the given wavelength (cannot be resolved by the incident light) and the gradient refractive index results in effective antireflection. This antireflective sample with $P = 210$ nm and a VO_2 thickness of 40 nm can offer an approximately 10% enhancement in T_{lum} and 24.5% increase in the solar modulation (Fig. 6.3c,d).

The iridescent colors of cicada wings are revealed to come from the periodic structures on the surface of the wings (Fig. 6.1c,d)

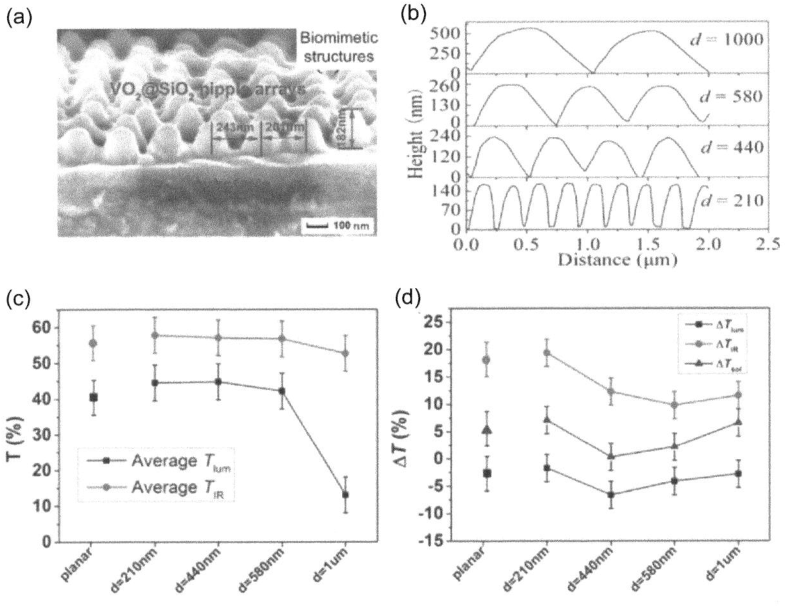

Figure 6.3 (a) SEM cross-sectional profile of 210 nm VO_2-SiO_2 nipple arrays. (b) AFM results of AR samples with different periods: 210 nm, 44 nm, 580 nm, and 1000 nm. (c) Effects of periodicity on T_{lum} and T_{IR}. (d) Effects of periodicity on ΔT_{lum}, ΔT_{IR}, and ΔT_{sol}. Reprinted from Ref. [6].

[1]. This phenomenon has been developed into the concept called the "photonic crystal" [7–9]. The photonic crystal is one of the most promising structures that display conspicuous structural color due to their photonic bandgaps generated by coherent optical diffraction.

In 2016, Ke et al. successfully introduced photonic crystals into VO_2-based thermochromic smart windows (Fig. 6.4) [10]. In their report, they proposed that a smart window with photonic crystals can selectively reflect incident light of a certain wavelength, achieving different colors. Specifically, smart windows can appear red, blue, or green by selectively reflecting the red, blue, or green incident light (Fig. 6.4a). Meanwhile, smart windows are also equipped with near-infrared (NIR) and solar energy modulation ability due to the phase change of VO_2. Ke et al. proposed 2D SiO_2-VO_2 core–shell photonic crystals, where a 20 nm VO_2 film homogeneously coats a SiO_2 nanosphere (Fig. 6.4b). Simulation results demonstrated that such smart windows can achieve diameter-dependent colors, demonstrated as an adjustable transmittance valley in the visible region (Fig. 6.4c). Also, a good thermochromic performance is achieved in such designs, indicated as tunable NIR transmittance—the calculated values of T_{lum} and ΔT_{sol} are 49.6% and 11%, respectively (Fig. 6.4c). The tunable photonic bandgaps in the visible range and solar modulation were further verified by experiments. Tunable colors were observable on the samples, where colors change from yellowish brown to greenish blue by changing the silica sphere size from 200 nm to 400 nm, 600 nm, and 700 nm. However, the ΔT_{sol} was largely sacrificed and could only achieve a value of 3.1% with T_{lum} at 44.2%. Such a difference between the simulation and experimental results is mainly due to the sol-gel method employed in an experiment that could not produce perfect core–shell structures. Though the solar modulation property in experimental results is not high, they suggested other methods, such as atomic layer deposition and chemical vapor deposition, to solve such issues.

Though progress has been made, application of biomimetic structures in VO_2-based smart windows is still in its early stages. The introduction of bioinspired structures and materials is expected to not only further improve the thermochromic performance of smart windows but also make them multifunctional. Moreover, processable bioinspired structures rely highly on the facile and controllable fabrication of monoclinic VO_2, while its preparation in the sub-50 nm scale is still a challenge [10, 11].

6.2 Grid Structures

Films based on VO_2 with grid structures are attracting increasing interest due to the great potential for structure-induced thermochromic enhancements, mainly including enhanced transmittance, antireflection, increased effective solar modulation volume, and localized surface plasmon resonance (LSPR) [12]. Transmittance enhancement is due to void formation in the film, decreasing the affected area of VO_2. The antireflection property can be

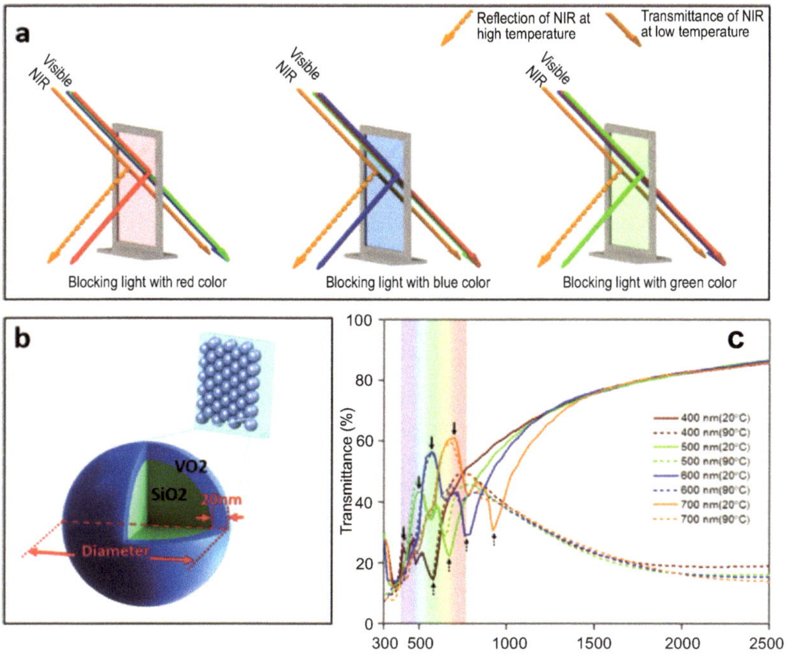

Figure 6.4 (a) Illustration of how a color-changed thermochromic smart window works. (b) Illustration of designed structures for simulation. (c) Calculated transmittance spectrum. The colorful background in (c) denotes the visible spectrum from 370 nm to 770 nm. Reprinted from Ref. [10].

attributed to the formation of a layer with a gradient optical constant along the vertical direction of the interfacial plane, compared to the planar VO_2 film with a fixed optical constant. The grid structure, in principle, also increases the effective volume for solar modulation. Moreover, the LSPR phenomenon in the nanostructure is able to tune the VO_2 absorption of the incident light with a shorter wavelength (around 1250 nm), which contains much denser solar energy compared to the light with a longer wavelength.

In this section, we start with the simulation work regarding the grid VO_2 film for enhanced thermochromic performance, followed by a discussion of the experiment works. We summarize previous studies, including simulation and experiment work.

On the simulation side, optical properties of nanogridded VO_2 films were investigated by Long's group using a 3D FDTD [13]. They proposed three different grid VO_2 nanostructures: structures with square holes, structures with circular holes, and structures with circular holes in a hexagonal arrangement (Fig. 6.5a). They numerically optimized a number of parameters, including structural models, cavity size, periodicity, film thickness, and fill factor. The thermochromic performance of square holes in square lattices is plotted in Fig. 6.5b in relation with the periodicity and the square hole length. The T_{lum} and ΔT_{sol} are indicated by the black and red dashed lines, respectively. It can be concluded that an increase in the square hole length induces a decrease in the ΔT_{sol} while increasing the T_{lum}. Also, an increase in the periodicity results in a decrease in the T_{lum}, accompanied by a slight decrease in the ΔT_{sol}. Optimized thermochromic performance ($T_{lum} = 76.5\%$ and $\Delta T_{sol} = 14\%$) was found on hexagonal-packed circular cells of VO_2 with a radius of 80 nm, periodicity of 160 nm, and thickness of 300 nm. The method demonstrated that the nanogrid structure is an effective way to enhance the T_{lum} without reducing the ΔT_{sol}, compared to the planar film.

The successful fabrication of grid VO_2 films in experiments is relatively more complex compared to fabrication of disordered or randomly porous films. For fabrication in or above the microscale, methods such as mesh printing and electrodeposition are demonstrated to be efficient. However, for fabrication in the nanoscale, more advanced and sophisticated methods are required, for example, nanosphere lithography.

Micropatterned VO_2/Si-Al gel composites were produced by Lu et al. using the mesh printing method (Fig. 6.6a) [14]. Besides achieving a competitive thermochromic performance, the fabrication method is facile, efficient, and controllable. In the experiment, commercialized quasi-sphere VO_2 particles with an average diameter of around 100 nm were dispersed in a Si-Al gel matrix. The grid structures of the produced films are consistent with the patterns of the meshes applied (Fig. 6.6b), demonstrating the effectiveness of the mesh printing method in preparing micropatterned structures. Factors such as film thickness, VO_2 concentration, and grid styles are optimized to enhance

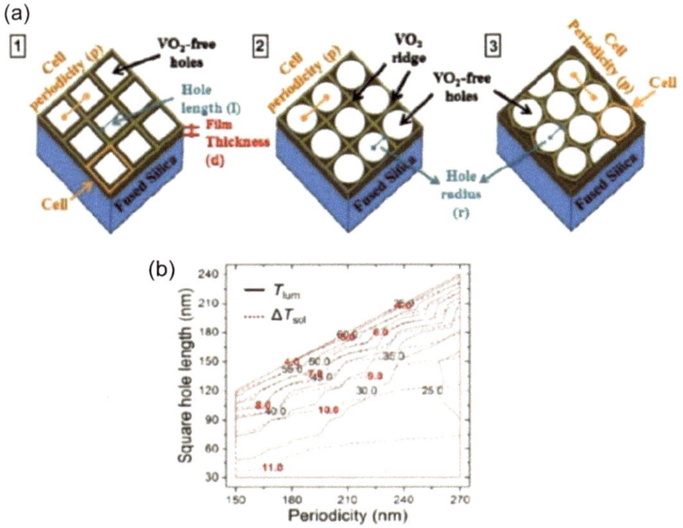

Figure 6.5 (a) Simulation models of grid films with square holes (a1) and circular holes (a2) in square lattices as well as circular holes in a hexagonal arrangement (a3). (b) Summary of the calculated T_{lum} and ΔT_{sol} of the model with square holes in square lattices under various periodicities and square hole lengths. Reprinted from Ref. [13].

the thermochromic performance. Rationally, a decreasing T_{lum} and an increasing ΔT_{sol} are observed with an increasing film thickness or VO_2 concentration. According to the report, optimized performance is demonstrated by the samples using meshes of sizes 230 μm and 55 μm, showing ΔT_{sol} of 15 and 14.9, respectively, accompanying by T_{lum} of 40.1 and 43.3, respectively. The achieved performance is comparable to the best results achieved through other approaches, such as antireflection overcoating, biomimetic design, nanothermochromism, and porosity.

The electrodeposition method has been demonstrated by Liu et al. as an effective method for fabricating patterned VO_2 films above the microscale [15]. In the electrodeposition method, an ionic strength is applied to facilitate the deposition process of suspended VO_2 nanocrystals from the solution to the predesigned conductive mesh (Fig. 6.7a). The fabricated films inherit the patterning from the conductive mesh (Fig. 6.7b), demonstrating

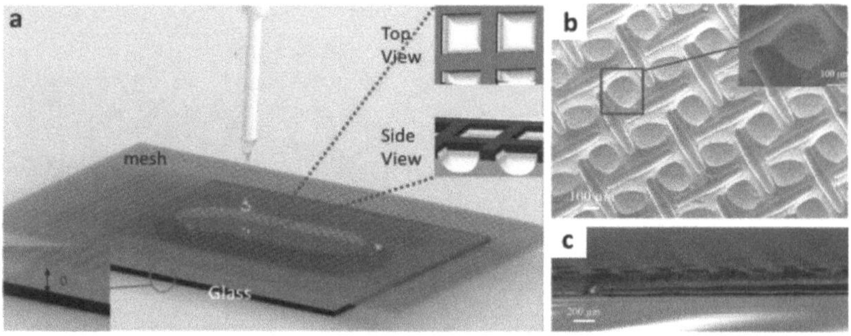

Figure 6.6 Illustration of the mesh printing process. (b) The SEM image of the patterned VO_2 composite film. (c) The cross-section optical image of the patterned sample. Reprinted from Ref. [14].

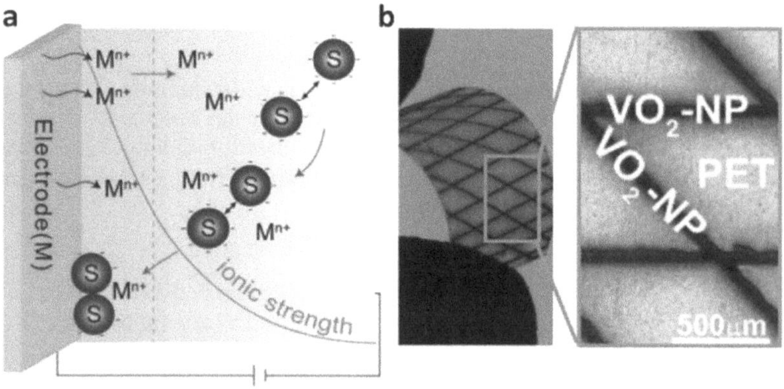

Figure 6.7 Schematic of the electrodeposition process via ionic strength. (b) Optical images of the prepared sample and the magnified optical image of the grid structures. Reprinted from Ref. [15].

the effectiveness of the morphology control of the method. Flexible substrates, such as polyethylene terephthalate, can be applied in this approach (Fig. 6.7b). Liu et al. further measured the thermochromic performance of the produced films and demonstrated a better performance than that of the continuous VO_2 film in terms of both T_{lum} and ΔT_{sol}. Besides VO_2 nanocrystals, other commonly used materials can also be successfully deposited via this method, including 0D Au nanoparticles, 1D multiwalled carbon nanotubes (CNTs), and 2D graphene oxide nanosheets.

Preparation of the grid VO_2 film in the nanoscale is relatively much more sophisticated than that in or above the microscale. The preparation method requires better precision control. Both template-free and template–assisted methods are used for grid VO_2 film though generally the template-assisted methods enable much better morphology control compared to the template-free methods.

Liu et al. [16] used the template-free methods called "on spontaneous self-template and assembly process during dual-phase transformation" to successfully prepare ordered VO_2 honeycomb structures. These processes are shown in Fig. 6.8a–d, and the corresponding results are demonstrated in Fig. 6.8e–h. In the experiment, an aqueous vanadium precursor containing vanadyl dichloride (VCl_2), hydrazine (N_2H_4), hydrochloric acid (HCl), and polyvinylpyrrolidone (PVP) was spin-coated on a quartz substrate and then dried in nitrogen. The formation of hydrous colloids was promoted by N_2H_4 but impeded by HCl. During the evaporation process, the promotion effect of N_2H_4 was enhanced but the quenching effect of HCl was weakened, which helped the homogeneous precursor to develop into hydrous colloidal spheres. These spheres self-assembled into a close-packed structure on the colloid-substrate interface when a saturated level was reached in the hydrous colloids. Then, hollow sphere structures developed at 300°C and collapsed and crystallized into VO_2 on annealing at 500°C. These processes are demonstrated experimentally. By optimizing the experimental conditions, a high transparency (95.4% at 700 nm) was achieved accompanied by $\Delta T_{sol} = 5.5\%$ on the honeycomb VO_2 film with a thickness of 65 nm. They further calculated the transmittance and reflectance spectra of periodic honeycomb and nonperiodic VO_2 films using the 3D FDTD method. From both experimental and experimental sides, they demonstrated that the honeycomb-structured VO_2 coating displayed higher transmittance than the nonperiodic VO_2 film.

Using a process different from self-assembly relying on dual-phase transformation, Zhang et al. prepared the self-assembling VO_2 nanonet on a (001) single-crystal sapphire substrate by

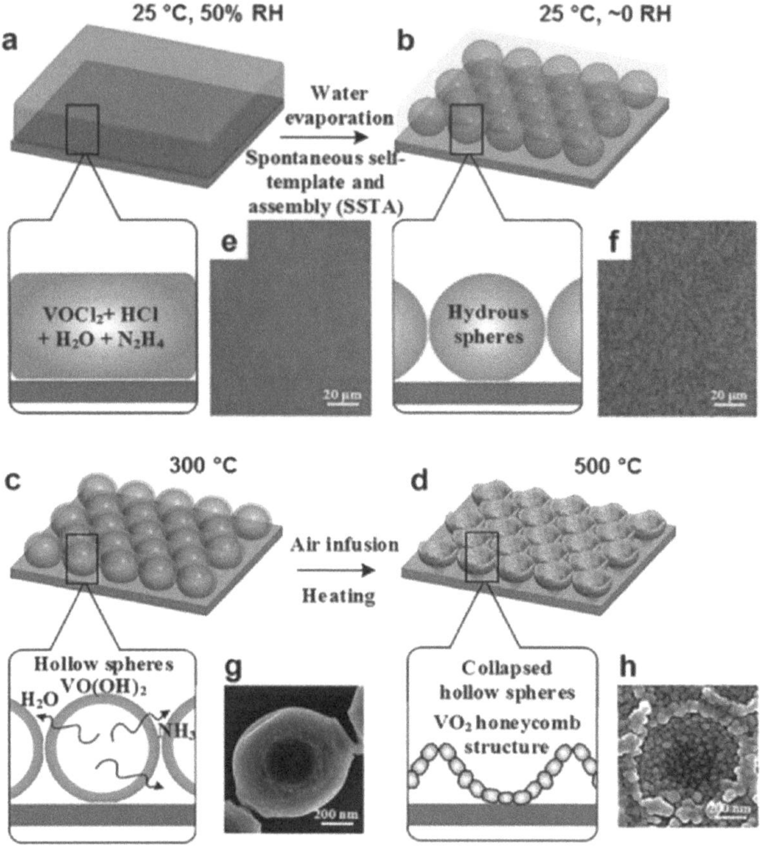

Figure 6.8 (a–d) Illustrations of the spontaneous self-template and assembly process during dual-phase transformation processes. (e–h) Optical/SEM images of the corresponding steps in (a–d). Reprinted from Ref. [16].

synchronously controlling the growth direction and crystallographic orientation [17]. The key is lattice matching between the low surface energy plane of VO_2 and the threefold symmetric (001) plane of sapphire, which induced the growth of $(020)_M$ VO_2 on (001) sapphire substrates followed by the growth of $(001)_S$ VO_2 to suspended nanorods in three equivalent directions. The films were produced in the waferscale and displayed dramatic fatigue endurance.

Template-assisted methods are relatively more effective in controlling the structures and reducing the defects, such as grain boundaries, point defects, and unit size distribution. Also, template-assisted methods help control the grid structures on a relatively smaller scale, which is demonstrated by a periodicity down to around 200 nm, which is around 800 nm for template-free methods. Though these unique properties make template-assisted methods relatively more popular than the others, shortages remain, for example, the price of the templates is commonly high and the size of the produced sample is usually not large due to the limited size of the templates.

Nanosphere lithography has been recognized as a facile, inexpensive, productive, and flexible technique for preparing functional 2D patterned nanostructures [18–20]. This method relies on the template made by a close-packed nanosphere monolayer. The monolayer is well developed and easily prepared by an interfacial self-assembly method.

The group of Xie pioneered monolayer colloidal crystal (MCC) templates made of PS nanospheres to prepare periodic porous VO_2 films [21]. They illustrated the process as shown in Fig. 6.9a. More specifically, the MCC templates are fabricated via air-water interfacial self-assembly of PS nanospheres. Then, the templates are first immersed in a $VOSO_4$ solution and then in a NH_4HCO_3 solution. After that, a high-temperature annealing process is applied to remove the organic template and to assist in the crystallization of VO_2 crystals. In the produced samples, hexagonal patterned arrays are observed, such as the nanonet structure (Fig. 6.9b). As in the case of most of the other template-assisted methods, the morphologies of the produced samples are inconsistent with those of the templates, the reason for which is illustrated in Fig. 6.9c. Multilayer structures are achievable by repeating the process shown in Fig. 6.9a. Using nanospheres with fixed diameters of around 460 nm, Xie and group further measured the optical performance of the produced samples and demonstrated the enhancement of T_{lum} on the grid VO_2 structures.

More systematically, the group of Long explored the 2D-patterned VO_2 films with tunable periodicities and diverse nanostructures, including nanoparticle, nanonet, and nanodome arrays, by

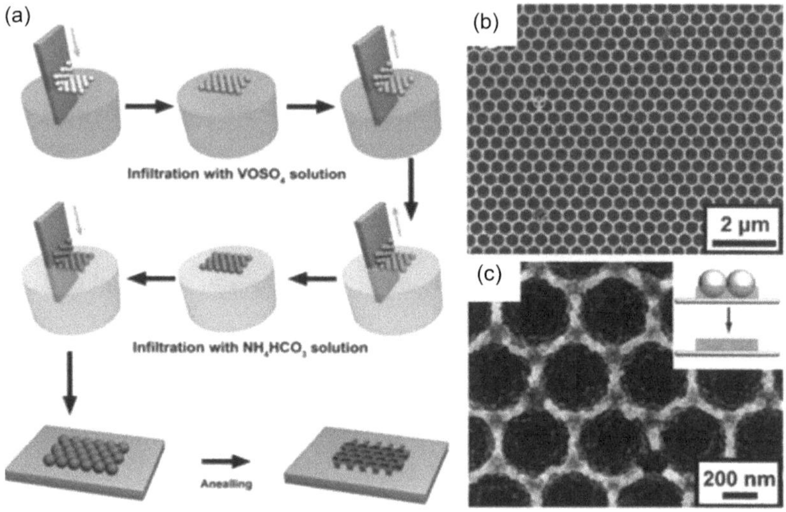

Figure 6.9 (a) Diagram of the solution-based lithography method utilized to fabricate a periodic VO_2 film. (b, c) SEM images of the periodic VO_2 film produced by the colloidal lithography method in (c) low and (b) high magnifications and its proposed mechanism as the inset of (c). Reprinted from Ref. [21].

the nanosphere lithography technique [22]. The fabrication process is quite flexible because the plasma etching (PE) duration and the precursor viscosity can be controlled. They concluded the synthesis routes shown in Fig. 6.10. When PE of short duration is applied, nanoparticle and nanodome arrays are produced using low-viscosity (Route 1) and high-viscosity (Route 2) precursors, respectively. Nanonet arrays are fabricated by prolonging the PE duration and using a low-viscosity precursor (Route 3). Besides, the periodicity of the pattern structures is controlled from 200 to 800 nm using nanospheres with corresponding diameters.

The produced 2D patterned VO_2 arrays are highly uniform and can be precisely controlled. For example, samples are produced in a hexagonally patterned VO_2 nanoparticle array with an average diameter of 60 nm and periodicity of 160 nm. The successful preparation of the three structures is shown as the SEM images in the three lines of Fig. 6.11. The hexagonal patterns are indicated as the yellow hexagons in Fig. 6.11a–c. The hollow structures below

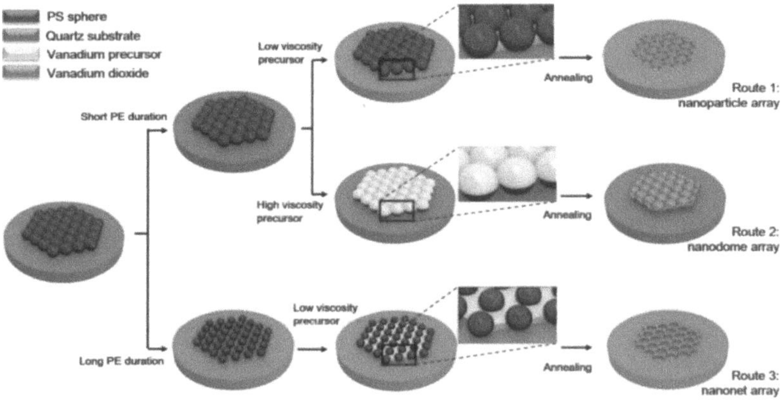

Figure 6.10 Schematic of the fabrication routes for the nanoparticle, nanodome, and nanonet arrays. Reprinted from Ref. [22].

the top of the domes are demonstrated on nanodome arrays (inset of Fig. 6.11h). In addition, the periodicity of the nanostructures is successfully controlled using different templates with nanospheres of sizes 160 nm, 490 nm, and 830 nm. These results are presented as the SEM images in the three columns of Fig. 6.11. These results can verify the feasibility of the three preparation routes in Fig. 6.10. On the basis of these results, it is observed that the morphological control is better on samples with relatively large periodicities (800 and 490 nm) than on samples with small periodicities (160 nm) (Fig. 6.11a–f). For example, more grain boundaries are observed in nanonet arrays with a periodicity of ~160 nm than those with periodicities of ~490 and ~830 nm. Another more distinguish example is the nanodome samples. The nanodome structures are well maintained on 490 and 830 nm samples (Fig. 6.11h,i) but are almost unobservable on the 160 nm samples (Fig. 6.11g).

The group of Long further explored the optical properties of these films. It is of great interest that the nanoparticle arrays exhibit transmittance valleys in the NIR range with a peak position around 1200–1500 nm, which cannot be observed on continuous VO_2 films. These transmittance valleys are due to the phenomenon of LSPR, commonly observed on metal nanocrystals. The nanoparticle array gives rise to a tunable peak position and intensity of the LSPR at

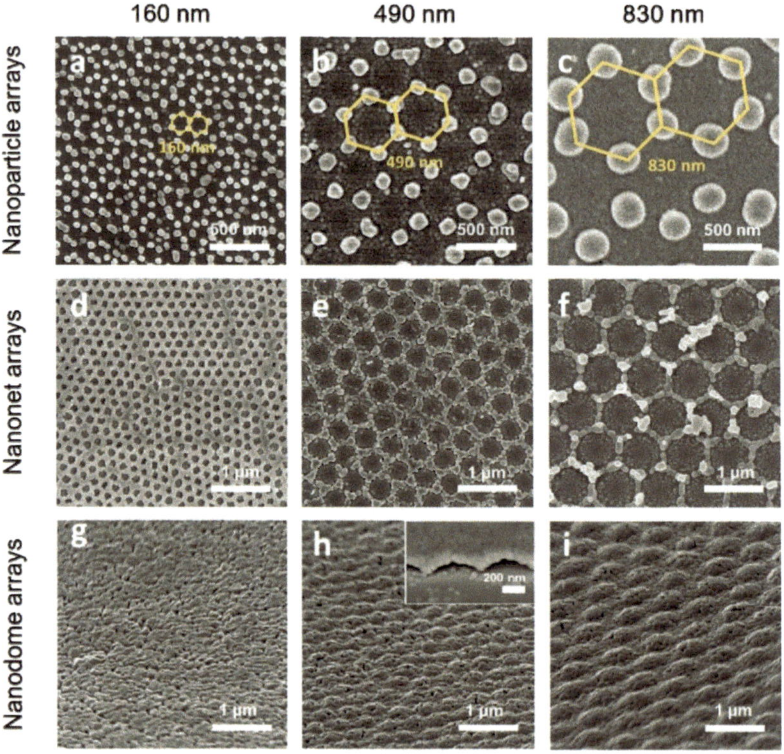

Figure 6.11 SEM images of VO_2 crystal arrays with (a–c) the nanoparticle, (d–f) nanonet, and (g–i) nanodome morphologies. Different patch sizes are prepared using 160 nm, 490 nm, and 830 nm PS spheres. (d–f) A tiled view. The inset of (h) is a high-magnitude image of (h) on the sample edge. Reprinted from Ref. [22].

different temperatures. LSPR happens when the VO_2 nanoparticles change from semiconductor to metallic phases, releasing the free carriers to support the plasmon resonance on the crystal surfaces. LSPR was also found to red-shift with an increase in the particle size and media reflective index, and these results fit well with the tendency calculated using 3D FDTD. Besides the fact that a decent thermochromic performance (up to $\Delta T_{sol} = 13.2\%$ and $T_{lum} = 46\%$) has been achieved, 2D patterned VO_2 films have been demonstrated as efficient smart thermal radiation filters to enable remote control of the lower critical solution temperature (LCST) behavior of a poly(*N*-isopropylacrylamide) (PNIPAm) hydrogel. Compared with the template-free method, periodic films produced by the nanosphere lithography technique offer more uniform periodicity (fewer periodic defects) as well as a smaller periodicity (as small as ~200 nm).

In conclusion, compared to planar VO_2 films, gridded structures are an efficient method to promote thermochromic performance, demonstrated through methods such as mesh printing, ionic strength–induced electrodeposition, dual-phase transformation, and nanosphere lithography. The successful preparation of grid structures in the nanoscale is relatively more challenging. Also template-assisted methods display relatively higher reliability than template-free methods in terms of morphology control. However, controllable preparation of grid VO_2 films in the nanoscale is still a challenge, which may be due to the harsh conditions for the crystallization of monoclinic/rutile VO_2 crystals. Further exploration of grid VO_2 films is one promising direction. Considering the diverse existing technologies for the fabrication of nanoscale materials and the interesting phenomenon of materials in the nanoscale, grid VO_2-based thermochromic films in the nano- and microscale should be further explored.

6.3 Hybridation

VO_2 has a higher refractive index below the critical temperature (T_c) than above, and a large luminous absorption in both states poses a veritable challenge to attain a large solar modulating ability (ΔT_{sol}) and a high luminous transmittance (T_{lum}) simultaneously. Combining VO_2 with other materials to form composites has been investigated to (i) enhance thermochromism in VO_2 (to enhance both ΔT_{sol} and T_{lum}), (ii) increase the stability of VO_2, and (iii) develop multifunctional VO_2-based smart windows.

6.3.1 Enhance Thermochromic Properties

To enhance thermochromic properties, numerous materials, including hydrogels, ionic liquids, and liquid crystals (LCs), have been

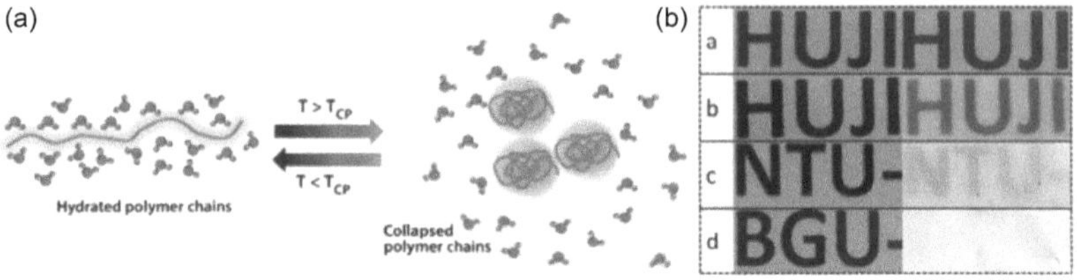

Figure 6.12 (a) Schematic of the polymer chains of a thermoresponsive hydrogel that undergoes hydrophilic-to-hydrophobic transition. Reproduced with permission from Ref. [36]. Copyright 2016, Elsevier. (b) PNIPAm hydrogel thin films at room temperature (left) and 35°C (right) with different thicknesses: (a) 26 μm, (b) 52 μm, (c) 78 μm, and (d) 200 μm. Reproduced with permission from Ref. [30]. Copyright 2014, RSC.

selected as a response matrix for fabricating VO_2-based composite devices.

6.3.1.1 VO_2/hydrogel composites

A hydrogel has a 3D network structure of hydrophilic polymer chains in which water is used as the dispersion agent. Due to the intrinsic properties, hydrogels have been applied in different research areas, including tissue engineering [23, 24], drug delivery [25, 26], cell culture [27, 28], biological implantation [29], and energy saving [30, 31].

Thermochromic hydrogels include polyampholyte hydrogels [32], PNIPAm [30, 33, 34], and hydroxypropyl cellulose (HPC) [35] and have been applied in thermochromic smart windows applications. Significantly, thermochromic hydrogels undergo a hydrophilic-to-hydrophobic transition [36] (Fig. 6.12). In other words, when the temperature is lower than the LCST, hydrogels are transparent and when the temperature is above the LCST, hydrogels turn translucent and opaque [37–42].

To enhance the visible light modulating ability, thermochromic hydrogels were selected as host materials for VO_2-based composites [33, 35]. The first incorporation of pure VO_2 particles with a PNIPAm hydrogel was achieved by Zhou et al. on a hybrid of the hydrogel with inorganic VO_2 particles [33]. With the inherited

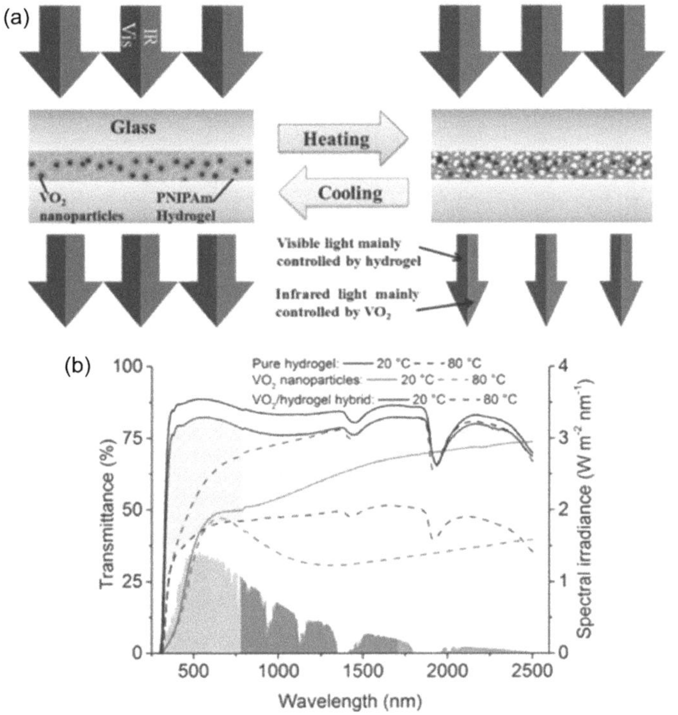

Figure 6.13 (a) Schematic of the sandwiched structure of a PNIPAm/VO_2 hybrid smart window. (b) Transmittance spectra of a PNIPAm hydrogel, VO_2 nanoparticles, and a PNIPAm/VO_2 hybrid at 20°C and 80°C from 250 to 2500 nm. Reproduced with permission from Ref. [33]. Copyright 2015, The Royal Society of Chemistry.

structure, the PNIPAm hydrogel/VO_2 hybrid thin film is sandwiched between two pieces of glass substrates (Fig. 6.13a). Figure 6.13b shows the transmittance spectra of a pure PNIPAm hydrogel, pure VO_2 nanoparticles, and a PNIPAm/VO_2 hybrid at 20°C and 80°C. Compared with the pure PNIPAm hydrogel, the ΔT_{sol} of the hybrid material has been significantly improved, especially in the NIR region, with a giant contrast of 22.2%. The PNIPAm/VO_2 hybrid film also shows relatively high durability because after it is heated and cooled between 20°C and 80°C for 20 cycles, the changes in both T_{lum} and ΔT_{sol} are negligible. Compared with a pure VO_2 nanoparticles sample, the PNIPAm/VO_2 hybrid hydrogel shows a higher T_{lum} at both room and elevated temperatures. Zhou et al. also developed an HPC tungsten–doped VO_2 hybrid microgel, responsive at near room temperature [35].

6.3.1.2 Ion liquids/VO_2 composites

An ion liquid is a kind of salt in a liquid state without any vaporization or decomposition. An ion liquid is made up of ions and short-lived ion pairs instead of neutrally charged molecules, which is recognized as the solvent of the future [43, 44]. It has been ascribed with diverse potential applications, including nuclear fuel processing [45], cellulose processing [46, 47], solar thermal energy system [48, 49], and batteries [50, 51].

A mixture of a transition metal compound and an ionic liquid can be applied to energy saving smart windows due to its solar modulating ability, which originates from solar radiation absorption instead of reflection or scattering. Currently, energy saving smart windows applications are mainly based on the octahedral-tetrahedral configuration change of transition metal complexes assisted by the interaction with the donor solvent molecules (Fig. 6.14a). The ion liquid–based smart windows display high sensitivity to and efficiency in temperature change.

Zhu et al. reported a composite film for smart windows applications by combining IL-Ni-Cl ion liquid complexes and VO_2 nanoparticles [52]. It is observed that a pure IL-Ni-Cl ion liquid film displays a significant transmittance change mainly in the 500–800 nm spectrum range, leading to a limited modulation ability because this range is relatively narrow compared to the broad solar energy spectrum (250 to 2500 nm) (Fig. 6.14b). The composite film of VO_2 nanoparticles and IL-Ni-Cl complexes significantly broadens the modulating spectrum and promotes ΔT_{sol}, contributing to the combination of tunable absorption of the IL-Ni-Cl to visible light (500 to 800 nm) as well as dominated modulation of the VO_2 nanoparticles to NIR light (700 to 2500 nm) (Fig. 6.14b). As shown in Fig. 6.14b, IL-Ni-Cl/VO_2-hybridized materials demonstrate a significantly widened solar regulation spectrum (500–2500 nm).

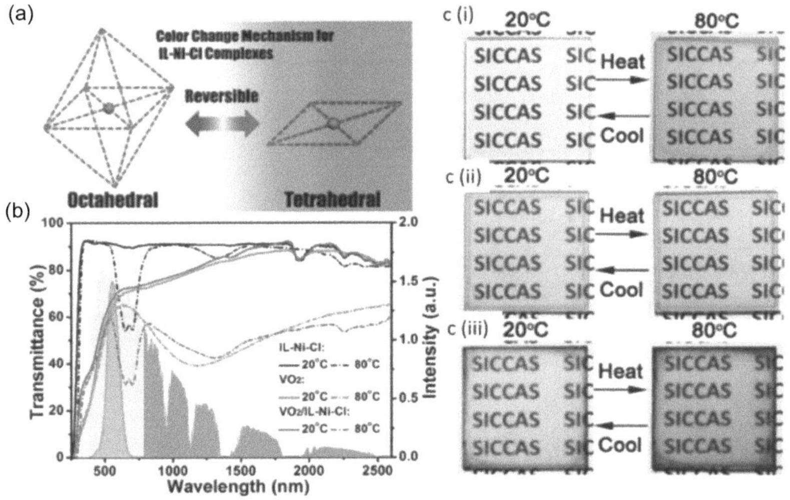

Figure 6.14 (a) Schematic of octahedral-to-tetrahedral configuration of an ionic liquid–based complex system. (b) Transmittance spectra of a pure IL-Ni-Cl film, VO_2 nanoparticles, and a VO_2/IL-Ni-Cl hybrid film at 20°C and 80°C from 250 to 2500 nm. (c) Photographs of a pure IL-Ni-Cl film, a VO_2 nanoparticle film, and a VO_2/IL-Ni-Cl hybrid film at both 20°C (left) and 80°C (right). Reproduced with permission from Ref. [52]. Copyright 2016, American Chemical Society.

Solar regulation efficiency up to 26.5% is achieved while the composite system is heated up to 80°C. Meanwhile, the luminous transmittance is maintained at 43.9%, which is a moderate level. An appearance change is observed on the composite as the color of the film changes from light brown (20°) to dark green (80°), which contributes to the combined effect of color changes of IL-Ni-Cl (transparent to blue) and VO_2 (yellowish brown at all time) (Fig. 6.14c). Another system of mixing VO_2 nanoparticles and NLETS (nickel (II)–based ligand exchange thermochromic system) has been developed by this group [53]. It demonstrates an excellent ΔT_{sol} of 18.19% while a relatively high illumination intensity of 68.71% is maintained. VO_2 nanoparticles hybridized with an ion liquid–based complex system have been recognized as a promising candidate for energy saving smart windows applications.

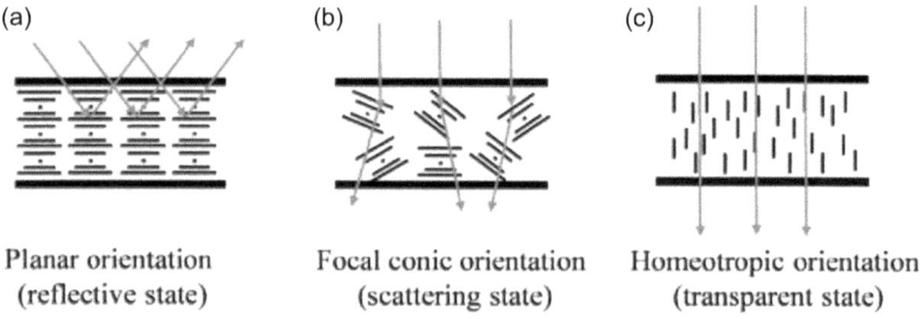

Figure 6.15 Schematic of the orientation-dependent optical behavior of cholesteric liquid crystalline materials. Reproduced with permission from [59], Copyright © 2017 John Wiley & Sons.

6.3.1.3 Liquid crystals/VO_2 composites

LCs are an intermediate material between conventional liquids and crystalline solids [54]. Most LCs are made of organic compounds, which are characterized by the anisotropy of electrical, optical, and mechanical properties [55]. These properties of LCs are responsive to multiple external stimuli, including electric field, temperature, and light, which make LCs materials of interest across a broad range of applications [56–58].

Thermochromic LCs can tune the orientation of anisotropic molecules with temperature changing because the optical properties of LCs are decided by the orientation of anisotropic molecules [59]. Normally, LC-based smart windows have a sandwich structure (LC molecules sandwiched in between two pieces of glasses) and can be tuned to different orientations. As shown in Fig. 6.15a, in a reflective state, the molecules are oriented parallel to glass slides and assembled in helical structures. In the scattering state, the helical structures are randomly tilted in a focal conic orientation (Fig. 6.15b). In a homeotropic state, the molecules are oriented perpendicular to the glass slides or the LC smart window (Fig. 6.15c). In scattering and reflective states, the light incident angle, the pitch of the LCs, and the materials' refractive index decide the reflection band.

Liang et al. developed an LC/VO_2-hybrid thermochromic smart window [60]. In this work, as shown in Fig. 6.16a, tungsten-doped VO_2 (W-VO_2) nanoparticles and LCs were dispersed in the phase-

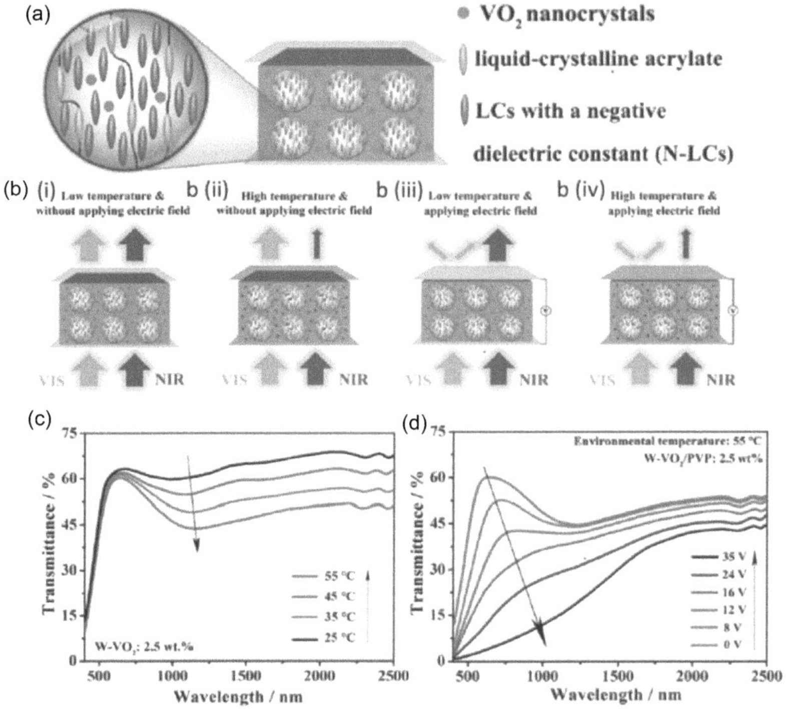

Figure 6.16 (a) Schematic of the LC/VO_2 hybrid film, (b) LC/VO_2 composite film with four optical states under different stimuli, and (c) temperature-dependent spectra of the composite film containing 2.5 wt.% W-doped VO_2. (d) Electric field dependence of the transmittance spectra of the composite film containing 2.5wt.% W-doped VO_2 at 55°C. Reproduced with permission from Ref. [60]. Copyright 2017, The Royal Society of Chemistry.

separated polymer matrix homogeneously. The LCs and W-VO_2 nanoparticles work independently, and both electrochromic and thermochromic properties of LCs and W-VO_2 are retained after they combine. Thus, as both LCs and W-VO_2 nanoparticles can work individually, this responsive hybrid film displays four states under different stimuli, which show different optical properties that help control both visible and infrared (IR) lights individually (Fig. 6.16b). For example, as shown in Fig. 6.16c, as the temperature increased from 25°C to 55°C, the transmittance of the pure W-VO_2 film (2.5 wt.%) decreased, especially in the NIR spectrum range. The transmittance (400–2500 nm) was further decreased significantly in the visible range when a voltage of 0–35 V was applied (Fig. 6.16d), where a ΔT_{sol} of up to 34.6% was achieved.

6.3.2 Increasing Stability

Numerous materials, including polyurethane (PU), SiO_2, and other stable transparent materials, have been selected as the host in order to fabricate VO_2-based composites with enhanced stability, for example, in terms of antioxidation. At the same time, such composites can also have better thermochromic properties.

6.3.2.1 Transparent host material/VO_2 composite

Transparent, nonresponsive materials, like PU, have been used as the matrix for VO_2-based thermochromic composites [61–63]. Chen et al. combined pure VO_2 nanocrystals with PU, which showed excellent thermochromic properties [61]. The VO_2 nanoparticles (26 mm) were well dispersed in the PU matrix with the assistance of PVP (Fig. 6.17a,b) to form a visible homogenous film (Fig. 6.17c). Reflectance of this PU/VO_2 composite film (below 8%) is much lower than that of pure VO_2 coated on bare glass. Due to the low reflectance and small size of the VO_2 nanoparticles (high crystallinity), a remarkable ΔT_{sol} of 22% was achieved. At the same time, the average T_{lum} of 42.8% is also acceptable (Fig. 6.17d). The thermochromic properties of this PU/VO_2 composite thin film are very close to that shown in the best simulation results (ΔT_{sol} = 23.7%).

To further block IR light, Gao et al. introduced Sb-doped SnO_2 (ATO) into a PU/VO_2 composite film, where ATO is able to block IR light and maintain high visible transparency at the same time [63]. The new ATO- PU/VO_2 composite achieved thermochromic properties comparable to those of the pure PU/VO_2 composite film (Fig. 6.18a,b). The thermochromic properties were further tested in two model houses (Fig. 6.18c) and demonstrated that an ATO-PU/VO_2 composite was more effective at shielding against IR transmittance than the PU/VO_2 composite.

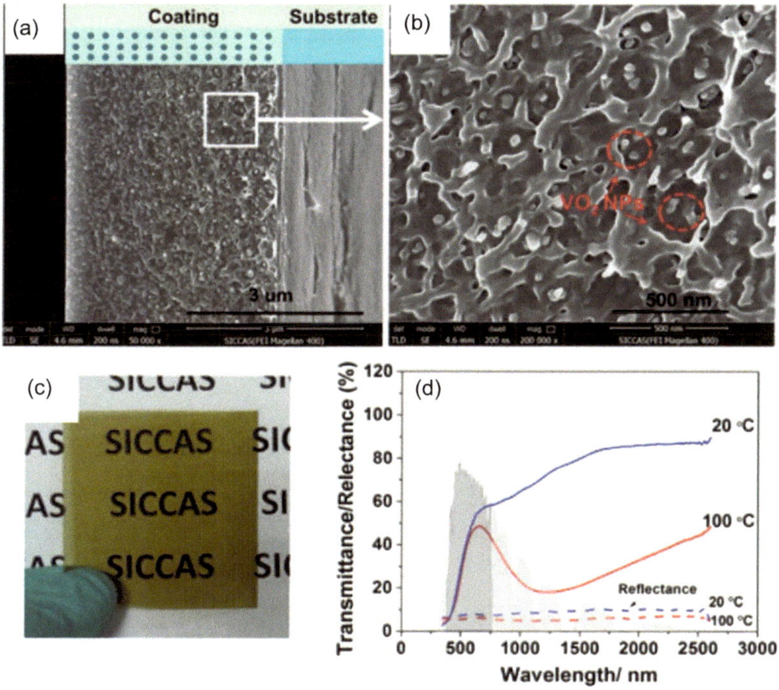

Figure 6.17 (a) Cross-sectional SEM of the PU/VO_2 composite coating and (b) a partial enlargement of the image in (a). (c) Intrinsic yellowish-green color of a PU/VO_2 composite film, and (d) transmittance and reflectance of a PU/VO_2 composite film at 20°C and 100°C [61]. Copyright 2014, The Royal Society of Chemistry.

A new transparent Si-Al gel was developed by Long's group as the host material [64]. As VO_2 nanoparticles were well dispersed in the Si-Al gel, an optimized thermochromic performance was achieved by a 3 μm thick film (10 wt.% VO_2).

6.3.2.2 VO_2/SiO_2 core–shell composite

To prevent VO_2 nanoparticles from oxidation, core–shell structures were developed to act as a protective layer to restrain oxygen diffusion, which will protect thermodynamically unstable VO_2 from oxidation. At the same time, the nanothermochromic properties of VO_2 nanoparticles were maintained. Gao et al. successfully

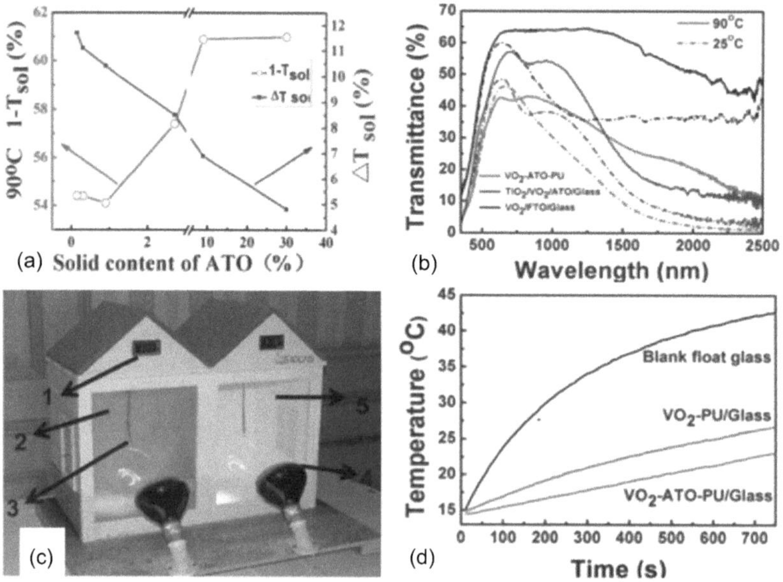

Figure 6.18 (a) $1 - T_{sol}$ (90°C) and ΔT_{sol} change with the solid content of ATO (a higher $1 - T_{sol}$ indicates a better solar heat shielding ability), (b) transmittance comparison of a VO_2-ATO-PU film with an ATO solid content of 2.7% with a reported VO_2/FTO and a TiO_2/VO_2/FTO multilayer film, (c) photographic illustration of the testing system (1) temperature monitor, (2, 5) test windows, (3) temperature probe, and (4) infrared lamps. (d) Temperature dependence on irradiation time for blank float glass, PU/VO_2 glass, and VO_2-ATO-PU/glass (ATO solid content of 30%) [63]. Copyright 2012, The Royal Society of Chemistry.

fabricated flexible thermochromic foils by introducing VO_2-SiO_2 core–shell nanocrystals into a PU matrix [65]. As shown in Fig. 6.19a, in order to obtain a SiO_2 shell on VO_2 nanoparticles, VO_2 was first treated by PVP and then by tetraethyl orthosilicate. VO_2-SiO_2 core–shell nanocrystals were further treated by a silane coupler to promote their dispersion in PU. Figure 6.19b,c shows the transmission electron microscopy (TEM) images of silica-coated VO_2 nanoparticles with PVP pretreatment. As shown in Fig. 6.19d,e, the films display excellent flexibility.

Jin's group developed VO_2@SiO_2 core–shell nanorod structures and W-doped VO_2-SiO_2 core–shell structures [66, 67]. Both

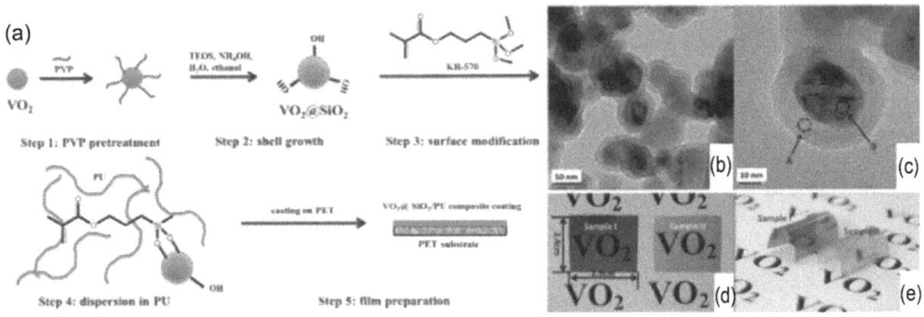

Figure 6.19 (a) Scheme of the preparation procedure for a $VO_2@SiO_2/PU$ composite, starting from VO_2 nanoparticles. (b, c) TEM images of silica-coated VO_2 particles after functionalization with PVP. (d, e) Photographs of flexible thermochromic films based on $VO_2@SiO_2/PU$ composites with high (Sample I) and low (Sample II) VO_2 contents [65]. Copyright 2012, The Royal Society of Chemistry.

approaches achieved high thermochromic properties as well as better stability compared with those of pure VO_2 nanocrystals. The decent thermochromic performance benefited from surface plasmon resonance.

6.3.3 Multifunction

Combining some advanced functions, such as self-cleaning and energy conservation and generation, into one thermochromic smart window device is a highly attractive scientific undertaking. To generate multifunctional energy saving smart windows, some functional materials, like TiO_2, fiber-optic strain, and poly(methyl methacrylate), or devices like solar cells (SCs) and transparent heaters have been applied on VO_2-based smart windows.

6.3.3.1 Self-cleaning and wettability smart windows

In daily life, windows always come in contact with water or pollutants, which may lower there transparency and, at the same time, increase the cleaning cost. To address these problems and achieve a multifunctional device, some highly transparent materials with wettability and self-cleaning properties have been integrated with VO_2-based smart windows as a coating [68].

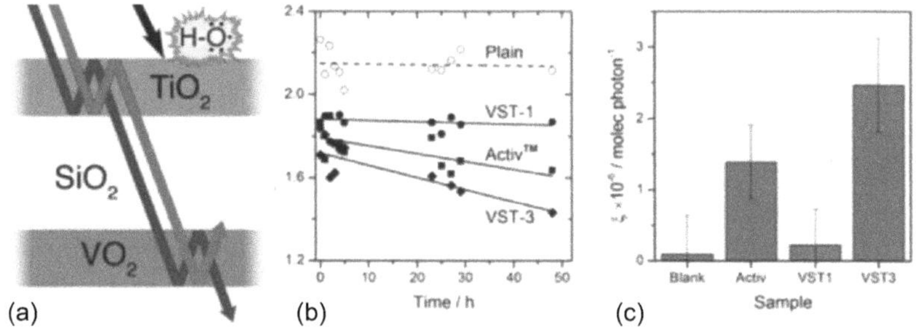

Figure 6.20 (a) Illustration of the $VO_2/SiO_2/TiO_2$(A) multilayer for self-cleaning and energy-saving coatings. (b, c) Degradation of stearic acid under UV illumination ($I = 4$ mW cm^{-2}) on $VO_2/SiO_2/TiO_2$(R) (VST-1), $VO_2/SiO_2/TiO_2$(A) (VST-3), plain (blank), and Pilkington Activ glass (Activ): (b) time-dependent integrated area of stearic acid and (c) the calculated corresponding formal quantum efficiency (ξ), presenting the molecular quantity of degraded stearic acid per incident photon. Reproduced with permission from Ref. [73]. Copyright 2014, American Chemical Society.

TiO_2(A) (anatase) enables effective photocatalytic degradation of organic contaminants, has antireflection properties, and can be coated on a VO_2 surface as a highly transparent thin film [69–72]. Parkin's group produced a $TiO_2/SiO_2/VO_2$ composite film with a multilayer structure that has the dual function of energy saving and self-cleaning (Fig. 6.20a) [73]. A TiO_2(A)-coated film was able to effectively remove adherent stearic acid over time under ultraviolet (UV) illumination ($I = 4$ mW cm^{-2}), which was comparable to the performance of commercial Pilkington Activ glass, as shown in Fig. 6.20b. The self-cleaning character of TiO_2(A) arises from its photocatalytic property, as indicated by a comparison of the molecular quantity of degraded stearic acid per incident photon (ξ) on these four samples (Fig. 6.20c). The sandwiched SiO_2 layer works as a barrier against Ti^{4+} ion diffusion as well as enhances the antireflection performance.

Instead of applying a sandwiched SiO_2 layer, TiO_2(A) and TiO_2(R) were also directly deposited on each side of a planar VO_2 layer [74]. As UV illumination duration increased, the wettability of the film changed from hydrophobic to hydrophilic, while the water contact

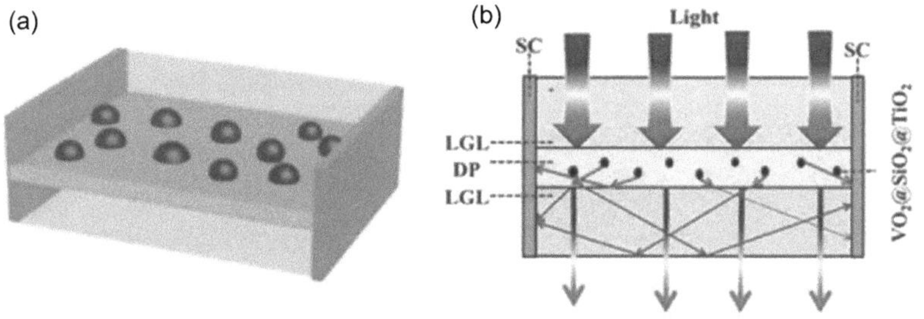

Figure 6.21 (a) 3D structure of the prepared VO_2-based smart window. The solar cells are assembled in a manner that surrounds the module. Here, three sides are shown. (b) Illustration of the integrated SC and VO_2-based thermochromic film utilizing light scattering. The light guide layer (LGL) and the low reflective index medium (DP) are made of polycarbonate (PC) and a VO_2/polyurethane (PU) composite, respectively. Some of the incident light is scattered to the SCs. Reproduced with permission from Ref. [78], Copyright © 2013, Nature Publishing Group.

angle decreased from 86.4° to 3.7°. Furthermore, a VO_2 thin film with ZnO coating also showed enhanced durability and antibacterial and biocompatible properties [75–77].

6.3.3.2 Energy-generating VO_2 smart windows

Smart windows that both conserve and generate energy are a highly attractive scientific undertaking. VO_2-based smart windows with both these properties have been achieved by integrating SCs with VO_2-based thermochromic layers [78, 79].

Zhou et al. fabricated an SC smart window by using the light scattering effect of VO_2 nanoparticles [78]. In this approach, the SCs were attached to the four sides of the VO_2-based thermochromic bulk, vertical to the VO_2 film (Fig. 6.21a). This thermochromic bulk has a sandwich structure—a VO_2@SiO_2@TiO_2 nanoparticle/PU composite layer was sandwiched between two light guide layers (LGLs) (Fig. 6.21b). The incident light was scattered by the VO_2@SiO_2@TiO_2 nanoparticles to the SCs on the four sides, facilitating electricity generation (Fig. 6.10a). For this SCs/VO_2 smart

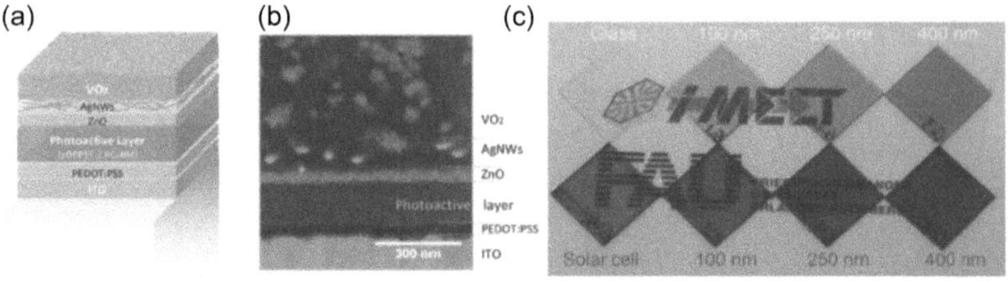

Figure 6.22 (a) Illustration of the layered integration of SC and VO_2-based thermochromic films utilizing light transmittance. pDPP5T-2:PC_{60}BM and PEDOT:PSS refer to the diketopyrrolopyrrole-quinquethiophene alternating with copolymer: [6,6]-phenyl-C_{60}-butyric acid methyl ester and poly(3,4-ethylenedioxythiophene):polystyrene sulfonate, respectively. (b) The cross-sectional SEM image of the multilayer configuration corresponding to the illustration of (a). (c) Photographs of the VO_2-coated glass samples (top) and the prepared smart windows (bottom). For comparison, a piece of glass and a semitransparent OPV cell without a VO_2 coating are placed in parallel. Reproduced with permission from [79], Copyright © 2015 John Wiley & Sons.

window, a ΔT_{sol} of 7.5% accompanied by a power conversion efficiency of 0.52% was achieved.

Another approach was to fabricate a semitransparent SC on a VO_2-based thermochromic film [79]. Figure 6.22a shows a layered structure of this VO_2-SC smart window, which consists of a VO_2 thermochromic film, silver nanowires (Ag NWs), ZnO, one photoactive layer, poly(3,4-ethylene dioxythiophene):polystyrene sulfonate (PEDOT:PSS), and an indium tin oxide (ITO) layer in sequence. The photoactive layer consists of diketopyrrolopyrrole-quinquethiophene alternating with copolymer: [6,6]-phenyl-C_{60}-butyric acid methyl ester (pDPP5T-2: PC_{60}BM). The VO_2-based thermochromic film was made of a composite of VO_2 nanoparticles and PVP. Distinct layer boundaries are observed in Fig. 6.22b through SEM. Figure 6.22c shows the image of pure VO_2 smart windows and VO_2-SC smart windows. The VO_2-SC smart window showed the same ΔT_{sol} (7.5%) but with a much higher power conversion efficiency (3.1%) compared with the scattering-based smart window. However, the T_{lum} of both configurations decreased considerably to 28.2%, which is due to the semitransparency of the SC.

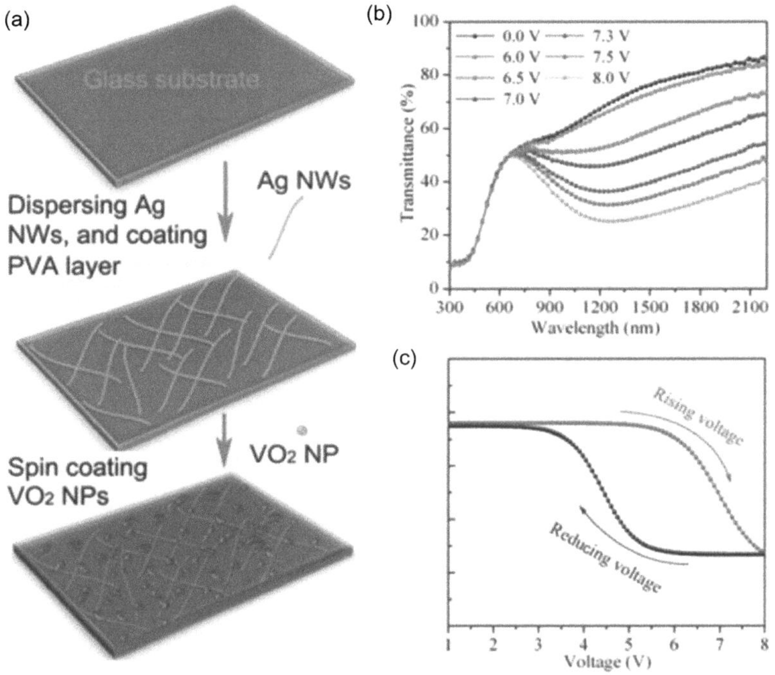

Figure 6.23 (a) Schematic representation of a VO_2/Ag-based electro-thermochromic device and its fabrication process. (b) UV-vis-NIR transmittance spectrum of the VO_2/Ag-based electro-thermochromic device at various voltages from 0 V to 8 V. (c) Transmittance hysteresis loop of the VO_2/Ag-based electro-thermochromic device as a function of the applied voltage at a wavelength of 1.5 μm.

6.3.3.3 Dual-response electrothermal VO_2 smart window

Most thermochromic smart windows are passive as they just respond to the surrounding temperature automatically. Recently, researchers have focused on developing transparent conductive films through Ag nanowires, CNTs, etc. These transparent conductive films composed of VO_2-based films can generate heat under an applied voltage, which can be used for electro-thermo control thermochromic devices. The working mechanism of the dual-responsive devices is based on joule heating.

Electro-thermochromic devices are created by depositing VO_2 nanoparticles on a transparent conductive film, such as Ag NWs or

ITO glass [80]. Once sufficient voltage is applied to the transparent electrode, joule heat is generated and transferred to the VO_2 crystal layer. Once the temperature increases to a value above the T_c, VO_2 crystals will go through a metal- to insulator-phase transition. Li et al. developed a double-layered VO_2-based electro-thermochromic device (Fig. 6.23a) [81]. In this VO_2/Ag NW system, when the applied voltage increased from 0 V to 6 V, the IR transmittance of the system decreased gradually. Once the applied voltage was further increased to 6.5 V, the transmittance at 1200 nm decreased significantly, from ~70% to ~30% (Fig. 6.23b). Figure 6.23c shows the transmittance hysteresis loop of the VO_2-based electro-thermochromic device as a function of applied voltage at 1500 nm, which indicates high reversibility and stability.

In an additional approach, Chen et al. prepared monoclinic crystalline muscovite mica as the substrate [82]. The VO_2 film was deposited on a substrate through radio frequency–plasma-assisted oxide molecular beam epitaxy and a layered, single-walled CNT film was deposited by the chemical vapor method and employed as the electrode.

6.4 Conclusions

Due to the large surface areas of windows available in the modern city, it is of great significance to provide multifunctionality to the windows, especially energy conversion. Meanwhile, for energy conservation, to continue the trend of enhancing thermochromic properties, hybridation and active control have so far offered the best results compared with other conventional approaches. This direction is worth more attention for further development as well as commercialization of multifunctional smart windows.

References

1. Vukusic, P. and Sambles, J. R. (2003). Photonic structures in biology, *Nature*, **424**, p. 852.
2. Li, Y., Zhang, J. and Yang, B. (2010). Antireflective surfaces based on biomimetic nanopillared arrays, *Nano Today*, **5**(2), pp. 117–127.

3. Cai, J. and Qi, L. (2015). Recent advances in antireflective surfaces based on nanostructure arrays, *Mater. Horiz.*, **2**(1), pp. 37–53.
4. Li, Y. F., et al. (2009). Biomimetic surfaces for high-performance optics, *Adv. Mater.*, **21**(46), pp. 4731–4734.
5. Taylor, A., et al. (2013). A bioinspired solution for spectrally selective thermochromic VO_2 coated intelligent glazing, *Opt. Express*, **21**(S5), pp. A750–A764.
6. Qian, X., et al. (2014). Bioinspired multifunctional vanadium dioxide: improved thermochromism and hydrophobicity, *Langmuir*, **30**(35), pp. 10766–10771.
7. Ge, J. and Yin, Y. (2011). Responsive photonic crystals, *Angew. Chem. Int. Ed. Engl.*, **50**(7), pp. 1492–1522.
8. González-Urbina, L., et al. (2012). Linear and nonlinear optical properties of colloidal photonic crystals, *Chem. Rev.*, **112**(4), pp. 2268–2285.
9. Zhang, J., Sun, Z. and Yang, B. (2009). Self-assembly of photonic crystals from polymer colloids, *Curr. Opin. Colloid Interface Sci.*, **14**(2), pp. 103–114.
10. Ke, Y., et al. (2016). Two-dimensional SiO_2/VO_2 photonic crystals with statically visible and dynamically infrared modulated for smart window deployment, *ACS Appl. Mater. Interfaces*, **8**(48), pp. 33112–33120.
11. Ke, Y., et al. (2018). Vanadium dioxide: the multistimuli responsive material and its applications, *Small*, **14**, p. e1802025.
12. Cui, Y., et al. (2018). Thermochromic VO_2 for energy-efficient smart windows, *Joule*, **2**, pp. 1707–1746.
13. Liu, C., et al. (2015). Vanadium dioxide nanogrid films for high transparency smart architectural window applications, *Opt. Express*, **23**(3), pp. A124–A132.
14. Lu, Q., et al. (2016). Periodic micro-patterned VO_2 thermochromic films by mesh printing, *J. Mater. Chem. C*, **4**(36), pp. 8385–8391.
15. Liu, C., et al. (2017). Ionic strength induced electrodeposition: a universal approach for nanomaterial deposition at selective areas, *Nanoscale*, **9**, pp. 485–490.
16. Liu, M., et al. (2017). Dual-phase transformation: spontaneous self-template surface-patterning strategy for ultra-transparent VO_2 solar modulating coatings, *ACS Nano*, **11**(1), pp. 407–415.
17. Zhang, J., et al. (2015). Self-assembling VO_2 nanonet with high switching performance at wafer-scale, *Chem. Mater.*, **27**(21), pp. 7419–7424.

18. Zhang, J., et al. (2010). Colloidal self-assembly meets nanofabrication: from two-dimensional colloidal crystals to nanostructure arrays, *Adv. Mater.*, **22**(38), pp. 4249–4269.
19. Li, Y., et al. (2013). Physical processes-aided periodic micro/nano-structured arrays by colloidal template technique: fabrication and applications, *Chem. Soc. Rev.*, **42**(8), pp. 3614–3627.
20. Ye, X. and Qi, L. (2011). Two-dimensionally patterned nanostructures based on monolayer colloidal crystals: controllable fabrication, assembly, and applications, *Nano Today*, **6**(6), pp. 608–631.
21. Zhou, M., et al. (2013). Periodic porous thermochromic VO_2(M) films with enhanced visible transmittance, *Chem. Commun.*, **49**(54), pp. 6021–6023.
22. Ke, Y., et al. (2017). Controllable fabrication of two-dimensional patterned VO_2 nanoparticle, nanodome, and nanonet arrays with tunable temperature-dependent localized surface plasmon resonance, *ACS Nano*, **11**(7), pp. 7542–7551.
23. Drury, J. L. and Mooney, D. J. (2003). Hydrogels for tissue engineering: scaffold design variables and applications, *Biomaterials*, **24**(24), pp. 4337–4351.
24. Yang, W., et al. (2017). High-throughput fabrication and modular assembly of 3D heterogeneous microscale tissues, *Small*, **13**(5), p. 1602769.
25. Hamidi, M., Azadi, A. and Rafiei, P. (2008). Hydrogel nanoparticles in drug delivery, *Adv. Drug Deliv. Rev.*, **60**(15), pp. 1638–1649.
26. Choi, S., et al. (2017). Hydrogels: supertough hybrid hydrogels consisting of a polymer double-network and mesoporous silica microrods for mechanically stimulated on-demand drug delivery (Adv. Funct. Mater. 42/2017), *Adv. Funct. Mater.*, **27**(42).
27. Tibbitt, M. W. and Anseth, K. S. (2009). Hydrogels as extracellular matrix mimics for 3D cell culture, *Biotechnol. Bioeng.*, **103**(4), pp. 655–663.
28. Omidinia-Anarkoli, A., et al. (2017). An injectable hybrid hydrogel with oriented short fibers induces unidirectional growth of functional nerve cells, *Small*, **13**(36), p. 1702207.
29. Adams, T. S. T., Crook, T. and Cadier, M. A. M. (2007). A late complication following the insertion of hydrogel breast implants, *J. Plast. Reconstr. Aes. Surg.*, **60**(2), pp. 210–212.

30. Zhou, Y., et al. (2014). Temperature-responsive hydrogel with ultra-large solar modulation and high luminous transmission for "smart window" applications, *J. Mater. Chem. A*, **2**(33), pp. 13550–13555.
31. Zhou, Y., et al. (2018). Fully printed flexible smart hybrid hydrogels, *Adv. Funct. Mater.*, **28**(9), p. 1705365.
32. La, T.-G., et al. (2017). Highly flexible, multipixelated thermosensitive smart windows made of tough hydrogels, *ACS Appl. Mater. Interfaces*, **9**(38), pp. 33100–33106.
33. Zhou, Y., et al. (2015). VO_2/hydrogel hybrid nanothermochromic material with ultra-high solar modulation and luminous transmission, *J. Mater. Chem. A*, **3**(3), pp. 1121–1126.
34. Lee, H. Y., et al. (2017). A dual-responsive nanocomposite toward climate-adaptable solar modulation for energy-saving smart windows, *ACS Appl. Mater. Interfaces*, **9**(7), pp. 6054–6063.
35. Yang, Y.-S., et al. (2017). Tungsten doped VO_2/microgels hybrid thermochromic material and its smart window application, *RSC Adv.*, **7**(13), pp. 7758–7762.
36. de la Rosa, V. R., Woisel, P. and Hoogenboom, R. (2016). Supramolecular control over thermoresponsive polymers, *Mater. Today*, **19**(1), pp. 44–55.
37. Schild, H. G. and Tirrell, D. A. (1990). Microcalorimetric detection of lower critical solution temperatures in aqueous polymer solutions, *J. Phys. Chem.*, **94**(10), pp. 4352–4356.
38. Pennadam, S. S., et al. (2004). Protein-polymer nano-machines. Towards synthetic control of biological processes, *J. Nanobiotechnol.*, **2**, p. 8.
39. Jain, K., et al. (2015). Tunable LCST behavior of poly(N-isopropylacrylamide/ionic liquid) copolymers, *Polym. Chem.*, **6**(38), pp. 6819–6825.
40. Wu, C. and Zhou, S. (1995). Laser light scattering study of the phase transition of poly(N-isopropylacrylamide) in water. 1. Single chain, *Macromolecules*, **28**(24), pp. 8381–8387.
41. Wang, X. and Wu, C. (1999). Light-scattering study of coil-to-globule transition of a poly(N-isopropylacrylamide) chain in deuterated water, *Macromolecules*, **32**(13), pp. 4299–4301.
42. Brijitta, J., Tata, B. V. R. and Kaliyappan, T. (2009). Phase behavior of poly(N-isopropylacrylamide) nanogel dispersions: temperature dependent particle size and interactions, *J. Nanosci. Nanotechnol.*, **9**(9), pp. 5323–5328.

43. Ma, Y., et al. (2017). Soft elastomers with ionic liquid-filled cavities as strain isolating substrates for wearable electronics, *Small*, **13**(9), p. 1602954.
44. Walden, P. (1914). Molecular weights and electrical conductivity of several fused salts, *Bull. Russ. Acad. Sci.*, pp. 405–422.
45. Sun, X., Luo, H. and Dai, S. (2012). Ionic liquids-based extraction: a promising strategy for the advanced nuclear fuel cycle, *Chem. Rev.*, **112**(4), pp. 2100–2128.
46. Swatloski, R. P., et al. (2002). Dissolution of cellose with ionic liquids, *J. Am. Chem. Soc.*, **124**(18), pp. 4974–4975.
47. Wang, H., Gurau, G. and Rogers, R. D. (2012). Ionic liquid processing of cellulose, *Chem. Soc. Rev.*, **41**(4), pp. 1519–1537.
48. Valkenburg, M. E. V., et al. (2005). Thermochemistry of ionic liquid heat-transfer fluids, *Thermochim. Acta*, **425**(1), pp. 181–188.
49. Wang, P., et al. (2003). Gelation of ionic liquid-based electrolytes with silica nanoparticles for quasi-solid-state dye-sensitized solar cells, *J. Am. Chem. Soc.*, **125**(5), pp. 1166–1167.
50. Nam, K. T., et al. (2006). Virus-enabled synthesis and assembly of nanowires for lithium ion battery electrodes, *Science*, **312**(5775), pp. 885–888.
51. Armand, M., et al. (2009). Ionic-liquid materials for the electrochemical challenges of the future, *Nat. Mater.*, **8**, p. 621.
52. Zhu, J., et al. (2016). Composite film of vanadium dioxide nanoparticles and ionic liquid–nickel–chlorine complexes with excellent visible thermochromic performance, *ACS Appl. Mater. Interfaces*, **8**(43), pp. 29742–29748.
53. Zhu, J., et al. (2017). Hybrid films of VO_2 nanoparticles and a nickel(II)-based ligand exchange thermochromic system: excellent optical performance with a temperature responsive colour change, *New J. Chem.*, **41**(2), pp. 830–835.
54. Binnemans, K. (2005). Ionic liquid crystals, *Chem. Rev.*, **105**(11), pp. 4148–4204.
55. Stephen, M. J. and Straley, J. P. (1974). Physics of liquid crystals, *Rev. Mod. Phys.*, **46**(4), pp. 617–704.
56. Yan, J., et al. (2017). Chiroptical resolution and thermal switching of chirality in conjugated polymer luminescence via selective reflection using a double-layered cell of chiral nematic liquid crystal, *Adv. Funct. Mater.*, **27**(2), p. 1604529.

57. Liu, Q. and Smalyukh, I. I. (2017). Liquid crystalline cellulose-based nematogels, *Sci. Adv.*, **3**(8).
58. Liu, H., et al. (2017). A lyotropic liquid-crystal-based assembly avenue toward highly oriented vanadium pentoxide/graphene films for flexible energy storage, *Adv. Funct. Mater.*, **27**(12), p. 1606269.
59. Khandelwal, H., Schenning, A. P. H. J. and Debije, M. G. (2017). Infrared regulating smart window based on organic materials, *Adv. Energy Mater.*, **7**(14), p. 1602209.
60. Liang, X., et al. (2017). Dual-band modulation of visible and near-infrared light transmittance in an all-solution-processed hybrid micro–nano composite film, *ACS Appl. Mater. Interfaces*, **9**(46), pp. 40810–40819.
61. Chen, Z., et al. (2014). Fine crystalline VO_2 nanoparticles: synthesis, abnormal phase transition temperatures and excellent optical properties of a derived VO_2 nanocomposite foil, *J. Mater. Chem. A*, **2**(8), pp. 2718–2727.
62. Shen, N., et al. (2015). Solid-state-reaction synthesis of VO_2 nanoparticles with low phase transition temperature, enhanced chemical stability and excellent thermochromic properties, *RSC Adv.*, **5**(130), pp. 108015–108022.
63. Gao, Y., et al. (2012). VO_2–Sb:SnO_2 composite thermochromic smart glass foil, *Energy Environ. Sci.*, **5**(8), pp. 8234–8237.
64. Liu, C., et al. (2014). VO_2/Si–Al gel nanocomposite thermochromic smart foils: largely enhanced luminous transmittance and solar modulation, *J. Colloid Interface Sci.*, **427**, pp. 49–53.
65. Gao, Y., et al. (2012). Enhanced chemical stability of VO_2 nanoparticles by the formation of SiO_2/VO_2 core/shell structures and the application to transparent and flexible VO_2-based composite foils with excellent thermochromic properties for solar heat control, *Energy Environ. Sci.*, **5**(3), pp. 6104–6110.
66. Zhou, Y., et al. (2013). Surface plasmon resonance induced excellent solar control for VO_2@SiO_2 nanorods-based thermochromic foils, *Nanoscale*, **5**(19), pp. 9208–9213.
67. Zhu, J., et al. (2015). Vanadium dioxide nanoparticle-based thermochromic smart coating: high luminous transmittance, excellent solar regulation efficiency, and near room temperature phase transition, *ACS Appl. Mater. Interfaces*, **7**(50), pp. 27796–27803.

68. Zheng, S., et al. (2016). Superhydrophilic coating induced temporary conductivity for low-cost coating and patterning of insulating surfaces, *Adv. Funct. Mater.*, **26**(48), pp. 9018–9025.
69. Chen, Z., et al. (2014). Crystallised mesoporous TiO_2(A)–VO_2(M/R) nanocomposite films with self-cleaning and excellent thermochromic properties, *J. Mater. Chem. A*, **2**(30), pp. 11874–11884.
70. Li, Y., et al. (2013). Core-shell VO_2@TiO_2 nanorods that combine thermochromic and photocatalytic properties for application as energy-saving smart coatings, *Sci. Rep.*, **3**, p. 1370.
71. Lu, X., et al. (2017). A multi-functional textile that combines self-cleaning, water-proofing and VO_2-based temperature-responsive thermoregulating, *Sol. Energy Mater. Sol. Cells*, **159**, pp. 102–111.
72. Zhang, Z., et al. (2011). Solution-based fabrication of vanadium dioxide on F:SnO_2 substrates with largely enhanced thermochromism and low-emissivity for energy-saving applications, *Energy Environ. Sci.*, **4**(10), p. 4290.
73. Powell, M. J., et al. (2016). Intelligent multifunctional VO_2/SiO_2/TiO_2 coatings for self-cleaning, energy-saving window panels, *Chem. Mater.*, **28**(5), pp. 1369–1376.
74. Zheng, J., Bao, S. and Jin, P. (2015). TiO_2(R)/VO_2(M)/TiO_2(A) multilayer film as smart window: combination of energy-saving, antifogging and self-cleaning functions, *Nano Energy*, **11**, pp. 136–145.
75. Zhou, H., et al. (2016). Use of ZnO as antireflective, protective, antibacterial, and biocompatible multifunction nanolayer of thermochromic VO_2 nanofilm for intelligent windows, *Appl. Surf. Sci.*, **363**, pp. 532–542.
76. Chen, Y., et al. (2017). High performance and enhanced durability of thermochromic films using VO_2@ZnO core-shell nanoparticles, *ACS Appl. Mater. Interfaces*, **9**(33), pp. 27784–27791.
77. Kang, L., et al. (2011). Thermochromic properties and low emissivity of ZnO:Al/VO_2 double-layered films with a lowered phase transition temperature, *Sol. Energy Mater. Sol. Cells*, **95**(12), pp. 3189–3194.
78. Zhou, J., et al. (2013). VO_2 thermochromic smart window for energy savings and generation, *Sci. Rep.*, **3**, p. 3029.
79. Guo, F., et al. (2015). Printed smart photovoltaic window integrated with an energy-saving thermochromic layer, *Adv. Opt. Mater.*, **3**(11), pp. 1524–1529.
80. Li, M., et al. (2016). Active and dynamic infrared switching of VO_2 (M) nanoparticle film on ITO glass, *J. Mater. Chem. C*, **4**(8), pp. 1579–1583.

81. Li, M., et al. (2014). Infrared response of self-heating VO_2 nanoparticles film based on Ag nanowires heater, *J. Mater. Chem. A*, **2**(48), pp. 20470–20473.
82. Chen, Y., et al. (2017). Free-standing SWNTs/VO_2/Mica hierarchical films for high-performance thermochromic devices, *Nano Energy*, **31**, pp. 144–151.

Chapter 7

Hydrothermal Synthesis of Thermochromic VO_2 for Energy-Efficient Windows

Ming Li,[a,b] Guanghai Li,[a] and Yi Long[b]

[a] *Key Laboratory of Materials Physics, Anhui Key Laboratory of Nanomaterials and Nanotechnology, Institute of Solid State Physics, Chinese Academy of Sciences, Hefei 230031, China*

[b] *School of Materials Science and Engineering, Nanyang Technological University, 50 Nanyang Avenue, Singapore 639798, Singapore*

liming@issp.ac.cn

7.1 Introduction

A pioneer hydrothermal study performed by a British geologist, Sir Roderick Murchison, in the middle of the 19th century described the action of water at elevated temperatures and pressures in bringing about changes in the earth's crust and leading to the formation of various rocks and minerals. Much of the early hydrothermal syntheses were concerned with the development of an autoclave that can realize high-temperature and high-pressure environments and the growth of particular minerals similar to natural ones in

Vanadium Dioxide-Based Thermochromic Smart Windows
Edited by Yi Long and Yanfeng Gao

ISBN 978-981-4877-06-0 (Hardcover), 978-1-003-17731-9 (eBook)
www.jennystanford.com

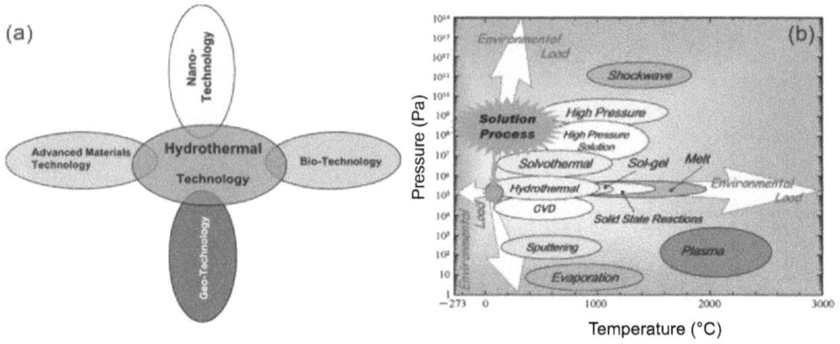

Figure 7.1 (a) Branches of hydrothermal technology in the 21st century; (b) Pressure-temperature map of materials processing techniques [3]. Republished with permission of Elsevier.

a laboratory. Hydrothermal synthesis has been widely accepted since the 1960s, and almost all inorganic species can be obtained by using this method. Nowadays, hydrothermal technology has found its unique place in different branches of science and technology, involving highly interdisciplinary research fields, such as geotechnology, biotechnology, nanotechnology, and advanced materials technology, as schematically illustrated in Fig. 7.1a [1].

Compared with other processing methods for the synthesis of advanced materials, like physical vapor deposition, sol-gel, laser ablation, plasma synthesis, and microwave techniques, hydrothermal processing facilitates the fabrication of even the toughest or the most complex materials with desired physicochemical properties. The hydrothermal method has several advantages over other conventional ones, for example, it enables energy saving, is simple and cost effective, allows better nucleation control, is pollution free, involves higher dispersion and a higher rate of reaction, allows better shape control, and involves a relative lower operation temperature in an appropriate solvent. Hydrothermal processing involves a slightly longer reaction time compared with the vapor deposition process or milling, but it provides highly crystalline products with better controlled sizes and shapes [2]. Figure 7.1b shows the pressure-temperature map of various material processing techniques, in which hydrothermal processing can be considered as environmentally benign [3].

The principle of hydrothermal synthesis involves a process of dissolution-supersaturation and subsequent crystallization in which the temperature, pressure, and time are the three most common parameters to be considered. Temperature plays an important role in the kinetics of product formation as well as on the thermodynamic stability of the resultant phase. Pressure is essential for the solubility, the supersaturation range directing the crystallization process, as well as the thermodynamic stability of the resulting phase. Time is also an important parameter because the synthesis of kinetically stable phases is favored in a short-term process while the thermodynamically stable phases are generally formed in long-term experiments within a given temperature- pressure regime [4]. Rational design of a hydrothermal product requires understanding of the nucleation and crystal growth process under isothermal and isobar conditions [5, 6]. The fluid consisting only of water with solid precursor materials at the beginning of the hydrothermal reaction (point I in Fig. 7.2) will become denser with time and even above the solubility limit of the precursor materials (point II, supersaturation). A glassy precursor can be dissolved at a faster rate than a crystalline material with the same composition. The supersaturation zone is wider with a glassy starting material than that with a crystalline one, which is pressure dependent. At a certain level of supersaturation (marked as point III), spontaneous crystallization will finally take place, leading to a decrease in the concentration of the hydrothermal fluid as the experiment continues [4]. The hydrothermal technique has been popularly utilized for the crystallization of nanomaterials, and the growth mechanism follows either Ostwald ripening (OR) or oriented attachment (OA) or both. The OR mechanism describes the growth of large particles at the expense of smaller ones due to the lowering of the surface energy. On the other hand, for the OA mechanism, as the surface energy is low or the solubility of the material is weak, the adjacent primary particles spontaneously self-organize into chains of nanoparticles (NPs) with a common crystallographic orientation, followed by joining of these particles via the fusion process. It is worth noting that the OA mechanism was first found in the synthesis of TiO_2 particles by using the hydrothermal process [7].

Hydrothermal technology occupies a unique place in the fabrication of vanadium dioxide (VO_2) nanostructured materials owing

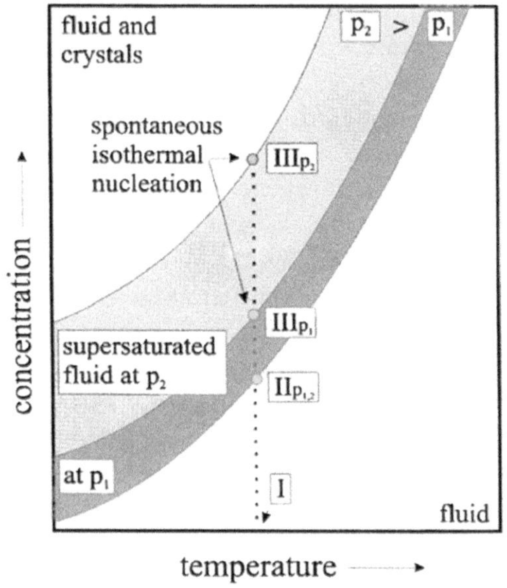

Figure 7.2 Isothermal-isobar hydrothermal processing. At (I), only water and precursor materials are present. Between (I) and (II), time-dependent precursor material dissolution occurs. Note that the region of supersaturated fluid is often extended at higher pressures. Nucleation occurs spontaneously if point (III) is achieved. A subsequent crystal growth process may only take place in the field of supersaturated solution [4]. Republished with permission of Springer.

to its advantages over other conventional ones mentioned above. In this chapter, we describe the hydrothermal synthesis of VO_2 polymorphs and the thermochromic property of VO_2 with various nanostructures, like NPs, nanowires, and nanosheets. Note that the term "hydrothermal" throughout this chapter refers to any chemical reaction in supercritical or near-supercritical conditions, whether it is aqueous or nonaqueous.

7.2 Hydrothermal Synthesis of VO_2 Polymorphs

VO_2 has many polymorphs, like VO_2(B), VO_2(A), VO_2(D), VO_2(M/R), and VO_2(P). Hydrothermal processing is the only way to selectively synthesize VO_2 polymorphs with controlled sizes, morphologies, and crystal structures by adjusting the reaction parameters, such as temperature, precursor concentration and composition, pressure, reaction time, pH value, and autoclave filling ratio. For the precursor, V_2O_5, NH_4VO_3, and $VOSO_4$ are widely used as the V source; $H_2C_2O_4$, PEG, and N_2H_4 are the common reducing agents; and HCl, HNO_3, H_2SO_4, and $CO(NH_2)$ act as the pH regulator. For most hydrothermal conditions, the temperature is between 180°C and 260°C and the reaction time ranges from a few hours to a few days depending on the target product [8].

VO_2(B): VO_2(B) is the most common phase prepared by hydrothermal processing. It is a metastable phase and adopts a monoclinic structure with $a = 1.203$, $b = 0.3693$, $c = 0.642$ nm, and $\beta = 106.6°$. VO_2(B) undergoes a hysteretic first-order metal–insulator transition (MIT) at about 290 K without change in the monoclinic space group across the phase transition [9, 10] and has a double-layered crystal structure, in which the distorted VO_6 octahedra share both corners and edges (Fig. 7.3a). The resulting VO_2(B) usually has a nanorod [9] or nanobelt (Fig. 7.3b) [11, 12] morphology in hydrothermal synthesis. There are also some other specific morphologies, like nanoring (Fig. 7.3c) [13], sphere-like (Fig. 7.3d) [14, 15], and flower-like structures [16].

VO_2(B) is of great interest for its promising application as material in lithium batteries due to its high theoretical capacity (323 mA h g^{-1}) and good rate capability. The morphology, crystalline, and purity are important factors in optimizing the electrochemical performance [17]. VO_2(B) also displays field emission (FE) properties with a turn-on field of 1.4 V/μm and a threshold field of about 5.38 V/μm, which are competitive data for VO_2(B) to be used as cathode material in FE devices [18].

VO_2(A): VO_2(A) is another metastable phase of VO_2, adopting a tetragonal structure with $a = b = 0.885$ nm and $c = 0.7686$ nm, which was first synthesized by a hydrothermal reaction in the system of V_2O_3-V_2O_5-H_2O [19]. VO_2(A) usually has a 1D nanostructure (Figs. 7.4a and 7.4b), for example, it may be in the shape of a nanorod, a nanobelt, or a nanowire [20]. Theoretical and experimental studies reveal that VO_2(A) is a thermodynamically stable phase and has a lower formation energy compared with

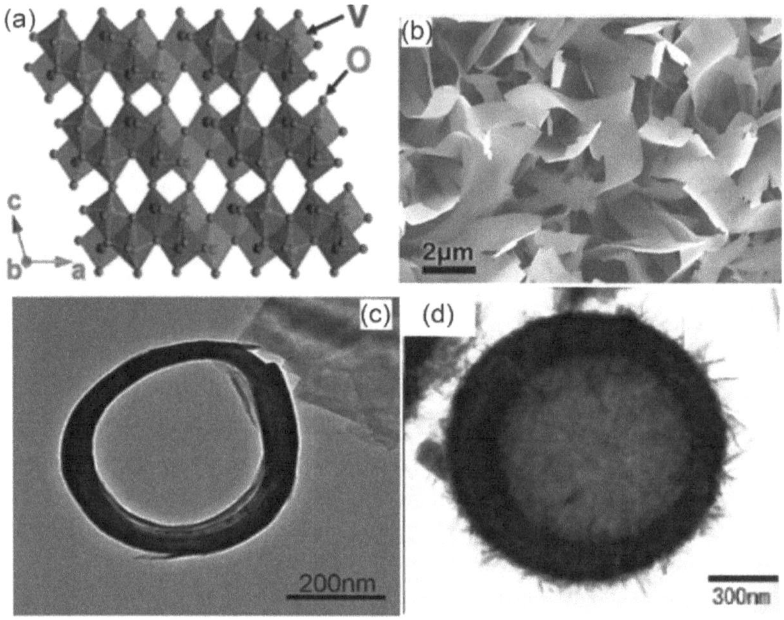

Figure 7.3 (a) Lattice structure illustrating the double-layered crystal structure of VO_2(B). Crystal directions *a*, *b*, and *c* represent [100], [010], and [001] directions, respectively; (b) SEM image of VO_2(B) nanobelts [12]; (c) TEM image of a VO_2(B) nanoring [13]; (d) typical TEM image of an individual VO_2(B) hollow microsphere [14]. Republished with permission of Royal Society of Chemistry.

the metastable VO_2(B). Figure 7.4c shows that a high reaction temperature plays a more crucial role than the reaction time for the formation of VO_2(A) rather than VO_2(B) in the hydrothermal reaction systems [21]. Oka et al. observed the crystallographic transition of VO_2(A) between a low-temperature tetragonal phase (A_L, space group P4/ncc) and a high-temperature body-center tetragonal phase (A_H, space group I4/m) at about 162°C [22]. The magnetic and electrical properties of VO_2(A) have been investigated for a long time, but no thermal effect during the phase transition has been measured from A_H to A_L [23]. Recently, Jin et al. [24] and Li et al. [25] detected the phase transition exothermic peak in VO_2(A) during differential scanning calorimetry (DSC) measurement (Fig. 7.4d). Both size and surface roughness affect the phase transition

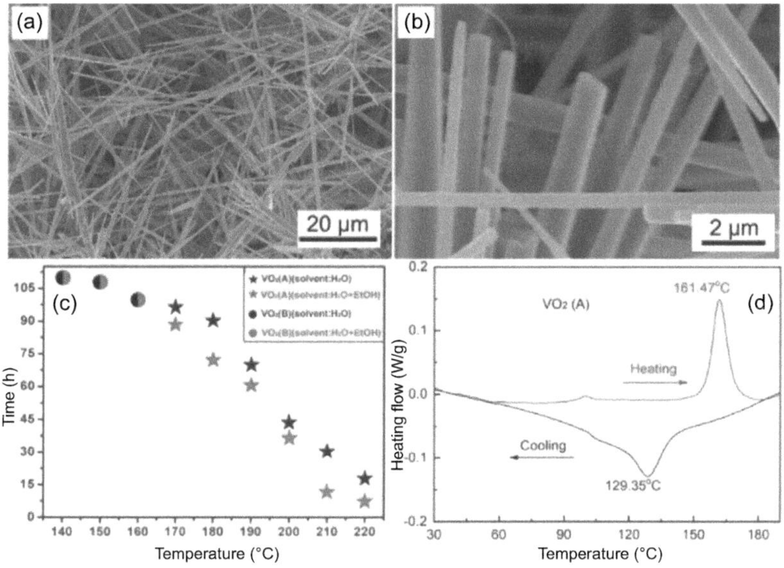

Figure 7.4 (a, b) SEM images of a typical 1D VO_2(A) nanostructure [25]; (c) reaction time vs. reaction temperature for the formation of VO_2(A) and VO_2(B) [21]; (d) DSC curve of VO_2(A) nanobelts with a scan rate of 10°C min^{-1} [25]. Republished with permission of Royal Society of Chemistry.

characteristics of VO_2(A), and the phase transition in VO_2(A) even coverts from a first-order to a high-order continuous phase transition as the nanowire width decreases to a critical size of 72 nm [26].

VO_2(A) has been used in Li-ion batteries [27], field electron emission [25, 28], and optical switch devices [29]. The FE behavior exhibits a stable and high emission current at a low voltage in the air environment, which makes VO_2(A) a candidate for application in new types of temperature-controlled field emitters [28]. Recently, the performance of VO_2(A) as a battery cathode material has received increasing attention due to its unique layered structure [27, 30]. The discharge capacity is 109.3 mA h g^{-1} in the initial cycle and 79 mA h g^{-1} after 100 cycles [31]. The infrared (IR) transmittance of VO_2(A) could change with temperature at different vibratory absorption bands, which makes VO_2(A) a candidate for optical switching devices [32].

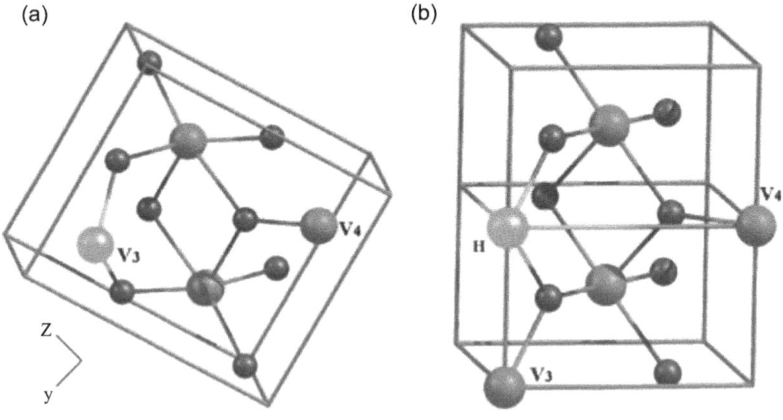

Figure 7.5 The inequivalent ions in (a) the unit cell of VO_2(D) and (b) the $1 \times 1 \times 2$ supercell of VO_2(R). The V and O ions are represented by the large and small balls, and the site denoted by H does not represent any ion at all [34]. Republished with permission of Elsevier.

VO_2(D): VO_2(D) is the latest reported polymorph of VO_2, by Xie et al. in 2012 [33]. The crystal structure of VO_2(D) resembles that of monoclinic $NiWO_4$, and the lattice constants are $a = 0.4597$ nm, $b = 0.5684$ nm, $c = 0.49133$ nm, and $\beta = 89.39°$. Theoretical studies reveal that the local bonding features, including the vibrational modes and the strength of the V–O bonds, have a great influence on the electronic and magnetism properties of VO_2(D) [34]. As displayed in Fig. 7.5, four inequivalent V ions and eight inequivalent O ions are contained in each unit cell of VO_2(D) and VO_2(R). In Fig. 7.5b, the light ball marked with H does not represent any ion at all, and the structural transition from VO_2(D) to VO_2(R) involves a 2.8 Å migration of the V_3 ion (Fig. 7.5a) toward the H site along the [011] direction, while there are no significant changes in the atomic positions of the other atoms. Thus, the formation energy of VO_2(D) shows a value very close to that of VO_2(R), which ensures that the structural transformation from VO_2(D) to VO_2(R) occurs in mild annealing conditions, such as at exposure to 165°C temperature for only a few minutes [35]. VO_2(D) nanostructures with flower [35, 36], NP [37], and micronanostructural morphologies [38, 39] have been reported. VO_2(D) is mainly used as an intermediate phase to obtain

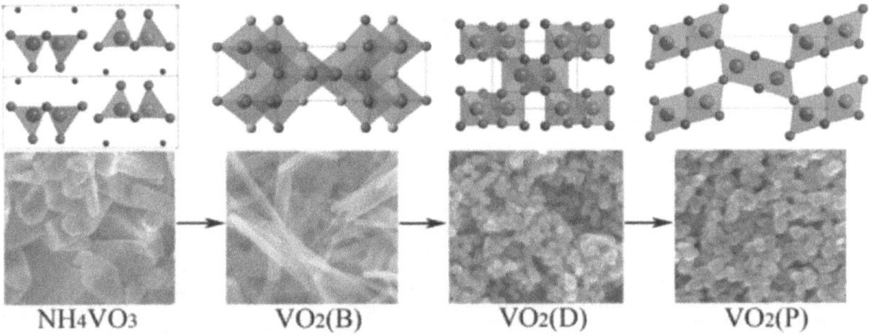

Figure 7.6 A distinct phase evolution of VO_2 polymorphs has been demonstrated in the presence of AOT. VO_2(B) nanorods first nucleated and grew and then dissolved and recrystallized to form VO_2(D) nanoparticles. Finally, VO_2(D) transformed to VO_2(P) nanoparticles via a recrystallization process [45]. Republished with permission of American Chemical Society.

VO_2(M), and currently there are few studies on the application of VO_2(D).

VO_2(P): VO_2(P) has a paramontroseite structure with lattice constants of $a = 0.489$ nm, $b = 0.939$ nm, and $c = 0.293$ nm, first noted by Evans et al. in 1955 [40]. It was obtained only from minerals during the early-stage studies. In 2008, Wu et al. reported a simple hydrothermal route to synthesize VO_2(P) with walnut-like microstructures in the laboratory [41]. Besides hydrothermal processing, VO_2(P) nanocrystals have been prepared by the vapor-transport method by Lugstein et al. in 2010 [42]. Structure analysis and first-principle calculation reveal the metallic behavior VO_2(P) displays by sharing d electrons along the c axis because the nearest V^{4+}–V^{4+} distance is smaller than the critical value of about 0.293 nm [43]. Similar to VO_2(D), VO_2(P) can also be transformed to VO_2(M) by being exposed to mild annealing conditions, for example, to 400°C for 60 s under N_2 or open space atmosphere, and the obtained VO_2(M) has nearly the same morphology as VO_2(P) [44]. Recently, a definite phase evolution process of $(NH_4)_2V_6O_{16} \cdot 1.5H_2O \rightarrow NH_4V_4O_{10} \rightarrow VO_2(B) \rightarrow VO_2(B+D) \rightarrow VO_2(D) \rightarrow VO_2(D+P) \rightarrow VO_2(P)$ was disclosed during hydrothermal [45], as shown in Fig. 7.6. VO_2(P) also has been examined as a potential cathode material for battery applications [46, 47].

VO_2(M/R): VO_2(M/R) undergoes a reversible semiconductor-to metal-phase transition at $T_c = 68°C$. Below T_c, VO_2(M) is a monoclinic structure with lattice constants of $a = 0.5752$ nm, $b = 0.4538$ nm, $c = 0.538$ nm, and $\beta = 122.64°$and above T_c, VO_2(R) is a tetragonal structure with $a = b = 0.4554$ nm and $c = 0.2856$ nm. VO_2(M/R) is the most thermodynamically stable phase, and all of the metastable VO_2 polymorphs can be transformed to VO_2(M/R) via an appropriate thermal treatment process. For example, the phase transition from VO_2(B) or VO_2(A) to VO_2(M) is irreversible and the coarsening phenomenon is inevitable due to the high-temperature annealing (>500°C), while VO_2(D) or VO_2(P) can be transformed to VO_2(M) at a lower temperature within a short time, and the nanostructure can be thus preserved after annealing treatment. A one-step hydrothermal method is cost effective and facile to directly obtain VO_2(M/R), but it is still subjected to many limitations, including phase purity, high hydrothermal temperature, long reaction time, and less rich nanostructures. And thus, in most cases, hydrothermal processing combined with subsequent annealing treatment is commonly used to prepare VO_2(M/R) nanostructures that can be used for building energy-efficient thermochromic smart windows (Fig. 7.7). We will discuss hydrothermal-related preparation and thermochromic performance of VO_2(M/R) nanostructures later.

Other VO_2 polymorphs, like VO_2(C) [48], VO_2(T) [49, 50], $VO_2(M_2)$ [51–54], and $VO_2(M_3)$ [51], receive less attention due to their harsh synthesis processes and rare application fields.

7.3 Hydrothermal Synthesis of VO_2 Powders

7.3.1 0D VO_2 Nanoparticles

0D VO_2 NPs with sizes much smaller than the wavelength of incident light can provide superior luminous transmittance and solar energy transmittance modulation. The nanothermochromic concept was thus proposed to indicate VO_2 NPs in a dielectric host, which opens new avenues in energy-efficient fenestration [55]. High-quality VO_2 NPs are essential to improve thermochromic performance.

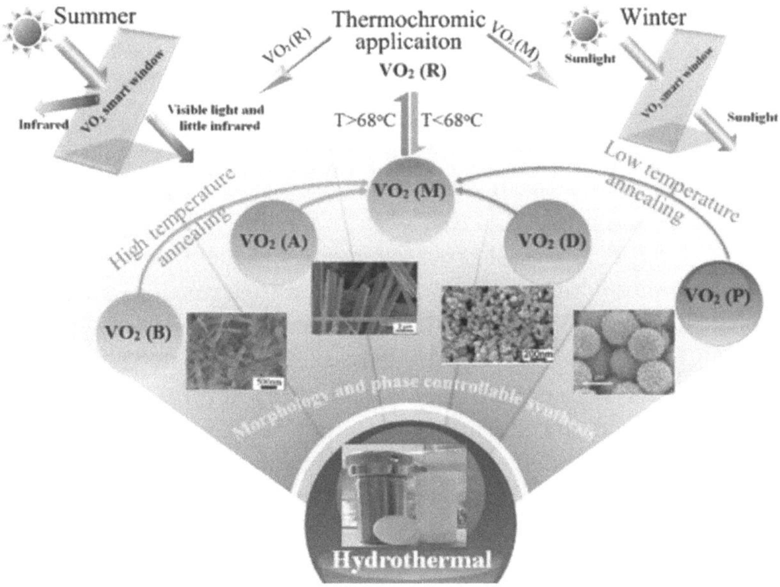

Figure 7.7 Controllable hydrothermal synthesis of VO_2 polymorphs and their transformation to thermochromic VO_2 for the application of energy-saving windows [8]. Republished with permission of Wiley.

7.3.1.1 One-step hydrothermal method

Cao et al. synthesized a mixture of VO_2(M) and VO_2(B) NPs by a hydrothermal reaction in the V_2O_5-H_2SO_4-$N_2H_4{\cdot}H_2O$ system, and the obtained NPs had diameters ranging from 20 to 50 nm at pH = 10. In the experiment, they believe that lower concentrations of the vanadium source and hydrazine are necessary for the formation of pure VO_2 [56]. VO_2 NPs with an average size less than 20 nm were obtained by Jin et al. by the hydrothermal treatment of a $V_2O_5 \cdot nH_2O$ sol. at 260°C for 24 h [57]. In this experiment, N_2H_4 performed as both reductant and pH controller and the molar ratio of V_2O_5 to N_2H_4 was found to be a critical factor for the preparation of VO_2 NPs. In addition, a hydrothermal temperature lower than 260°C will lead to poor crystallinity of VO_2. Gao et al. reported a new phase of $(NH_4)_2V_4O_9$ that acts as an intermediate phase for the formation of VO_2, and a self-assembly decomposition nucleation growth process was proposed to explain the growth mechanism [58]. Lowering systematic energy is the driving force for the self-assembly of $(NH_4)_2V_4O_9$, which can be decomposed at an elevated hydrothermal temperature (300°C), and then bud-like VO_2 nuclei are generated at the expense of $(NH_4)_2V_4O_9$. Note that the common metastable phase of either VO_2(A) or VO_2(B) was not found in the hydrothermal process. Meanwhile, Jin et al. prepared Sb-doped VO_2 NPs with a tunable size (8–30 nm) by a one-pot hydrothermal processing, and the metal–semiconductor transformation takes place at temperatures ranging from 55°C to 68°C [59]. A convenient hydrothermal-assisted homogeneous precipitation approach was developed to tune the size of VO_2 from several tens to hundreds of nanometers by controlling the initial vanadium source concentration [60]. The as-prepared VO_2 NPs exhibit size-dependent phase transition property, and the decreased dimension of VO_2 particles led to an improvement in the optical properties along with a slightly increased transition temperature and wider hysteresis. Long et al. reported one-step hydrothermal synthesis of rare earth-/W-codoped VO_2 NPs with a reduced phase transition temperature and improved thermochromic properties compared with those of W-doped VO_2 [61]. High-quality crystalline VO_2 NPs with diameters of 25–45 nm were synthesized by Gao et al. [62], and the obtained NPs have the highest latent heat, 43 Jg^{-1}, which is close to that of bulk VO_2, 51 Jg^{-1}. After being dispersed in polyurethane and coated on polyethylene terephthalate to form a flexible film, the VO_2 shows a luminous transmittance (T_{lum}) of 45.6% and a solar energy regulate ability (T_{sol}) of 22.3%.

Most recently, Jin et al. reported an original one-step hydrothermal approach by controlling the oxidizing atmosphere of the reaction with hydrogen peroxide to prepare ultrafine VO_2 NPs [63]. Figure 7.8a shows the schematic of the hydrothermal synthesis process, and the evolution of the synthetic route is shown below. Figure 7.8b displays the SEM image of the obtained NPs with high dispersity and an average particle size of about 30 nm. Figure 7.8c is the corresponding scheme image of the VO_2 flexible film. The transmittance spectra of flexible VO_2 films of different thicknesses are measured at 30°C and 90°C to examine the thermochromic properties, as shown in Fig. 7.8d. The flexible VO_2 film shows excellent thermochromic performance, with a T_{lum} of 54.26% and a

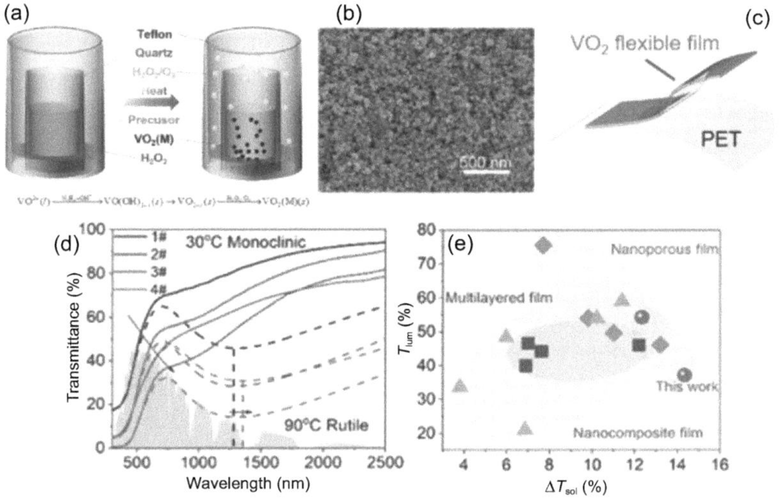

Figure 7.8 (a) Schematic of one-step hydrothermal synthesis assisted by self-released oxidizing atmosphere; (b) SEM image of the as-prepared VO_2 NPs; (c) scheme of the VO_2 flexible film; (d) transmittance spectra of VO_2 flexible films with VO_2 NPs synthesized with 0.3 mL H_2O_2 (1#, 2#, 3#, and 4# with thicknesses of about 1.1, 2.3, 4, and 2.7 μm, respectively.). The real lines denote transmittance spectra measured at 30°C, and the dashed lines are measured at 90°C; (e) comparison of thermochromic properties in this work with those of reported ones [63]. Republished with permission of American Chemical Society.

T_{sol} of 12.34% (Fig. 7.8e). Li et al. developed a TiO_2 seed–assistant epitaxial growth of Mo-doped VO_2(M)/TiO_2 composite NPs by a one-step hydrothermal method [64]. They found that the size of the composite NPs can be adjusted from 100 nm to about 20 nm, the T_c can be tuned to room temperature (about 25°C), and the photochromic properties can be optimized by changing the content of the TiO_2 seed.

7.3.1.2 Hydrothermal method combined with annealing

The strategy of combining the hydrothermal method with annealing involves hydrothermal synthesis of the metastable phase of nanostructured VO_2 and a subsequent annealing treatment. Some newly reported metastable phases of VO_2, such as VO_2(D) and VO_2(P), can be transformed to VO_2(M/R) by being exposed to annealing at a low temperature for a short time. The resulting VO_2 has nearly the same morphology as VO_2(D) or VO_2(P). Having been exposed to a lower annealing temperature for a shorter thermal treatment time, the obtained VO_2 nanostructures show remarkably different phase transition and thermochromic performance.

Chung et al. synthesized organically functionalized highly monodisperse VO_x nanocrystals 7–10 nm in size. After undergoing thermal annealing at 350°C under an air/N_2 flow for 5 min, the VO_2 films exhibited good thermochromic properties (T_{lum} = 49%–66% and T_{sol} = 7.4%–10.7%) [65]. Li et al. reported the fabrication of VO_2 NPs with an average diameter of about 80 nm by mildly annealing VO_2(D) NPs, as shown in Fig. 7.9a.

They found that VO_2 NPs can be obtained at an annealing temperature range of 300°C–450°C (Fig. 7.9c). The annealing conditions have a great effect on the T_c: with increasing annealing temperature, annealing time, and vacuum conditions, the T_c increases but the effect of the annealing temperature is the most obvious (Fig. 7.9d). After mixing with poly(4-vinylphenol) (PVP) in alcohol, a stable VO_2 NP dispersion is made, which can be spin-coated on glass or flexible substrates to form a film. The VO_2 NP film shows an excellent thermochromic performance, with high visible transmittance and IR modulation (Fig. 7.9b). Li et al. also found that after mixing with epoxy resin, VO_2 NP films have not only high stability at a high temperature and in an acid environment but also improved transmission in visible and modulation properties in the IR region [66]. On the basis of the VO_2 NP film, an active and dynamic IR switching device was built by integrating VO_2 with some transparent conductive substrates, such as silver nanowire and indium tin oxide [67, 68]. These devices utilize the advantages of the thermal-sensitive performance of VO_2 and the joule heat produced by a transparent electrode to achieve both thermochromic and optical switching performances and provide a new insight into the applications of transparent conductive electrodes in the thermochromic field. High-quality single-domain VO_2 nanocrystals with precisely tailored sizes were obtained within an ultrashort annealing time of 40 s at 400°C in air [44]. The author suggested that the protection of surface-adsorbed oleylamine helps to realize VO_2

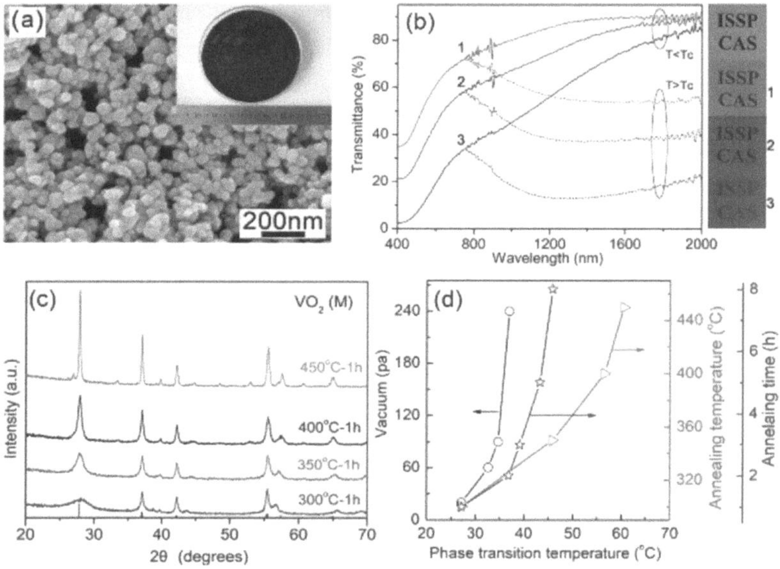

Figure 7.9 (a) FESEM image of the VO_2 NPs; the inset is a photograph of the large-scale production of NPs; (b) optical transmittance and the corresponding photographs (the topmost photo is a bare glass substrate) of the VO_2(M) NP films with thicknesses of 463 nm (curve 1), 730 nm (curve 2), and 1175 nm (curve 3); (c) XRD patterns of VO_2 NPs obtained at different annealing temperatures; (d) dependence of the phase transition temperature of VO_2NPs on annealing conditions [37]. Republished with permission of Royal Society of Chemistry.

nanocrystals without destroying their original size and appearance. Meanwhile, the increase of particle size from 10 to 30 nm leads to a higher transition temperature and remarkably broadened phase transition hysteresis, which is related to the defect density in VO_2.

7.3.2 1D VO_2 Nanowires

1D VO_2 nanostructures, including nanorods, nanowires/micro-beams, and nanobelts, have a single-domain phase transition behavior, which is especially attractive for various device applications as well as for mechanism investigation. Because of two confined directions, 1D VO_2 nanostructures are expected to exhibit significantly different optical, especially thermochromic, properties.

The hydrothermal approach is a simple, low-cost but effective approach to synthesize high-quality single-crystalline 1D VO_2 with a high yield [69]. An optical simulation of VO_2 nanowire polymer composite films was conducted by Naoi and Amano, and it was found that VO_2 nanowires with diameters of 50 nm and lengths of 100 nm show the optimum thermochromic property [70].

Anisotropic freestanding VO_2 nanobelts were prepared by the hydrothermal cleavage of bulk VO_2 particles in the presence of aliphatic alcohols. These nanosheets are highly crystalline and show a strong finite size effect on the phase transition of VO_2 by depressing the T_c to as low as 32°C from the 67°C observed in the bulk one [71]. H-doped VO_2 nanowires with different sizes were achieved by one-step hydrothermal processing, and the hydrogen doping concentration of as-prepared VO_2 nanowires can be adjusted by modifying the filling ratio and reductive agent concentration. The promotion of T_c and the reduction of the hysteresis width can be simultaneously realized by properly increasing the annealing temperature or time [72]. Single-crystalline W-doped VO_2 nanobeams with the widths and lengths of 90–100 nm and 400–800 nm, respectively, were prepared by Long et al. (Fig. 7.10a). Figure 7.10b shows the high-resolution high-angle annular dark field (HAADF) of a single VO_2 nanobeam, and the energy-dispersive X-ray (EDX) maps in Fig. 7.10c–e indicate that the W atoms are dispersed uniformly in the nanobeam. Two distinct surface plasmon resonance peaks, situated at 1344 nm and 619 nm, as a result of the polarization along their transverse and longitudinal axes are the benefit of increasing the solar modulating abilities of VO_2, which has the optimal T_{lum} and T_{sol} of 26% and 5%, respectively (Fig. 7.10f). As shown in Fig. 7.10g, the temperature-dependent hysteresis loop was recorded at a wavelength of 2500 nm, and the reduction of the T_c should be ascribed to the generated free electron carriers as well as the extra strains resulting from the atom replacement of V^{4+} by W^{6+} in VO_2 [73]. W-doped VO_2 nanorods were prepared using a one-step hydrothermal synthesis, and the T_c of VO_2 nanorods is noted to be depressed by as much as 103°C per at.% of W. The obtained 0.5 at.% W-doped VO_2 nanorod composite film shows superior thermochromic performance with an impressive T_{lum} of 48.4% and T_{sol} of 10.3% with a low phase transformation temperature [74].

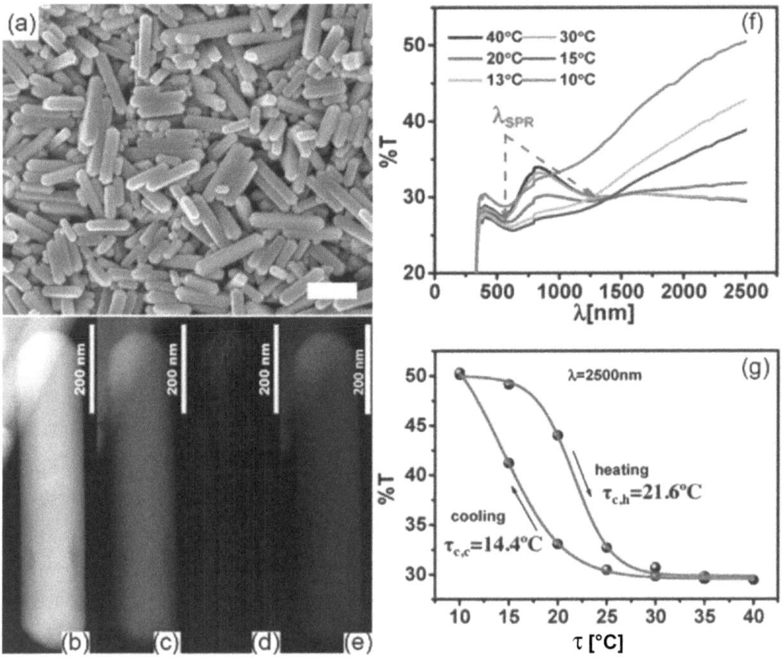

Figure 7.10 (a) FESEM image of VO_2 nanobeams; (b) high-angle annular dark field (HAADF) STEM and EDX maps of (c) V, (d) W, and (e) O; (f) UV-vis-NIR spectra recorded in the temperature from 10°C to 40°C; (g) temperature-dependent transmittance (%T) hysteresis loop recorded at a wavelength of 2500 nm. The solid line represents a best-fitting result [73]. Republished with permission of Wiley.

Often, the metastable phase of VO_2(B) or VO_2(A) has a 1D morphology, which can be used as raw materials to synthesize VO_2(M) by postannealing treatment [75]. The core–shell VO_2@SiO_2 nanowire structure was introduced to retain the 1D morphology and avoid thermal oxidation during heat treatment [76, 77].

A stepwise scalable hydrothermal and annealing route for obtaining VO_2 nanowires was reported by Banerjee et al., and the obtained nanowires exhibit high quality and a controlled stoichiometry [78]. The high-yield freestanding nanowires can be dispersed on arbitrary substrates or incorporated within composite materials. VO_2 nanorods with lengths of more than a few micrometers were synthesized by annealing the hydrate precursor obtained by the hydrothermal reaction of KOH, V_2O_5, and hydrazine in a sealed autoclave at a low temperature [79], in which the VO_2 nanorods are self-assembled by nanocrystalline particles along the rod axis and have the same morphology as their hydrate precursor. Pollet et al. reported rapid hydrothermal synthesis of VO_2(B) at 180°C for 2 h or at 220°C for 1 h, which can be thermal transformed to thermochromic VO_2(M) at 550°C for 1 h while keeping the platelet morphology [80].

7.3.3 2D VO_2 Nanosheets

Ultrathin—exceptionally thin—2D nanostructures that have a quantum confinement effect have received extensive research interest because their fascinating physical and chemical properties differ from those of their bulk counterparts. Xie et al. proposed a structural phase transition–assisted synthetic strategy to achieve single-crystalline ultrathin VO_2 nanosheets with a thickness of about 3 nm (Fig. 7.11a) [81]. The as-obtained ultrathin VO_2 nanosheets show a unique electronic state of half-metallic character induced by surface structure distortion and quantum confinement (Fig. 7.11b). Notably, the nanosheets show a higher IR transmittance compared with their bulk analogue (Figs. 7.11c and 7.11d). As of now, the growth of high-quality 2D VO_2 ultrathin nanosheets and the construction of the corresponding film devices is still essential but difficult.

7.3.4 3D VO_2 Microstructure

Gao et al. reported for the first time the synthesis of W-doped VO_2 snowflake powders by using V_2O_5 and oxalic acid mingled with a small amount of tungstic acid by one-step hydrothermal reactions at 240°C for 7 days. The adulterating agent tungstic acid plays a key role in the synthesis of rutile-phase VO_2, and the W-doped VO_2 shows a reduced T_c [82]. VO_2 with both high-crystallinity-star-shape and hollow-sphere-shape morphologies was produced via direct hydrothermal techniques by Ban et al., and the VO_2 films displayed significant thermal stability (about 300°C) and strong optical modulation in the IR region [83]. The star-shaped VO_2(D)

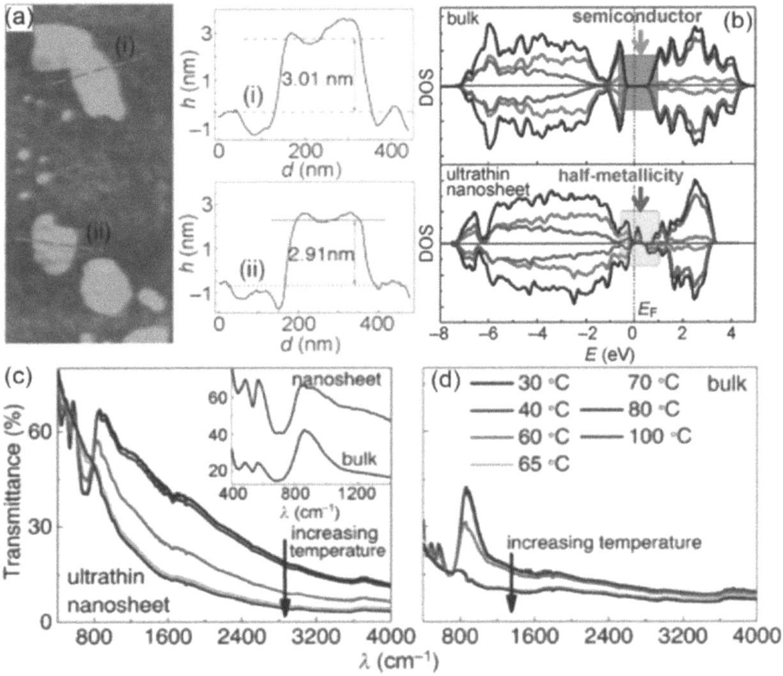

Figure 7.11 (a) AFM image and the corresponding height profiles of VO_2 ultrathin nanosheets; (b) the calculated band structures of bulk VO_2 and 2D ultrathin nanosheets of VO_2; variable-temperature far-IR spectra of (c) ultrathin VO_2 nanosheets and (d) bulk VO_2. The insert compares the far-IR spectrum of a nanosheet with that of the bulk at room temperature [81]. Republished with permission of Wiley.

was synthesized by the hydrothermal treatment of ammonium metavanadate and formic acid, which can be further transformed into VO_2(M) through mild annealing treatment. The IR modulation amplitude, phase transition temperature, and hysteresis width of VO_2 can be regulated by changing the annealing conditions [36]. By similar a hydrothermal-calcination method, Gao et al. prepared VO_2(D) with various micronanostructures that can be converted to VO_2(M). The obtained VO_2(M) has a low T_c (about 41°C) and a small hysteresis width (6.6°C), and can be a suitable candidate for smart windows [38].

7.3.5 VO_2 Smart Windows with Hydrothermal Powders

Generally, hydrothermal products are powders that cannot be used directly for smart windows. To be more practical, a VO_2 powder should be dispersed in an inorganic host matrix and then coated on glass or a flexible substrate to form a thermochromic film by spray coating or spin coating. PVP has been identified to be an optimum polymer matrix that can prevent the agglomeration and deactivation of VO_2. The space distribution of VO_2 particles and their mean size play an important role in determining their thermochromic property [8]. In contrast to many other methods that have been undertaken to deposit VO_2 thin film coatings, such as chemical vapor deposition, pulsed laser deposition, magnetic sputtering, sol-gel, and polymer-assisted deposition method, this strategy offers several advantages. Firstly, it is a convenient method to prepare VO_2 thin films on a large scale at a low cost. Secondly, it is a feasible way to prepare a flexible film that can be used on various nonflat substrates. Thirdly, the polymer matrix possesses a protective property that enhances the stability and thermochromic property of the VO_2 coating. Finally, it is readily to obtain the doped-VO_2 with reduced T_c.

Several strategies have been proposed to further improve the thermochromic performance of VO_2 smart windows, such as introducing periodical patterned VO_2 arrays, enabling multifunctional VO_2 films, adding active stimuli on VO_2 films, and fabricating inorganic-organic composite films.

A periodical patterned VO_2 NPs film was constructed by Long et al. using the screen printing method, and its optimum T_{lum} and T_{sol} are 43.3% and 14.9%, respectively, which is comparable to some of the best reported results using other approaches [84]. To achieve a multifunctional smart window, VO_2 was incorporated with other functional materials that enable the window to not only adjust solar energy radiation but also harvest power, like solar cells [85]. Gao et al. fabricated TiO_2/VO_2 nanocomposite films with electrochromism and thermochromism dual-function performance, and the optimum sample showed a good thermochromic performance, with T_{sol} of 17.1% and T_{lum} of 44.7% [86]. A VO_2/FTO double-layered film was reported that possesses both low emissivity and enhanced thermochromic properties [87]. For a 65 nm thick VO_2/FTO double-layered film, low emissivity values of 0.19 and 0.27 before and after the metal–insulator phase transition are obtained, while a solar transmittance modulation efficiency of 4.9% can be achieved.

A transparent conductive electrode (TCE) heater, as a surface heating source, is an important kind of temperature adjuster, while VO_2 is sensitive to ambient temperature change. Therefore, the combination of VO_2 with TCE materials can provide a new insight in developing an advanced VO_2 electro-thermochromic device and extending the application of TCE materials in the thermochromic field at the same time. Moreover, this combination can avoid the high T_c limit and provide a dynamic means to control the thermochromic property of VO_2 [8, 68, 88].

Long and coworkers introduced an inorganic-organic VO_2/hydrogel hybrid film to increase both T_{lum} and T_{sol}, which was the first report on integrating inorganic VO_2 with organic thermochromic materials to enhance the thermochromic property [89]. Figure 7.12a shows the solar modulation mechanism of the laminated VO_2/hydrogel hybrid thin film. The excellent performance is a result of the hybrid structure that can modulate the IR as well as the visible spectra, which gives a high T_{sol}, of about 34.7%, while maintaining a moderate T_{lum}, of about 62.6%, as shown in Fig. 7.12b.

7.4 Hydrothermal Synthesis of VO_2 Films

Besides nanopowders, hydrothermal synthesis also allows for manufacturing a film structure on a substrate directly, which enables high-quality thin film deposition with controlled porosity, morphology, and structure. But to the best of our knowledge, there are only a few studies about the production of VO_2 films by hydrothermal synthesis.

Jin et al. developed a facile hydrothermal method to enable a wafer-scale VO_2 interconnected nanonets film to self-assemble on a (001)-sapphire substrate. The VO_2 nanonet film can create a sharp resistance change that is comparable to that by a single-crystal VO_2 over the MIT. In addition, it exhibits excellent structural stability and good thermochromic properties [90]. The maxima of visible transmittance of the film at 30°C and 100°C are both above 65%, while the solar modulating efficiency is up to 10.7%,

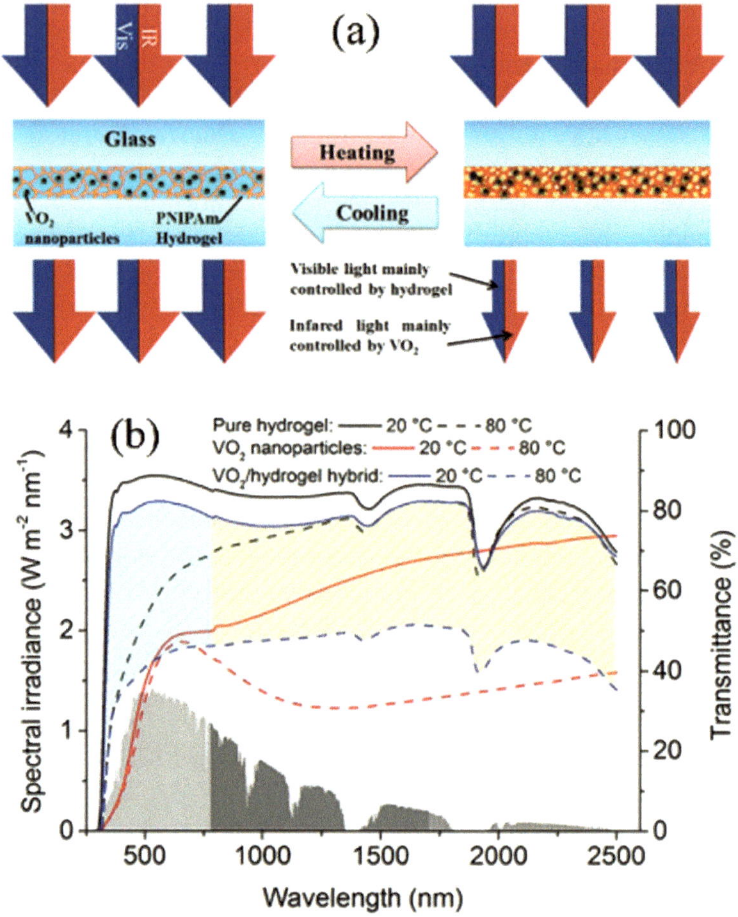

Figure 7.12 (a) Solar modulation mechanism of the VO_2/hydrogel hybrid film; (b) UV-vis spectra of a pure hydrogel (black), pure VO_2 (red), and a VO_2/hydrogel hybrid (blue) at low and high temperatures. The gray area indicates the normalized values of the visible (light) and IR (dark) spectral irradiance. The light-cyan area with sparse slashes is the luminous transmission difference, and the light-yellow area is the IR transmission difference of the VO_2/hydrogel hybrid schematic performance limits of various VO_2-based thermochromic glazings [89]. Republished with permission of Royal Society of Chemistry.

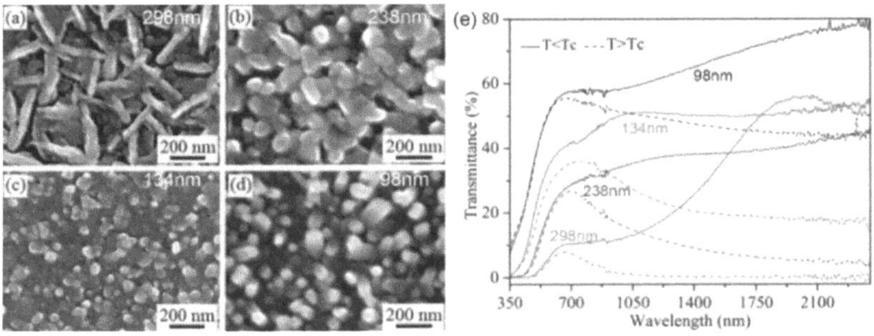

Figure 7.13 (a) FESEM images of annealed VO_2/TiO_2 composite films with a thickness of (a) 298 nm, (b) 238 nm, (c) 134 nm, and (d) 98 nm; (b) Optical transmittance change with temperature of VO_2/TiO_2 composite films of different thicknesses [92]. Republished with permission of Royal Society of Chemistry.

which shows thermochromic properties comparable to or better than those of periodic/aperiodic porous VO_2 films and multilayered films fabricated by complicated processes.

Compared to the costly sapphire substrate, the VO_2 film prepared by the hydrothermal method on glass is more economical. For example, a VO_2 nanoplate film was fabricated on a TiO_2-buffered glass substrate via a facile hydrothermal process. The VO_2 films composed of triangular plates standing on substrates exhibit a self-generated porous structure, which favors the transmission of solar light. The porosity of films is easily controlled by changing the concentration of precursor solutions. Excellent thermochromic properties are observed, with visible light transmittance as high as 70.3% and a solar modulating efficiency of up to 9.3%, in a VO_2 film with a porosity of about 35.9% [91]. Li et al. reported TiO_2 seed-assisted hydrothermal growth of VO_2 films on a quartz substrate. The rutile TiO_2 seed layer can promote the oriented growth of VO_2 at the initial stage of the hydrothermal reaction, and a pure VO_2 thin film can grow on the TiO_2 seed layer on a quartz substrate at a hydrothermal temperature as low as 200°C. Figure 7.13a–d shows the SEM images of $VO_2(M)/TiO_2$ composite films of different thicknesses and surface morphologies after the as-prepared $VO_2(D/M)/TiO_2$ composite films are annealed at 450°C for 2 h in vacuum. Figure 7.13e shows the optical transmittance of the VO_2(M)/TiO_2 composite films of different thicknesses before (20°C) and after phase transition (80°C). One can see that the optical transmittance in both visible and near-infrared (NIR) regions decreases with increasing film thickness and the transmittance in the NIR region of the film with a thickness of 298 nm drops to near zero in the metal state. The optimum values of T_{lum} and T_{sol} are about 30% and 13%, respectively [92].

7.5 Hydrothermal Derivation Technology

In this part, we describe the hydrothermal derivation technologies to synthesize VO_2, including microwave-assisted hydrothermal and hydrothermal flow methods, which involve the development of the hydrothermal method technique. The methods retain the hydrothermal functions and show certain advantages over pure hydrothermal processing, such as high yield and a short reaction time.

7.5.1 Microwave-Hydrothermal Method

The conventional hydrothermal synthesis is carried out by heating the precursor with an external heat source (e.g., oven) and generally requires the temperature of the reaction vessel to be higher than that of the reaction mixture. In contrast, for the microwave-assisted hydrothermal mode, the heating process proceeds rapidly and homogenously to the desired temperature by absorbing and transforming microwave energy into heat, which is a more efficient heating process to synthesize nanomaterials within less time compared to conventional processes [93, 94]. The microwave-assisted hydrothermal method opens up new options for time-, energy-, and cost-saving approaches toward the fabrication of VO_2 with respect to pure hydrothermal processing.

A typical microwave-assisted hydrothermal instrument consists of six main components—a magnetron, a waveguide, a microwave cavity, a mode stirrer, a circulator, and a turntable—as shown in Fig. 7.14a. The magnetron is a device for generating fixed-frequency microwaves, which is principally a thermionic diode with a heated

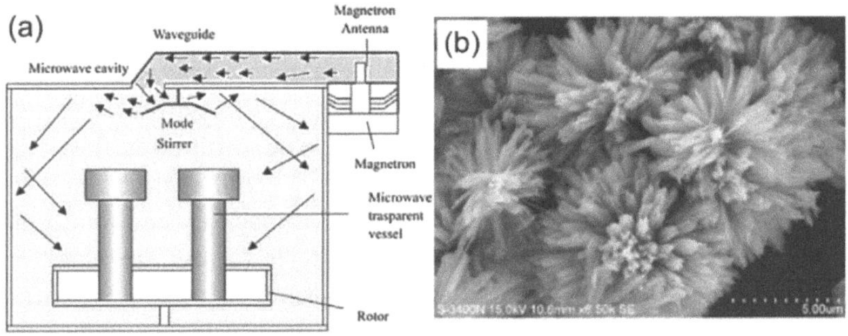

Figure 7.14 (a) Schematic illustration of the microwave equipment for the synthesis of inorganic materials [93]; (b) SEM image of Mo-W codoped VO_2 nanopowders [95]. Republished with permission of Elsevier.

cathode acting as a source of electrons. From the magnetrons, the microwaves are commonly directed toward a target placed in microwave cavities with the use of waveguides, where the mode stirrer homogeneously distributes the incoming microwave energy in various directions. The materials used for the construction of the primary components of microwave units are reflective materials, such as Teflon polymer [93].

Mo-W-codoped VO_2 nanopowders were synthesized by reducing V_2O_5 with oxalic acid using a microwave-assisted hydrothermal method (Fig. 7.14b). The codoping ions have little influence on the morphology of VO_2 but can effectively reduce the T_c. The Mo-W-codoped VO_2 nanopowders show thermochromic properties with luminous transmittance change from 50% to 70%, and the IR transmittance ranges from 85% to 40% at 293 K and 323 K, respectively [95]. VO_2 NPs were produced by the microwave-assisted hydrothermal method and were transferred onto the surface of ceramic glassy tiles by spray coating. This application of VO_2 aims to control the reflected IR radiation on roofs, thus improving the energy efficiency and reducing the carbon dioxide emissions. The best results obtained for this VO_2 coated tile showed a 2.4 times increase in reflectivity as compared with that of an uncoated tile. These superior thermochromic characteristics of VO_2 NPs in conjunction with this new application to a smart roof offer a great potential to regulate the energy in an intelligent way [96].

7.5.2 Continuous Hydrothermal Flow Method

Hydrothermal processing is generally conducted in a fixed-volume autoclave, but a continuous hydrothermal flow synthesis (CHFS) technique makes it possible to operate in a continuous flow-type apparatus [97]. M. J. Powell et al. built a CHFS system used for scalable production of thermochromic Nb-doped VO_2 nanomaterials [98, 99]. In which, nanomaterials are rapidly formed upon mixing of an ambient temperature stream of metal precursor salts with a stream of supercritical (or superheated) water within an engineered mixer which prevents blockages, and the large-scale production of VO_2 nanopowders is achievable [98]. The particles with size range from 50 to 200 nm are highly crystalline and show excellent thermochromic properties with a maximum of 25% change in reflectance during phase transition. The introduction of low Nb concentrations (1 at %) does not affect the T_c of VO_2, however this sample shows a relatively sharp transition within the range of 60–80°C and a larger change in transmittance compared to undoped VO_2. Relatively high Nb concentrations (5 at %) shifts the T_c to 50°C and gives a reduction of 3.6°C per at % Nb incorporated, but the 15 at % Nb-doped VO_2 shows no further decrease in the T_c [99]. Notably, the high yield of VO_2 NPs is beneficial for future industrial applications.

7.6 Conclusion and Perspectives

In summary, hydrothermal synthesis is an effective process to prepare VO_2 with controlled phase, size, morphology, crystallinity, and chemical stoichiometry by changing the hydrothermal conditions, such as precursors, reaction temperature and time, concentration, filling percentage in the autoclave, pressure, and pH value. Specifically, hydrothermal synthesis has been successful in the large-scale production of VO_2, with output in kilograms, which makes it closer to practical thermochromic applications. Some up-to-date strategies are presented and discussed, which are expected to enhance the thermochromic property and allow for multiple functionalities of VO_2 energy-efficient smart windows.

In spite of the above-mentioned achievements, there are still several challenges that need to be resolved in the materials fabrication and practical application of VO_2 smart windows. For instance, it could be interesting to perform one-step hydrothermal growth of high-quality VO_2 nanostructured materials at a low cost on a large scale. Meanwhile, technologies for coating VO_2 composite films based on hydrothermal powders also need to be developed to realize large-area, durable, and flexible coatings that can be readily integrated onto the existing glass production line. Further, it is important to identify ways to construct VO_2-based films with acceptable T_c, T_{lum}, and T_{sol} simultaneously for the application of VO_2 in energy-efficient smart windows. Finally, the products of hydrothermal methods are usually powders that have to be dispersed in a polymer host matrix to form a discontinuous VO_2 particles film, which leads to the nonconductive nature of the film, limiting its applications in certain optoelectronic and electro-thermochromic devices.

In the future, advanced hydrothermal synthesis and characterization of VO_2 should be performed in order to achieve maximum thermochromic energy efficiency. Much work needs to be done in order to favor the lab-to-market transition of optical VO_2 coatings for smart windows. Hydrothermal synthesis will still play an important role in promoting the practical application of VO_2 nanostructured materials in smart windows for improving the human living environment, saving energy, and reducing greenhouse gases in the future.

References

1. Byrappa, K. and Yoshimura, M. (2012). *Handbook of Hydrothermal Technology* (William Andrew).
2. Yoshimura, M. and Byrappa, K. (2008). Hydrothermal processing of materials: past, present and future, *J. Mater. Sci.*, **43**, pp. 2085–2103.
3. Byrappa, K. and Adschiri, T. (2007). Hydrothermal technology for nanotechnology, *Prog. Cryst. Growth Charact. Mater.*, **53**, pp. 117–166.
4. Knauth, P. and Schoonman, J. (2002). *Nanostructured Materials: Selected Synthesis Methods, Properties and Applications* (Kluwer Academic Publishers, Boston).

5. Dhanaraj, G., Byrappa, K., Prasad, V. and Dudley, M. (2010). *Springer Handbook of Crystal Growth* (Springer Science & Business Media).
6. Patzke, G. R., Zhou, Y., Kontic, R. and Conrad, F. (2011). Oxide nanomaterials: synthetic developments, mechanistic studies, and technological innovations, *Angew. Chem. Int. Ed.*, **50**, pp. 826–859.
7. Cöelfen, H. and Antonietti, M. (2008). *Mesocrystals and Nonclassical Crystallization* (John Wiley & Sons).
8. Li, M., Magdassi, S., Gao, Y. and Long, Y. (2017). Hydrothermal synthesis of VO_2 polymorphs: advantages, challenges and prospects for the application of energy efficient smart windows, *Small*, **13**, p. 1701147.
9. Corr, S. A., Grossman, M., Shi, Y. F., Heier, K. R., Stucky, G. D. and Seshadri, R. (2009). VO_2(B) Nanorods: solvothermal preparation, electrical properties, and conversion to rutile VO_2 and V_2O_3, *J. Mater. Chem.*, **19**, pp. 4362–4367.
10. Oka, Y., Yao, T., Yamamoto, N., Ueda, Y. and Hayashi, A. (1993). Phase transition and V^{4+}-V^{4+} pairing in VO_2(B), *J. Solid State Chem.*, **105**, pp. 271–278.
11. Mao, L. and Liu, C. (2008). A new route for synthesizing VO_2(B) nanoribbons and 1d vanadium-based nanostructures, *Mater. Res. Bull.*, **43**, pp. 1384–1392.
12. Ren, G., Hoque, M. N. F., Pan, X., Warzywoda, J. and Fan, Z. (2015). Vertically aligned VO_2(B) nanobelt forest and its three-dimensional structure on oriented graphene for energy storage, *J. Mater. Chem. A*, **3**, pp. 10787–10794.
13. Li, M., Kong, F. Y., Zhang, Y. X. and Li, G. H. (2011). Hydrothermal synthesis of VO_2(B) nanorings with inorganic V_2O_5 sol, *CrystEngComm*, **13**, pp. 2204–2207.
14. Liu, H. M., Wang, Y. G., Wang, K. X., Hosono, E. and Zhou, H. S. (2009). Design and synthesis of a novel nanothorn VO_2(B) hollow microsphere and their application in lithium-ion batteries, *J. Mater. Chem.*, **19**, pp. 2835–2840.
15. Li, R. and Liu, C. Y. (2010). VO_2(B) nanospheres: hydrothermal synthesis and electrochemical properties, *Mater. Res. Bull.*, **45**, pp. 688–692.
16. Fan, Z., Jia, L., Lin, C., Huang, X., Hu, X., Zhuang, N. and Chen, J. (2016). Enhanced electrochemical performance of vanadium dioxide (B) nanoflowers with graphene nanoribbons support, *RSC Adv.*, **6**, pp. 13297–13302.
17. Mai, L., Wei, Q., An, Q., Tian, X., Zhao, Y., Xu, X., Xu, L., Chang, L. and Zhang, Q. (2013). Nanoscroll buffered hybrid nanostructural VO_2(B) cathodes

for high-rate and long-life lithium storage, *Adv. Mater.*, **25**, pp. 2969–2973.

18. Yin, H., Yu, K., Zhang, Z. and Zhu, Z. (2011). Morphology-control of VO_2(B) nanostructures in hydrothermal synthesis and their field emission properties, *Appl. Surf. Sci.*, **257**, pp. 8840–8845.
19. Theobald, F. (1977). Hydrothermal study of VO_2-$VO_{2.5}$-H_2O system, *J. Less-Common Met.*, **53**, pp. 55–71.
20. Cheng, X. H., Xu, H. F., Wang, Z. Z., Zhu, K. R., Li, G. and Jin, S. (2013). Synthesis, characterization and formation mechanism of metastable phase VO_2(A) nanorods, *Mater. Res. Bull.*, **48**, pp. 3383–3388.
21. Zhang, S., Shang, B., Yang, J., Yan, W., Wei, S. and Xie, Y. (2011). From VO_2(B) to VO_2(A) nanobelts: first hydrothermal transformation, spectroscopic study and first principles calculation, *Phys. Chem. Chem. Phys.*, **13**, pp. 15873–15881.
22. Yao, T., Oka, Y. and Yamamoto, N. (1994). A structural study of the high-temperature phase of VO_2(A), *J. Solid State Chem.*, **112**, pp. 196–198.
23. Oka, Y., Ohtani, T., Yamamoto, N. and Takada, T. (1989). Phase transition and electrical properties of VO_2(A), *J. Ceram. Soc. Jpn.*, **97**, pp. 1134–1137.
24. Ji, S. D., Zhao, Y. G., Zhang, F. and Jin, P. (2010). Synthesis and phase transition behavior of W-doped VO_2(A) nanorods, *J. Ceram. Soc. Jpn.*, **118**, pp. 867–871.
25. Li, M., Kong, F. Y., Li, L., Zhang, Y. X., Chen, L., Yan, W. W. and Li, G. H. (2011). Synthesis, field-emission and electric properties of metastable phase VO_2(A) ultra-long nanobelts, *Dalton Trans.*, **40**, pp. 10961–10965.
26. Wang, C. Q., Shao, J., Liu, X. L., Chen, Y., Xiong, W. M., Zhang, X. Y. and Zheng, Y. (2016). Phase transition characteristics in the conductivity of VO_2(A) nanowires: size and surface effects, *Phys. Chem. Chem. Phys.*, **18**, pp. 10262–10269.
27. Dai, L., Gao, Y. F., Cao, C. X., Chen, Z., Luo, H. J., Kanehira, M., Jin, J. and Liu, Y. (2012). VO_2(A) nanostructures with controllable feature sizes and giant aspect ratios: one-step hydrothermal synthesis and lithium-ion battery performance, *RSC Adv.*, **2**, pp. 5265–5270.
28. Liu, M., Fu, W., Yang, Y., Li, T. and Wang, Y. (2018). Excellent field emission properties of VO_2(A) nanogap emitters in air, *Appl. Phys. Lett.*, **112**, p. 093104.
29. Liu, P., Zhu, K., Gao, Y., Wu, Q., Liu, J., Qiu, J., Gu, Q. and Zheng, H. (2013). Ultra-long VO_2(A) nanorods using the high-temperature

mixing method under hydrothermal conditions: synthesis, evolution and thermochromic properties, *CrystEngComm*, **15**, pp. 2753–2760.

30. Zheng, J., Zhang, Y., Wang, Q., Jiang, H., Liu, Y., Lv, T. and Meng, C. (2018). Hydrothermally encapsulating VO_2(A) nanorods into amorphous carbon by the carbonization of glucose for energy storage device, *Dalton Trans.*, **47**, pp. 452–464.
31. Li, L., Liu, P., Zhu, K., Wang, J., Liu, J. and Qiu, J. (2015). A general and simple method to synthesize well-crystallized nanostructured vanadium oxides for high performance Li-ion batteries, *J. Mater. Chem. A*, **3**, pp. 9385–9389.
32. Zhang, Y. F., Huang, Y. F., Zhang, J. C., Wu, W. B., Niu, F., Zhong, Y. L., Liu, X. H., Liu, X. and Huang, C. (2012). Facile synthesis, phase transition, optical switching and oxidation resistance properties of belt-like VO_2(A) and VO_2(M) with a rectangular cross section, *Mater. Res. Bull.*, **47**, pp. 1978–1986.
33. Liu, L., Cao, F., Yao, T., Xu, Y., Zhou, M., Qu, B., Pan, B., Wu, C., Wei, S. and Xie, Y. (2012). New-phase VO_2 micro/nanostructures: investigation of phase transformation and magnetic property, *New J. Chem.*, **36**, pp. 619–625.
34. Qu, B. Y., Liu, L., Xie, Y. and Pan, B. C. (2011). Theoretical study of the new compound VO_2(D), *Phys. Lett. A*, **375**, pp. 3474–3477.
35. Teixeira, D., Quesada Cabrera, R., Powell, M. J., Goh, G. K. L., Sankar, G., Parkin, I. P. and Palgrave, R. (2017). Particle size, morphology and phase transitions in hydrothermally produced VO_2(D), *New J. Chem.*, **41**, pp. 9216–9222.
36. Zhong, L., Li, M., Wang, H., Luo, Y. Y., Pan, J. and Li, G. H. (2015). Star-shaped VO_2(M) nanoparticle films with high thermochromic performance, *CrystEngComm*, **17**, pp. 5614–5619.
37. Li, M., Wu, X., Li, L., Wang, Y. X., Li, D. B., Pan, J., Li, S. J., Sun, L. T. and Li, G. H. (2014). Defect-mediated phase transition temperature of VO_2(M) nanoparticles with excellent thermochromic performance and low threshold voltage, *J. Mater. Chem. A*, **2**, pp. 4520–4523.
38. Song, Z., Zhang, L., Xia, F., Webster, N. A. S., Song, J., Liu, B., Luo, H. and Gao, Y. (2016). Controllable synthesis of VO_2(D) and their conversion to VO_2(M) nanostructures with thermochromic phase transition properties, *Inorg. Chem. Front.*, **3**, pp. 1035–1042.
39. Pan, A., Wu, H. B., Yu, L. and Lou, X. W. D. (2013). Template-free synthesis of VO_2 hollow microspheres with various interiors and their conversion into V_2O_5 for lithium-ion batteries, *Angew. Chem. Int. Ed.*, **125**, pp. 2282–2286.

40. Evans, H. and Mrose, M. E. (1955). A crystal chemical study of montroseite and paramontroseite, *Am. Mineral.*, **40**, pp. 861–875.

41. Wu, C., Hu, Z., Wang, W., Zhang, M., Yang, J. and Xie, Y. (2008). Synthetic paramontroseite VO_2 with good aqueous lithium-ion battery performance, *Chem. Commun.*, pp. 3891–3893.

42. Löffler, S., Auer, E., Weil, M., Lugstein, A. and Bertagnolli, E. (2010). Impact of growth temperature on the crystal habits, forms and structures of VO_2 nanocrystals, *Appl. Phys. A*, **102**, pp. 201–204.

43. Wu, C., Wei, H., Ning, B. and Xie, Y. (2010). New vanadium oxide nanostructures: controlled synthesis and their smart electrical switching properties, *Adv. Mater.*, **22**, pp. 1972–1976.

44. Sun, Y., Jiang, S., Bi, W., Long, R., Tan, X., Wu, C., Wei, S. and Xie, Y. (2011). New aspects of size-dependent metal-insulator transition in synthetic single-domain monoclinic vanadium dioxide nanocrystals, *Nanoscale*, **3**, pp. 4394–4401.

45. Zhong, L., Li, K. B., Luo, Y. Y., Li, M., Wang, H. and Li, G. H. (2017). Phase evolution of VO_2 polymorphs during hydrothermal treatment in the presence of AOT, *Cryst. Growth Des.*, **17**, pp. 5927–5934.

46. Xu, Y., Zheng, L. and Xie, Y. (2010). From synthetic montroseite vooh to topochemical paramontroseite VO_2 and their applications in aqueous lithium ion batteries, *Dalton Trans.*, **2010**, pp. 10729–10738.

47. Fei, H., Ding, X., Wei, M. and Wei, K. (2011). Facile synthesis of montroseite vooh, paramontroseite VO_2 and V_2O_3-VO_2 carbonaceous core-shell microspheres, *Solid State Sci.*, **13**, pp. 2049–2054.

48. Hagrman, D., Zubieta, J., Warren, C. J., Meyer, L. M., Treacy, M. M. J. and Haushalter, R. C. (1998). A new polymorph of VO_2 prepared by soft chemical methods, *J. Solid State Chem.*, **138**, pp. 178–182.

49. Eyert, V. (2011). VO_2: a novel view from band theory, *Phys. Rev. Lett.*, **107**, p. 016401.

50. Tomokuni, M. (1967). On the phase transformation of VO_2, *Jpn. J. Appl. Phys.*, **6**, pp. 1060.

51. Chamberland, B. (1973). New defect vanadium dioxide phases, *J. Solid State Chem.*, **7**, pp. 377–384.

52. Asayesh-Ardakani, H., Yao, W., Nie, A., Marley, P. M., Braham, E., Klie, R. F., Banerjee, S. and Shahbazian-Yassar, R. (2017). Direct evidence of M_2 phase during the monoclinic-tetragonal (rutile) phase transition of W-doped VO_2 nanowires, *Appl. Phys. Lett.*, **110**, p. 053107.

53. Lu, P., Zhou, J., Liu, X., Zhang, Z., Xu, F., Zhang, L., Mou, X., Feng, J., Gao, Y. and Zhao, J. (2014). Characterization of phase-transition-induced micro-domain structures in vanadium dioxide, *J. Appl. Crystallogr.*, **47**, pp. 732–738.
54. Dong, B., Shen, N., Cao, C., Chen, Z., Luo, H. and Gao, Y. (2016). An abnormal phase transition behavior in VO_2 nanoparticles induced by an M_1-M_2-R process: two anomalous high (>68°C) transition temperatures, *RSC Adv.*, **6**, pp. 50521–50528.
55. Li, S. Y., Niklasson, G. A. and Granqvist, C. G. (2010). Nanothermochromics: calculations for VO_2 nanoparticles in dielectric hosts show much improved luminous transmittance and solar energy transmittance modulation, *J. Appl. Phys.*, **108**, p. 063525.
56. Son, J. H., Wei, J., Cobden, D., Cao, G. and Xia, Y. (2010). Hydrothermal synthesis of monoclinic VO_2 micro- and nanocrystals in one step and their use in fabricating inverse opals, *Chem. Mater.*, **22**, pp. 3043–3050.
57. Ji, S., Zhang, F. and Jin, P. (2011). Preparation of high performance pure single phase VO_2 nanopowder by hydrothermally reducing the V_2O_5 gel, *Sol. Energy Mater. Sol. Cells*, **95**, pp. 3520–3526.
58. Dong, B., Shen, N., Cao, C., Chen, Z., Luo, H. and Gao, Y. (2016). An intermediate phase $(Nh_4)_2v_4o_9$ and its effects on the hydrothermal synthesis of VO_2 (M) nanoparticles, *CrystEngComm*, **18**, pp. 558–565.
59. Gao, Y., Cao, C., Dai, L., Luo, H., Kanehira, M., Ding, Y. and Wang, Z. L. (2012). Phase and shape controlled VO_2 nanostructures by antimony doping, *Energy Environ. Sci.*, **5**, pp. 8708–8715.
60. Li, W., Ji, S., Li, Y., Huang, A., Luo, H. and Jin, P. (2014). Synthesis of VO_2 nanoparticles by a hydrothermal-assisted homogeneous precipitation approach for thermochromic applications, *RSC Adv.*, **4**, p. 13026.
61. Wang, N., Goh, Q. S., Lee, P. L., Magdassi, S. and Long, Y. (2017). One-step hydrothermal synthesis of rare earth/W-codoped VO_2 nanoparticles: reduced phase transition temperature and improved thermochromic properties, *J. Alloys Compd.*, **711**, pp. 222–228.
62. Chen, Z., Gao, Y., Kang, L., Cao, C., Chen, S. and Luo, H. (2014). Fine crystalline VO_2 nanoparticles: synthesis, abnormal phase transition temperatures and excellent optical properties of a derived VO_2 nanocomposite foil, *J. Mater. Chem. A*, **2**, p. 2718.
63. Guo, D., Ling, C., Wang, C., Wang, D., Li, J., Zhao, Z., Wang, Z., Zhao, Y., Zhang, J. and Jin, H. (2018). Hydrothermal one-step synthesis of highly-dispersed M-phase VO_2 nanocrystals and application to flexible thermochromic film, *ACS Appl. Mater. Interfaces*, **10**, pp. 28627–28634.

64. Li, D. B., Li, M., Pan, J., Luo, Y. Y., Wu, H., Zhang, Y. X. and Li, G. H. (2014). Hydrothermal synthesis of Mo-doped VO_2/TiO_2 composite nanocrystals with enhanced thermochromic performance, *ACS Appl. Mater. Interfaces*, **6**, pp. 6555–6561.

65. Jung, Y.-H., Pack, S. P. and Chung, S. (2018). Solvothermal synthesis and characterization of highly monodisperse organically functionalized vanadium oxide nanocrystals for thermochromic applications, *Mater. Res. Bull.*, **101**, pp. 67–72.

66. Li, D. B., Li, M. and Li, G. H. Unpublished results.

67. Li, M., Wu, H., Zhong, L., Wang, H., Luo, Y. Y. and Li, G. H. (2016). Active and dynamic infrared switching of VO_2(M) nanoparticle film on ITO glass, *J. Mater. Chem. C*, **4**, pp. 1579–1583.

68. Li, M., Ji, S. L., Pan, J., Wu, H., Zhong, L., Wang, Q., Li, F. D. and Li, G. H. (2014). Infrared response of self-heating VO_2 nanoparticles film based on Ag nanowires heater, *J. Mater. Chem. A*, **2**, pp. 20470–20473.

69. Lu, J., Liu, H., Deng, S., Zheng, M., Wang, Y., van Kan, J. A., Tang, S. H., Zhang, X., Sow, C. H. and Mhaisalkar, S. G. (2014). Highly sensitive and multispectral responsive phototransistor using tungsten-doped VO_2 nanowires, *Nanoscale*, **6**, pp. 7619–7627.

70. Naoi, Y. and Amano, J. (2016). Optimization of VO_2 nanowire polymer composite thermochromic films by optical simulation, *J. Appl. Phys.*, **120**, p. 235301.

71. Whittaker, L., Jaye, C., Fu, Z. G., Fischer, D. A. and Banerjee, S. (2009). Depressed phase transition in solution-grown VO_2 nanostructures, *J. Am. Chem. Soc.*, **131**, pp. 8884–8894.

72. Shi, R., Wang, J., Cai, X., Zhang, L., Chen, P., Liu, S., Zhang, L., Ouyang, W., Wang, N. and Cheng, C. (2017). Axially modulation of metal-insulator phase transition of VO_2 nanowires by graded doping engineering for optically readable thermometers, *J. Phys. Chem. C*, **121**, pp. 24877–24885.

73. Wang, N., Duchamp, M., Xue, C., Dunin Borkowski, R. E., Liu, G. and Long, Y. (2016). Single-crystalline W-doped VO_2 nanobeams with highly reversible electrical and plasmonic responses near room temperature, *Adv. Mater. Interfaces*, **3**, p. 1600164.

74. Chen, R., Miao, L., Cheng, H., Nishibori, E., Liu, C., Asaka, T., Iwamoto, Y., Takata, M. and Tanemura, S. (2015). One-step hydrothermal synthesis of $V_{1-x}W_xO_2$(M/R) nanorods with superior doping efficiency and thermochromic properties, *J. Mater. Chem. A*, **3**, pp. 3726–3738.

75. Li, M., Li, D. B., Pan, J., Lin, J. C. and Li, G. H. (2013). Selective synthesis of vanadium oxides and investigation of the thermochromic properties of VO_2 by infrared spectroscopy, *Eur. J. Inorg. Chem.*, **2013**, pp. 1207–1212.
76. Li, S., Li, Y., Jiang, M., Ji, S., Luo, H., Gao, Y. and Jin, P. (2013). Preparation and characterization of self-supporting thermochromic films composed of VO_2(M)@SiO_2 nanofibers, *ACS Appl. Mater. Interfaces*, **5**, pp. 6453–6457.
77. Pelcher, K. E., Crawley, M. R. and Banerjee, S. (2014). Silica-shell encapsulation and adhesion of VO_2 nanowires to glass substrates: integrating solution-derived VO_2 nanowires within thermally responsive coatings, *Mater. Res. Express*, **1**, p. 035014.
78. Horrocks, G. A., Singh, S., Likely, M. F., Sambandamurthy, G. and Banerjee, S. (2014). Scalable hydrothermal synthesis of free-standing VO_2 nanowires in the M_1 phase, *ACS Appl. Mater. Interfaces*, **6**, pp. 15726–15732.
79. Gui, Z., Fan, R., Mo, W., Chen, X., Yang, L., Zhang, S., Hu, Y., Wang, Z. and Fan, W. (2002). Precursor morphology controlled formation of rutile VO_2 nanorods and their self-assembled structure, *Chem. Mater.*, **14**, pp. 5053–5056.
80. Popuri, S. R., Miclau, M., Artemenko, A., Labrugere, C., Villesuzanne, A. and Pollet, M. (2013). Rapid hydrothermal synthesis of VO_2(B) and its conversion to thermochromic $VO_2(M_1)$, *Inorg. Chem.*, **52**, pp. 4780–4785.
81. Yao, T., Liu, L., Xiao, C., Zhang, X., Liu, Q., Wei, S. and Xie, Y. (2013). Ultrathin nanosheets of half-metallic monoclinic vanadium dioxide with a thermally induced phase transition, *Angew. Chem. Int. Ed.*, **52**, pp. 7554–7558.
82. Cao, C., Gao, Y. and Luo, H. (2008). Pure single-crystal rutile vanadium dioxide powders: synthesis, mechanism and phase-transformation property, *J. Phys. Chem. C*, **112**, pp. 18810–18814.
83. Alie, D., Gedvilas, L., Wang, Z., Tenent, R., Engtrakul, C., Yan, Y., Shaheen, S. E., Dillon, A. C. and Ban, C. (2014). Direct synthesis of thermochromic VO_2 through hydrothermal reaction, *J. Solid State Chem.*, **212**, pp. 237–241.
84. Lu, Q., Liu, C., Wang, N., Magdassi, S., Mandler, D. and Long, Y. (2016). Periodic micro-patterned VO_2 thermochromic films by mesh printing, *J. Mater. Chem. C*, **4**, pp. 8385–8391.

85. Zhou, J., Gao, Y., Zhang, Z., Luo, H., Cao, C., Chen, Z., Dai, L. and Liu, X. (2013). VO_2 thermochromic smart window for energy savings and generation, *Sci. Rep.*, **3**, p. 3029.

86. Yang, F., Shi, X., Chen, Z., Ma, D., Wu, Y., Luo, H. and Gao, Y. (2016). Crystallized TiO_2(A)-VO_2(M/R) nanocomposite films with electrochromism-thermochromism dual-response properties, *RSC Adv.*, **6**, pp. 32176–32182.

87. Zhang, Z., Gao, Y., Luo, H., Kang, L., Chen, Z., Du, J., Kanehira, M., Zhang, Y. and Wang, Z. L. (2011). Solution-based fabrication of vanadium dioxide on F:SnO_2 substrates with largely enhanced thermochromism and low-emissivity for energy-saving applications, *Energy Environ. Sci.*, pp. 4290–4297.

88. Dalapati, G. K., Kushwaha, A. K., Sharma, M., Suresh, V., Shannigrahi, S., Zhuk, S. and Masudy-Panah, S. (2018). Transparent heat regulating (THR) materials and coatings for energy saving window applications: impact of materials design, micro-structural, and interface quality on the THR performance, *Prog. Mater. Sci.*, **95**, pp. 42–131.

89. Zhou, Y., Cai, Y., Hu, X. and Long, Y. (2015). VO_2/hydrogel hybrid nanothermochromic material with ultra-high solar modulation and luminous transmission, *J. Mater. Chem. A*, **3**, pp. 1121–1126.

90. Zhang, J., Jin, H., Chen, Z., Cao, M., Chen, P., Dou, Y., Zhao, Y. and Li, J. (2015). Self-assembling VO_2 nanonet with high switching performance at wafer-scale, *Chem. Mater.*, **27**, pp. 7419–7424.

91. Zhang, J., Li, J., Chen, P., Rehman, F., Jiang, Y., Cao, M., Zhao, Y. and Jin, H. (2016). Hydrothermal growth of VO_2 nanoplate thermochromic films on glass with high visible transmittance, *Sci. Rep.*, **6**, p. 27898.

92. Zhong, L., Luo, Y. Y., Li, M., Han, Y. Y., Wang, H., Xu, S. C. and Li, G. H. (2016). TiO_2 seed-assisted growth of VO_2(M) films and thermochromic performance, *CrystEngComm*, **18**, pp. 7140–7146.

93. Mirzaei, A. and Neri, G. (2016). Microwave-assisted synthesis of metal oxide nanostructures for gas sensing application: a review, *Sens. Actuators, B*, **237**, pp. 749–775.

94. Pan, J., Zhong, L., Li, M., Luo, Y. Y. and Li, G. H. (2016). Microwave-assisted solvothermal synthesis of VO_2 hollow spheres and their conversion into V_2O_5 hollow spheres with improved lithium storage capability, *Chem. Euro. J.*, **22**, pp. 1461–1466.

95. Lv, W., Huang, D., Chen, Y., Qiu, Q. and Luo, Z. (2014). Synthesis and characterization of Mo–W Co-doped VO_2(R) nano-powders by the

microwave-assisted hydrothermal method, *Ceram. Int.*, **40**, pp. 12661–12668.

96. Gonçalves, A., Resende, J., Marques, A. C., Pinto, J. V., Nunes, D., Marie, A., Goncalves, R., Pereira, L., Martins, R. and Fortunato, E. (2016). Smart optically active VO_2 nanostructured layers applied in roof-type ceramic tiles for energy efficiency, *Sol. Energy Mater. Sol. Cells*, **150**, pp. 1–9.
97. Darr, J. A., Zhang, J., Makwana, N. M. and Weng, X. (2017). Continuous hydrothermal synthesis of inorganic nanoparticles: applications and future directions, *Chem. Rev.*, **117**, pp. 11125–11238.
98. Powell, M. J., Marchand, P., Denis, C. J., Bear, J. C., Darr, J. A. and Parkin, I. P. (2015). Direct and continuous synthesis of VO_2 nanoparticles, *Nanoscale*, **7**, pp. 18686–18693.
99. Quesada Cabrera, R., Powell, M. J., Marchand, P., Denis, C. J., Maggio, F. D., Darr, J. A. and Parkin, I. P. (2016). Scalable production of thermochromic Nb-doped VO_2 nanomaterials using continuous hydrothermal flow synthesis, *J. Nanosci. Nanotechnol.*, **16**, pp. 10104–10111.

Chapter 8

Chemical Vapor Deposition and Its Application in VO_2 Synthesis

Shancheng Wang,[a] Dimitra Vernardou,[b] Charalampos Drosos,[c] and Yi Long[a,d,e]

[a]*School of Materials Science and Engineering, Nanyang Technological University, 50 Nanyang Avenue, Singapore 639798, Singapore*
[b]*Center of Materials Technology and Photonics, School of Applied Technology, Technological Educational Institute of Crete, 71004 Heraklion, Crete, Greece*
[c]*Delta Nano-Engineering Solutions Ltd., England, UK*
[d]*Singapore-HUJ Alliance for Research and Enterprise (SHARE), Nanomaterials for Energy and Energy-Water Nexus (NEW), Campus for Research Excellence and Technological Enterprise (CREATE), Singapore 138602, Singapore*
[e]*Sino-Singapore International Joint Research Institute (SSIJRI), GuangZhou 510000, China*
info@delta-nano.com, longyi@ntu.edu.sg

8.1 Introduction

This chapter focuses on a discussion of the chemical vapor deposition (CVD) technique and its application in synthesizing VO_2 thin films. Due to its unique vapor-phase chemical reaction and advantages such as good uniformity and high deposition speed, CVD is the film deposition technique widely employed in both academia

Vanadium Dioxide-Based Thermochromic Smart Windows
Edited by Yi Long and Yanfeng Gao

ISBN 978-981-4877-06-0 (Hardcover), 978-1-003-17731-9 (eBook)
www.jennystanford.com

and industry. Some types of films that can be synthesized via CVD are metal, ceramic, and semiconductor. In this chapter, the definition of CVD and its typical process is introduced, followed by some commonly used CVD methods. After the basis of CVD is reviewed, the factors that affect the synthesis of VO_2 in the CVD process are investigated through case studies. At the end of this chapter, the application of computer simulation to optimize the CVD process is presented.

8.2 Definition of Chemical Vapor Deposition

The concept of CVD can be easily understood by its name: the film is deposited on the substrate through a vapor-phase chemical reaction. The typical reaction process of CVD is illustrated in Fig. 8.1. The first step is the introduction of a precursor into the reaction chamber. Subsequently, the precursor diffuses from the main stream to the substrate surface (Step 2a) and a boundary layer is formed on the surface of the substrate. In the boundary layer, the velocity of gas gradually decreases from the main-stream gas flow rate to zero. The precursor in the boundary layer is then absorbed onto the surface

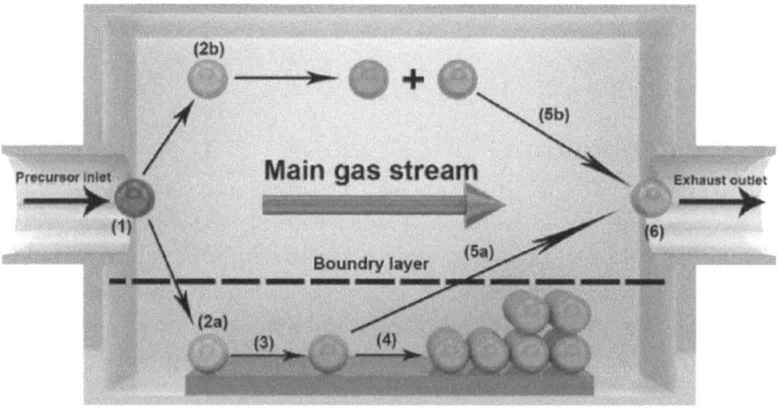

Figure 8.1 Schematic diagram of the CVD process: (1) introduction of the precursor; (2a) absorption of the precursor; (2b) homogeneous reaction; (3) heterogeneous reaction; (4) film forming and coalescence; (5a, 5b) by-product transportation; (6) exhausting. Adapted from Ref. [43].

of the substrate, and a chemical reaction occurs at the interface (Step 3), with the film subsequently forming on the substrate and coalescing (Step 4). Meanwhile, a CVD reaction can also happen in the gas phase instead of on the substrate surface, like the illustration of Step 2b. Commonly, the gas phase reaction is not preferred since this type of reaction will form particles, which will introduce defects, coarsen the film surface, and reduce adhesion [1]. The by-products and unreacted precursor are transported out from the chamber and exhausted (Steps 5 and 6).

8.3 Advantage and Limitation of CVD

Compared with other film deposition techniques, CVD is attractive in both academia and industry due to some unique advantages it offers, such as superb uniformity, a wide application range, and a relatively low requirement of vacuum. Physical vapor deposition processes, such as evaporation and sputtering, are all highly directional, which leads to poor step coverage. On the other hand, it is difficult to use solution-based deposition techniques, such as dip coating and spin coating, for creating a uniform thin film on a small geometry due to the surface tension of the liquid. In contrast, the substrate is submerged in a precursor cloud during the CVD process. As a result, it is possible to coat even a small geometry surface [2, 3] Meanwhile, because it offers the flexibility to apply a wide range of precursors, CVD can be employed to deposit metallic compounds, carbides, nitrides, sulfides, oxides, semiconductors, insulators, and polymers [4–6]. Moreover, it allows one to control the stoichiometries, morphologies, and crystal orientations of the deposited films by controlling the deposition parameters. Lastly, as the CVD system does not require ultrahigh vacuum, it can be compatible with high-volume manufacturing lines and batch production.

However, CVD has some limitations. Firstly, the CVD process usually uses a toxic precursor or has a toxic by-product [7]. Secondly, as the chemical reaction involves multiple gaseous species, intermediates, and by-products, the deposition process has to be carefully designed and controlled to minimize the introduction of impurities.

8.4 Commonly Used CVD Methods

8.4.1 Atmospheric Pressure Chemical Vapor Deposition

Atmospheric pressure chemical vapor deposition (APCVD) is a CVD technique that uses atmospheric pressure in a reaction chamber. Commonly, the air in the chamber is evacuated and an inert gas, such as nitrogen and argon, is introduced into the chamber to maintain the pressure and serve as a carrier gas. A typical system of APCVD is shown in Fig. 8.2. The typical APCVD system consists of several components: The mass flow controller or flow meter helps to control the concentration of the gaseous precursor by regulating the flow rate. The bubbler helps to vaporize the solid or liquid precursor. Depending on the properties and evaporation point of the precursor, the bubbler may require heating. The reaction chamber is usually combined with a furnace as most of the deposition process requires heating. APCVD is one of the most widely applied CVD methods due to its simplicity, fast deposition rate, and good repeatability.

8.4.2 Metal-Organic Chemical Vapor Deposition

As its name indicates, metal-organic chemical vapor deposition (MOCVD) is a CVD process that employs organometallic precursors. Different from APCVD, the pressure involved in MOCVD varies from atmospheric pressure by several pascals. Moreover, MOCVD has less restriction than APCVD. The precursor can be in vapor form or liquid form for MOCVD. Direct-liquid-injection MOCVD (DLI-

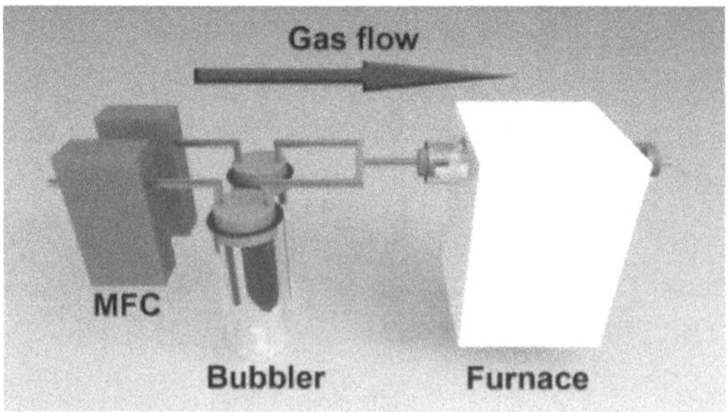

Figure 8.2 Schematic diagram for the APCVD system.

MOCVD) can help control film stoichiometry and morphology more precisely [8].

One of the most important advantages of MOCVD is the relatively low reaction temperature, attributed to the properties of the precursor. Some common types of MOCVD precursors are alkyls, alicyclic compounds, aryl compounds, and acetylacetonate compounds [9]. These compounds are able to decompose at moderate temperatures (100–500°C) and begin the chemical reaction.

Although MOCVD is good at producing epitaxial thin films at low temperatures, it has the disadvantage of involving toxic precursors and the potential risk of carbon contamination. Therefore, the reaction must be carefully designed.

8.4.3 Plasma-Enhanced Chemical Vapor Deposition

A plasma is a partially ionized gas. According to the temperature, plasmas can be categorized into "hot plasmas" and "cold plasmas" [10]. The hot plasma is produced simply by heating: when the temperature is high enough (usually higher than 5000 K), the electron will gain sufficient kinetic energy to escape from the atom and ionization will happen. On the other hand, a cold plasma is usually generated by an electrical field. As the mass of the electron is significantly lighter than that of the core, if the mean free path is small enough, the electron will not have a chance to exchange energy with the core. Therefore, a rapidly changed electrical field can "heat up" the electron without increasing the temperature of the gas. Because of the lower processing temperature, a cold plasma is preferred for the plasma-enhanced chemical vapor deposition (PECVD) process.

A PECVD system is categorized according to the frequency of the electrical field. Some commonly used PECVD systems are direct current PECVD, alternative current PECVD, radio frequency PECVD, and microwave PECVD. The choice of the PECVD system depends on the type of chemical reaction, the geometry of the reaction chamber, and the cost.

Although the PECVD system is more complex than the other types of CVD, PECVD is still considered a power tool for thin film deposition. As the gaseous precursor becomes more reactive after

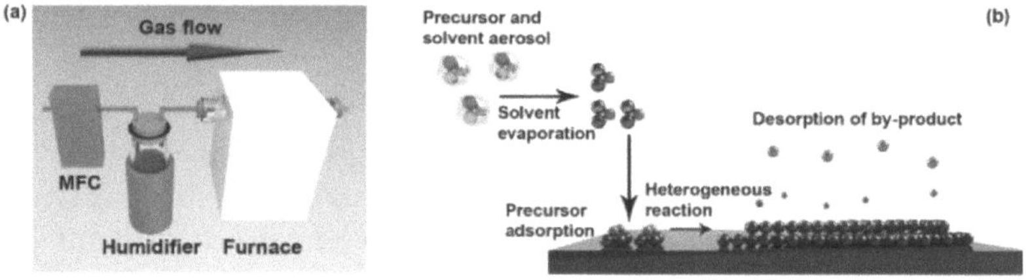

Figure 8.3 Mechanism and system configuration for AACVD. Adapted from Ref. [43].

the ionization process, PECVD is able to deposit a thin film, which normally cannot be achieved at a low temperature. Some other materials which can be synthesized include diamond, silica, and boron nitride. However, as the plasma may potentially damage the substrate and the deposited thin film, this method is not suitable for depositing low-defect-density or defect-free films.

8.4.4 Aerosol-Assisted Chemical Vapor Deposition

Instead of employing vapor, aerosol–assisted chemical vapor deposition (AACVD) utilizes an aerosol to deposit a thin film. Commonly, an ultrasonic humidifier or a nozzle is used to produce the aerosol. The deposition process of AACVD is described in Fig. 8.3: The precursors are usually dissolved in a solvent to form a solution or a slurry before use. The aerosol is first transported into the reaction chamber. Because of a high temperature in the chamber, the solvent evaporates, leaving behind precursor particles. The particles are then adsorbed onto the substrate and take part in a heterogeneous reaction. Finally, a film is formed on the substrate. Subsequently, the by-products are desorbed and transferred away from the substrate.

Different from other CVD techniques, AACVD does not require the precursor to be volatile. Therefore, the choice of the precursor becomes more flexible. Moreover, AACVD is able to deposit the film with a high speed.

However, the film deposited by AACVD has the problem of poor adhesion and relatively high defect density [11]. The problems are attributed to the high deposition speed. The deposited particles are not able to rearrange themselves and form a highly crystalline film as the formed film will be covered by a new layer of particles in a very short time. As a result, the interaction between substrate and film is weakened and defects such as pinholes are produced.

8.4.5 Atomic Layer Deposition

Atomic layer deposition (ALD) is consider a special type of CVD technique due to its unique character. Different from other CVD techniques, the precursor of ALD has to be introduced into the reaction chamber separately due to its self-limited reaction [12]. The self-limited growth is the core of ALD. Under this condition, the amount of thin film deposited on the substrate is fixed (theoretically one atomic layer per deposition). Because of its ability to precisely control the film thickness, ALD is able to form a very high-quality thin film. However, the reaction cycle must be repeated several times to deposit a thicker film, which makes ALD a slow and high-cost CVD technique.

A classic example of the self-limited growth process in ALD is the deposition of HfO_2 on SiO_2. In this process, $HfCl_4$ and H_2O are used as precursors. The first step is the purging of $HfCl_4$. Because of its own properties, only a monolayer of $HfCl_4$ will be absorbed by the surface –OH site on the SiO_2 and forms a "saturated" surface. In the second step, inert gas is purged into the reaction chamber to clean the excess $HfCl_4$ precursor and by-product. Subsequently, H_2O is introduced into the chamber to produce HfO_2 and create a new layer of –OH reactive sites. Finally, inert gas flows into the reaction chamber to clean the system and finish the whole cycle.

8.5 Application of CVD in VO_2 Deposition

8.5.1 Parameter Control in APCVD Growth of a VO_2 Thin Film

The first case of VO_2 thin film deposition was reported in 1968 [13]. In that experiment, vanadium oxytrichloride ($VOCl_3$) and a CO_2/CO mixture gas were used as precursors. The transition temperature for the deposited VO_2 film was 67°C, which was similar to the τ_c of the bulk VO_2. The commonly used vanadium sources for APCVD can be grouped into two categories: inorganic precursors and organic precursors. Examples of inorganic precursors are VCl_4 and $VOCl_3$. On the other hand, some commonly used organic precursors are vanadium acetylacetonate [$VO(acac)_2$], vanadium(III) acetylacetonate [$V(acac)_3$], and $VO(OC_3H_7)_3$. In this section, the parameters that influence the growth of VO_2 will be discussed according to the type of precursor.

8.5.1.1 Parameters that affect film growth in an APCVD system with an inorganic vanadium precursor

Theoretically, when the oxygen mole fraction is 0.66, VO_2 will be produced. With the oxygen mole fraction decreasing, triclinic Magnéli phases V_nO_{2n-1} (n = 4, 5, 6, 7, 8) would be formed. In contrast, V_2O_5 and other higher-valence vanadium oxides would be produced with excessive oxygen. Therefore, the reaction parameters must be carefully tuned to ensure the correct stoichiometry and phase.

Vernardou and coworkers conducted a series of experiments in 2006 [14] and 2011 [3] to show the effect of process parameters on the stoichiometry of a film. They proved that the growth temperature and mole fraction of V and O in the precursor had a significant influence on the stoichiometry of the film. In 2016, Gaskell et al. deposited VO_2(M) on a fluorine-doped-tin-oxide-precoated borosilicate substrate and found that the VO_2 thin film was produced at a lower temperature [15]. By analyzing the case reported by Vernardou et al. and Gaskell et al., a growth window was found for the APCVD system with VCl_4 and H_2O as the precursors and the total flow rate was fixed at 12 L/min (Fig. 8.4). When the temperature is higher than 450°C and the VCl_4-H_2O ratio is less than 0.5, V_2O_5 is formed. Meanwhile, V_2O_3 is formed when the temperature is 450°C and the VCl_4-H_2O ratio is 1. Below 400°C, no crystalline film was observed for all VCl_4-H_2O ratios except the ratio of 0.6. The growth window of VO_2 is T shaped: on SiO_2-precoated glass, VO_2 could be deposited at 450°C–475°C at a VCl_4-H_2O ratio of 0.55–0.8. Interestingly, VO_2 can be obtained on a fluorine-doped-tin-oxide-precoated borosilicate substrate at a lower temperature range (from 350°C) with a fixed VCl_4-H_2O ratio of 0.6. However,

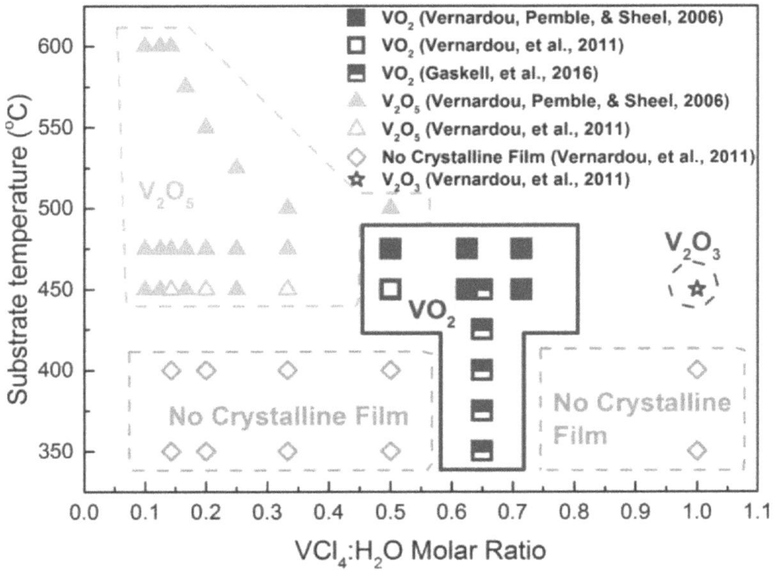

Figure 8.4 Binary diagram to show the experiment results for Vernardou's experiments [3, 14] and Gaskell's experiments [15]. Adapted from Ref. [43].

the possible reason for the reduction of the deposition temperature requires further investigation.

It was found that the total flow rate of gas is able to influence the stoichiometry of the film [14, 16]. The 3D-XYZ charts summarize the cases of VCl_4-H_2O and $VOCl_3$-H_2O systems and describe the relationship between film stoichiometry and growth parameters (Fig. 8.5). In the chart for the VCl_4-H_2O system, two growth windows can be observed. The first growth window is located in the moderate temperature range (350°C–475°C) and VCl_4-H_2O ratio (0.55–0.8) and at a high flow rate (12 L/min). And the second growth window has a higher temperature range (500°C–550°C), richer water content (VCl_4-H_2O ratio 0.05–0.2), and a significantly lower flow rate (~1 L/min). By increasing the substrate temperature and the concentration of the vanadium source, V_2O_5 could be formed with the flow rate fixed at 12 L/min.

In recent years, fewer studies involving $VOCl_3$ have been carried out. Manning and Parkin produced a VO_2 film with a small cluster

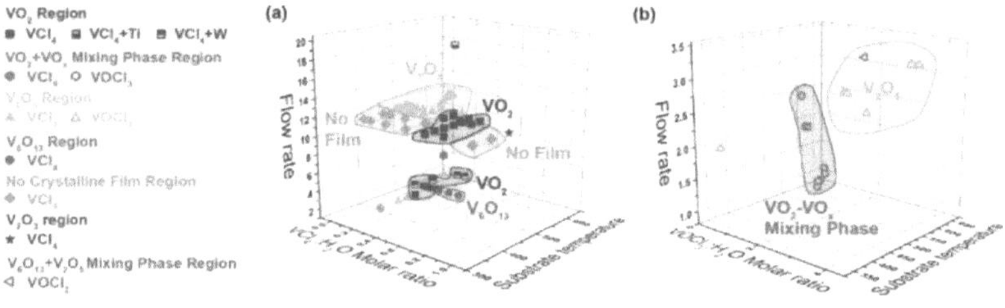

Figure 8.5 (a) 3D diagram for the VCl_4 system to show the phase produced. (b) 3D diagram for the $VOCl_3$ system to show the phase produced. The data of VCl_4 system are collected from Refs. [3, 14, 15, 17–19]. The data of the $VOCl_3$ system are collected from Ref. [16]. Adapted from Ref. [43].

of V_2O_5 or V_6O_{13} [16]. We suppose that the difficulty in controlling the stoichiometry might be due to the complex reaction mechanism of $VOCl_3$. The reaction that happens in the VCl_4-H_2O system is a simple hydrolysis reaction. In contrast, a reduction-oxidation (redox) reaction happens in the $VOCl_3$-H_2O system. The chart in Fig. 8.5b clearly shows that the $VOCl_3$-H_2O system produced V_2O_5 and the V_2O_5-V_6O_{13} mixing phase at the temperature range of 350°C–600°C, with the $VOCl_3$-H_2O ratio from 0.6 to 4 and the flow rate of 1.5–3.5 L/min. The V_2O_5 growth window for the $VOCl_3$-H_2O system was larger than for the VCl_4-H_2O system. Therefore, it indicates that the $VOCl_3$ precursor has a higher possibility to form V_2O_5. Moreover, only a VO_2-VO_x mixing phase was produced with the $VOCl_3$ precursor, which indicates that the VCl_4 system is easier to control than the $VOCl_3$ system and recommended for VO_2 growth.

8.5.1.2 Parameters that affect film growth in an APCVD system with an organic vanadium precursor

The film deposition of an APCVD system with an organic-vanadium precursor such as $VO(acac)_2$ is governed by growth temperature, the ratio of vanadium source to oxygen source, and the total flow rate of the carrier gas. Vernardou et al. [14] investigated the effect of changes in conditions on the stoichiometry in the $VO(acac)_2$-O_2 system and plotted a binary diagram (Fig. 8.6). As described in the diagram, the growth temperature of the $VO(acac)_2$ system is higher than that of the VCl_4 system. The different reaction routes

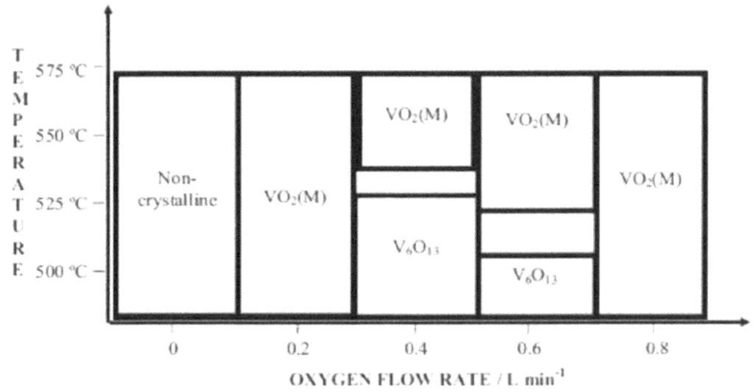

Figure 8.6 The temperature-oxygen flow rate and phase diagram at a fixed total flow rate of 12 L/min. Adapted from Ref. [14].

for $VO(acac)_2$ and VCl_4 may cause an increase in the reaction temperature. For $VO(acac)_2$, intramolecular decomposition is the major reaction. On the other hand, the reaction between VCl_4 and H_2O is a simpler intermolecular hydrolysis reaction. Moreover, it is worth noting that the $VO(acac)_2$ system has a wider growth window than the VCl_4 system when considering the oxygen flow rate.

There are several cases [20–23] on the synthesis of VO_2 with $VO(OC_3H_7)_3$ on SnO_2-precoated glass and a W/indium tin oxide-coated substrate. The growth temperature of these experiments was set in the range of 300°C–450°C, which was significantly lower than the growth temperature of the $VO(acac)_2$ system. Also, two cases [21, 23] evaluated the stoichiometry of the film produced with the carrier gas flowing through the precursor at different rates. The results showed that at the temperature range from 400°C to 450°C, VO_2 could be produced at a flow rate in the range of 2–4 L/min.

Deposition of VO_2 film with organometallic vanadium precursor is affected by the substrate. This phenomenon was observed with several kinds of precursors, such as $V(acac)_3$ [24] and $VO(OC_3H_7)_3$ [20, 22]. For example, attempts were made to produce VO_2 films on borosilicate glass, fused quartz, (1 0 0) MgO single crystal, (1 0 0) $SrTiO_3$ crystal, (1 0 0) Si wafer, and randomly oriented sapphire [20, 22, 24]. It was found that pure VO_2 was deposited only on the fused quartz and sapphire substrates. The film on other substrates had other phases, such as V_3O_7 or V_4O_9. It was supposed that the different results obtained for the substrate might be due to the lattice mismatch between VO_2 and substrate. However, a detailed explanation for this observation needs further experiment.

8.5.2 Parameter Control in MOCVD Growth of a VO_2 Thin Film

In the MOCVD system, two categories of precursors are commonly used for VO_2 deposition: (i) β-diketonates such as $VO(acac)_2$, $V(acac)_3$, and vanadyl-bis-hexafluoroacetylacetonate [$VO(hfa)_2$] and (ii) alkoxides such as $VO(OC_3H_7)_3$ and vanadylethoxide [$VO(OC_2H_5)_3$] [25]. However, the performance of precursors is different. Barreca et al. evaluated performances for four kinds of β-diketonate precursors: $VO(acac)_2$, $VO(hfa)_2(H_2O)$, $VO(dpm)_2$, and $VO(fod)_2$ (Hdpm: 2,2-6,6-tetra-methyl-3,5-heptanedione; Hfod: 2,2-dimethyl-6,6,7,7,8,8,8-hepta-fluoro-3,5-octanedione) [26]. VO $(acac)_2$, $VO(dpm)_2$, and $VO(fod)_2$ produced VO_2 in an oxygen atmosphere. On the other hand, $VO(hfa)_2(H_2O)$ produced V_2O_5 in oxygen but produced VO_2 in a nitrogen atmosphere. The group concluded that different decomposition mechanisms caused the product difference.

The film stoichiometry in MOCVD is affected by the temperature and fraction of the precursor [14]. The molar fraction in the MOCVD system is mainly controlled by the oxygen flow rate. As described in Fig. 8.7, the optimum growth temperature for DLI-MOCVD was 375°C–475°C, which was significantly lower than that for APCVD. Meanwhile, DLI-MOCVD seemed to be more sensitive to parameter changes in the precursor. When the oxygen flow rate was between 0.02 and 0.04 L/min, the VO_2(M) single phase was produced at temperatures from 450°C to 475°C. Furthermore, when the oxygen flow rate was within 0.04–0.08 L/min, the VO_2(M) and VO_2(B) mixing phase would be deposited at 425°C. However, if the temperature was lowered to 400°C, the VO_2(B) single phase would be produced. The region of VO_2 deposition in the MOCVD phase diagram is smaller than the region size shown in the diagram for the APCVD system with the same precursor.

The optimum temperature might be affected by chamber pressure. Spanò et al. [25] determined the influence of growth

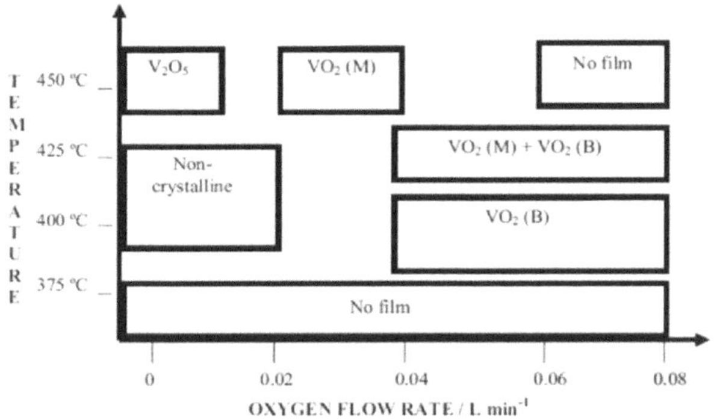

Figure 8.7 Influence of oxygen flow rate on film stoichiometry at a fixed total flow rate of 3 mL/h in an MOCVD system. Adapted from Ref. [14].

temperature on the film morphology and stoichiometry in a 3 torr MOCVD reactor. This group characterized the film growth at temperatures from 200°C to 750°C, with an oxygen flow rate of 150 sccm. VO_2 was produced at 200°C–350°C, which was lower than the value determined in Vernardou's experiment [14]. During the experiment, when the temperature increased to 400°C–500°C, V_6O_{13} was produced. Moreover, V_2O_5 deposition was observed if the temperature was higher than 500°C. In addition, the films grown at different temperatures varied in terms of morphology. The temperature influenced the roughness of the film. The film grown at 200°C consisted of 70–80 nm wide nanocolumns. If temperature increased to 300°C, the nanocolumns were replaced by plate-like structures. The surface of the film produced at 350°C was even flatter than that of the film produced at 300°C. This trend was also applicable to the film produced at 400°C–500°C [25].

8.5.3 Parameter Control in PECVD Growth of a VO_2 Thin Film

In 1998, Zhang and coworkers produced vanadium oxide for a battery cathode [27]. $VOCl_3$, O_2, and H_2 were used as precursors in this experiment. During the experiment, a radio-frequency power was applied to form the plasma. This experiment produced only a V_6O_{13} thin film. They found that the film had a high discharge capacity, of 408 mAh/g. At the same time, capacity fading due to repeated charging was negligible. The film also had an energy density of 960.3 Wh/kg. Regarding the outstanding electrochemical properties and the rapid deposition speed, the researchers concluded that PECVD was a technique with a high potential for future application. After reviewing the experiment procedure of this case, we suppose that V_6O_{13} was produced because O_2 was applied. The reactions that occurred during the experiment were recorded: In the experiment involving the deposition of VO_2, vanadium atoms in the precursor have to be reduced to the valence number of +4 to form VO_2. Hence, Zhang's experiment can be modified by the selection of a more suitable precursor or reducing the amount of oxygen pressure so that VO_2 can be produced.

It is worth mentioning that recently Wang et al. synthesized a VO_2/carbon hybrid material by applying the PECVD method and using the precursors $VO(acac)_2$ and O_2 [28]. In this paper, researchers investigated the factors that affect the deposition of VO_2 and pointed out that the growth window for high-crystallinity VO_2 is in the temperature range of 400°C to 450°C and at the O_2 flow rate of 10–20 sccm (Fig. 8.8). Moreover, the VO_2/carbon hybrid synthesized showed a decreased transition temperature (~45°C), superb emissivity switching property (~0.44), and good infrared regulation ability. Compared with the method reported previously by Zhang et al., this combination of precursors does not involve hydrogen as a reducer. Therefore, it simplifies the reaction process and reduces the risk of explosion.

8.5.4 Parameter Control in AACVD Growth of a VO_2 Thin Film

Commonly used precursors for AACVD are $VO(acac)_2$ [29], $V(acac)_3$ [30], and other kinds of organometallic precursors. Piccirillo et al. investigated the parameters that influence VO_2 deposition for two kinds of precursors, $VO(acac)_2$ and $V(acac)_3$ [30]. The factors they researched are the type of solvent and the temperature and flow rate of the carrier gas. The only combination that was able to produce a VO_2 thin film was a $VO(acac)_2$/ethanol combination at a carrier gas flow rate of 1.5 L/min in the temperature range of

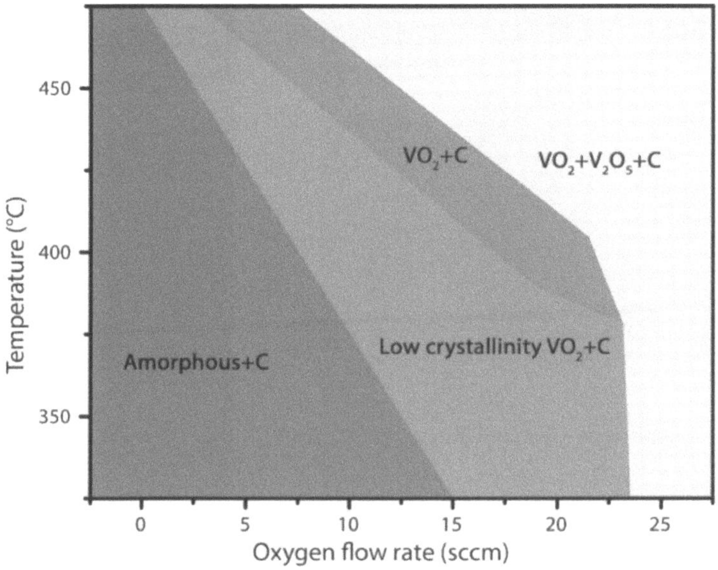

Figure 8.8 Phase diagram for the deposition of VO_2/carbon hybrid structures. Adapted from Ref. [28].

500°C–600°C. V_2O_3 was produced for $VO(acac)_2$/methanol/water and $V(acac)_3$/methanol/water combinations at the fixed flow rate of 1.5 L/min at 500°C–600°C. The $V(acac)_3$/ethanol combination deposited a mixing phase of VO_2 and other vanadium oxides when the carrier gas flow rate was set at 1.5 L/min. Lastly, when the carrier gas flow rate increased from 1.5 L/min to 3 L/min, vanadium oxides with a higher valence number were produced.

It is worth mentioning that Naik et al. produced VO_2 though AACVD without employing commonly used oxygen sources such as O_2 and water [11]. In this case, $VO(acac)_2$ was dissolved in ethanol. At the temperature of 450°C–600°C, VO_2 was formed with an island growth morphology.

8.5.5 Parameter Control in Hybrid AA/APCVD Growth of a VO_2 Thin Film

Hybrid AA/APCVD combines the advantages of both AACVD and APCVD. This system can deposit a high-quality, uniform film. At

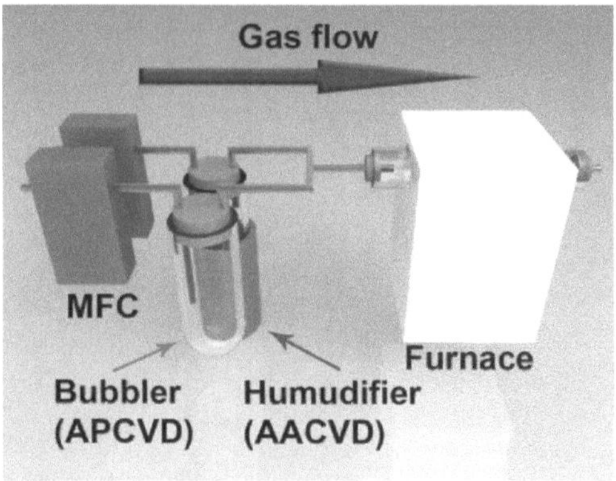

Figure 8.9 Schematic diagram of the hybrid AP/AACVD system.

the same time, it can add nanoparticles into the film by using a nanoparticle slurry aerosol. The hybrid CVD method is suitable for producing nanocomposites. A schematic illustration of the hybrid CVD system is shown in Fig. 8.9.

Warwick's group produced a VO_2 thin film, a VO_2-TiO_2 nanocomposite, and a VO_2-CeO_2 nanocomposite by hybrid CVD and compared their performance [31]. The VO_2 film was synthesized by the $VO(acac)_2$-O_2 APCVD system. At the same time, TiO_2 or CeO_2 nanoparticles were transported into the chamber by the AACVD system in the form of an aerosol. All three materials showed lowered τ_c compared with that of bulk VO_2, and the reflectance showed a significant change of 30% between 25°C and 80°C. Also, the films had photocatalytic properties similar to those of a titanium dioxide thin film. The group concluded that although the visible transmittance of the nanocomposite still needed to be improved, hybrid CVD was a promising way to produce thermochromic nanocomposites [1].

8.5.6 Parameter Control in ALD Growth of a VO_2 Thin Film

Tetrakis(ethylmethylamido)vanadium (TEMAV) [$V(NEtMe)_4$] is the vanadium-containing precursor commonly used in ALD [32–36].

Other precursors used in ALD are $VO(OC_3H_7)_3$ [37], and $VO(acac)_2$ [38]. The common oxygen sources used in ALD are water [37], ozone [36, 39], and oxygen plasma [40]. Compared with ozone, the system with water gives a higher deposition rate with a TEMAV precursor [32]. The enhancement of the growth rate is because the hydroxyl group dissociates from water [41]. The hydroxyl group serves as an intermediate for the metal compound to anchor on the substrate surface.

As discussed previously, the key to the ALD process is the self-terminating surface half-reaction. Therefore, the growth conditions, such as the growth temperature, the injection time, and the purging time, have to be carefully planned. To form a saturated monolayer, the binding energy between substrate and precursor should be larger than the binding energy between the monolayer and precursor particles above [42]. An adequate purging time is needed for the excess precursor to re-evaporate and leave the reaction chamber because the injected precursor has to be in excess to ensure a saturated layer is formed. The case reported by Rampelberg et al. [36] demonstrated a typical ALD process for VO_2(M) film deposition (Fig. 8.10). This group utilized TEMAV and O_3 as precursors. The pulsing scheme was [5-10-10-5]: TEMAV was injected into the chamber for 5 s, followed by 10 s of argon purging. Then O_3 was introduced into the chamber for 10 s Finally, another 5 s of argon purging was executed. The growth temperature was set at 150°C. During the TEMAV injection step, an excess amount of TEMAV was introduced into the system. The group has proven that the substrate would reach the TEMAV saturated state in 2 s. Therefore, excess TEMAV would accumulate on the monolayer and re-evaporate during the first purging step. After the first purging step, an excess amount of O_3 was introduced into the system and reacted with TEMAV. After the reaction, the unreacted O_3 was carried out from the chamber. An amorphous film was produced and then was annealed at 450°C in a He atmosphere for 30 min. The annealed sample showed a VO_2(M) phase. The τ_c for the sample was 67°C, and the hysteresis loop width was 12°C.

The optimum conditions for different CVD methods to synthesis VO_2 are summarized in Table 8.1. It should be pointed out that

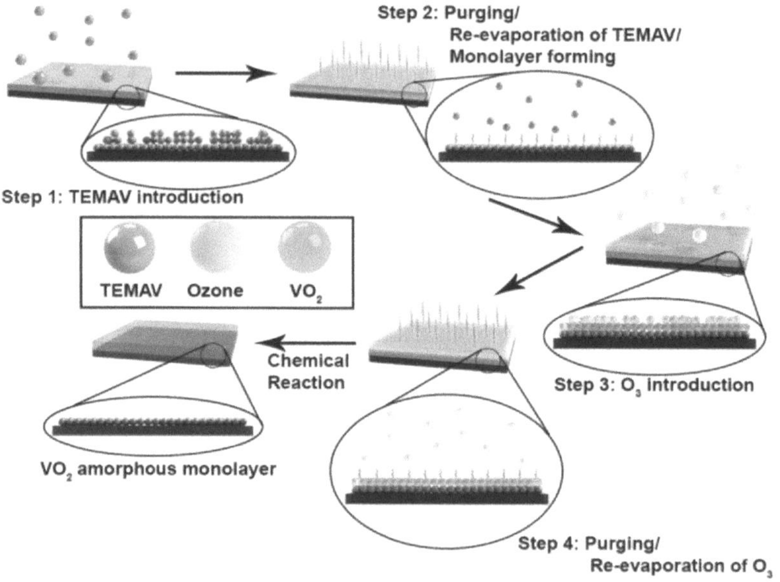

Figure 8.10 Schematic diagram to show the ALD process reported by Rampelberg et al. [36]. Adapted from Ref. [43].

an organovanadium precursor such as $VO(acac)_2$ and $VO(OC_3H_7)_3$ has the largest process window for APCVD, MOCVD, PECVD, and AACVD systems. Thus, organometallic precursors are fit for systems with rough condition controlling. The VCl_4-H_2O system is the most suitable system for the deposition of a high-purity VO_2 film. However, due to its toxicity, reactivity, and narrow growth window, the VCl_4-H_2O APCVD system must be able to control the growth parameters, such as H_2O concentration, temperature, and total flow rate, precisely. Although currently it is only used for ALD, TEMAV is the precursor with the lowest reaction temperature. The precursor has a high potential to be used in other CVD techniques, such as MOCVD and AACVD. $VOCl_3$ has been the least used precursor in recent years and is not recommended because of the difficulty to produce a VO_2 pure phase and the highest reaction temperature requirement among the precursors.

Table 8.1 Parameters for various CVD techniques to deposit pure-phase VO_2(M); pure-phase VO_2(B); and a VO_2(M), VO_2(B), and VO_2-VO_x mixing phase

Method	Precursor/ results	Controlling			Remarks
		Molar ratio	**T (°C)[a]**	**Flow rate (L/min)**	
	VCl_4-O_2^c	0.45–0.8	350–475	12	Is able to produce a pure VO_2 phase with little contamination
		0.05–0.2	500–550	0.5–1	Is recommended for a system with precise condition controlling
	$VOCl_3$-O_2/Mixing phase	0.5–1	600–650	0.5–2.5	Produces mixing-phase only Is not recommended
APCVD		**O_2 flow rate (L/min)**	**T (°C)**	**Total flow rate (L/min)**	
	VO(acac)$_2$-O_2/VO_2(M)	0.1–0.9	475–575	12	Has a wide process window Is recommended for a system with rough condition controlling
	VO(OC_3H_7)$_3$-O_2/VO_2(M)	0.6	300–450	0.7–0.8	Has the potential for carbon contamination, which may affect optical properties
		Precursor flow rate (L/min)	**T (°C)**	**Total flow rate (L/min)**	
	VO(OC_3H_7)$_3$/VO_2(M)	2–4	400–450	12	Is able to produce a VO_2 film with a single precursor Is suitable for an application with low purity requirement Has a relatively simple system configuration

(*Contd.*)

Table 8.1 (*Continued*)

MOCVD		**P (torr)[b]**	**T (°C)**	**Injection rate**	**O_2 flow rate (L/min)**	
	$VO(acac)_2$-O_2/VO_2(M)	760 3	430–475 300–350	3 mL/h 0.15 L/min	0.02–0.04 0.15	Has a wide process window Allows controlling the reaction conditions more precisely than APCVD
	$VO(acac)_2$-O_2/VO_2(M)+VO_2(B)	760	410–430	3 mL/h	0.04–0.08	
	$VO(acac)_2$-O_2/VO_2(B)	760 3	380–430 200–300	3 mL/h 0.15 L/min	0.04–0.08 0.15	
AACVD		**Solvent**	**T (°C)**	**Flow rate (L/min)**		
	$VO(acac)_2$/VO_2(M)	Ethanol	450–600	1.5		Has a high deposition rate but the product is of relatively poor quality
	$V(ACAC)_3$/Mixing phase	Ethanol	500–600	1.5		Is suitable for applications with low film quality requirement
PECVD		**Plasma power (W)**	**T (°C)**	**O_2 flow rate O_2 (sccm)**		Has a fast deposition rate (100–200 nm/min)
	$VO(acac)_2$-O_2/VO_2(M)	200	400–450	10–20		Creates a product of high crystallinity Has potential carbon contamination, which may affect optical properties

ALD		P (torr)	T (°C)	Pulse plan (s)[d]	Allows precise control of film thickness but is a complex system
	TEMAV-O_3/VO_2(M)	7.5×10^{-7}	150	2-25-5-15	Is suitable for an application with a high requirement of a specific film thickness
	VO(acac)$_2$-O_2/VO_2(M)	10	400–475	4-2-1-1	TEMAV especially suitable for applications requiring a low process temperature

[a]"T" stands for growth temperature.
[b]"P" stands for chamber pressure.
[c]Due to the lack of phase information in some of the references, we cannot differentiate the optimum conditions for VO_2(M) and VO_2(B) in the VCl_4-H_2O system.
[d]The pulse plan describes the purging step in ALD. The four values indicate the time for metal precursor injection, the first purging, the oxygen/ozone injection, and the second purging.

8.6 Application of Computer Simulation to the VO_2 Synthesis Process Optimization

As the CVD technique becomes more and more important in both industry and academia, it is necessary to predict the behavior of such elusive systems to understand the relationship between the system variables. The systems can be understood and predicted by using three tools: theory analysis, experiment, and simulation. However, due to the highly complex flow conditions in CVD reactors, it is not possible to manually calculate or even depict the properties of the flow, for example, the flow rates, the precursor concentrations, velocity profiles, temperature distributions. On the other hand, computational fluid dynamics (CFD) is the computational method that can potentially be used to improve the efficiency of existing operating systems as well as the design of new systems. It provides a qualitative and, occasionally, quantitative analysis of flowing fluids in relation to the fluids' surroundings. Using CFD, flow regimes in CVD reactors can be accurately simulated to give a more meticulous view of a reactor's regional behavior that is difficult to observe by experiment. Computer algorithms are being used to envisage flow features on the basis of a set of conservation equations for mass, momentum, and energy in a flow geometry of interest, such as 1D, 2D, or 3D, together with subsidiary sets of equations reflecting the problem in question.

Accompanied by the increase in computational power, software development, and distributed knowledge, CFD simulation becomes more and more important. Compared with the experiment, fluid dynamic phenomena simulation through CFD software has the advantage of low cost, less consumption of time, safer procedures, and better accuracy. Because APCVD is the most commonly used CVD system, the simulation and model for this CVD technique is the best established. Therefore, the CFD simulation of the APCVD system will be the focus in this section.

8.6.1 Governing Equations of Computational Fluid Dynamics Modeling

CFD simulations were performed in the software Ansys Fluent, which uses the finite volume method for solving equations in momentum, mass, and energy. Fluid motion within a fluid is described by the Navier–Stokes equations. More details of the numerical algorithm and procedures can be found in the user's guide of Ansys Fluent. The transport behavior of gas depends on the degree of rarefaction, expressed by the Knudsen number $\mathrm{Kn} = \lambda/L$, the ratio of the mean free path of molecules λ, and a typical length scale of the smallest detail of interest within the reactor, L. The mean free path, according to the kinetic gas theory, is described by Ref. [1].

$$\lambda = \frac{1}{\pi\sqrt{2}} \cdot \frac{k_B T}{p\sigma^2} \tag{8.1}$$

where T is the gas temperature, p is the reactor pressure, k_B is the Boltzmann constant, and σ is the collision diameter. In the case of the APCVD VO_2 deposition system, all tryouts were used at atmospheric pressure and temperatures between 300°C and 750°C with nitrogen as the carrier gas ($\sigma \cong 3.8$ Å). Under these conditions, the mean free path is approximately 0.2 μm. Depending on the pressure, the temperature, and the reactor's architecture, the flow in the reactor may be treated as continuous, free molecular, or transitional. For $\mathrm{Kn} < 0.01$, the gas flow can be simulated using the theory of continuum mechanics since the reactor's characteristic length scales of interest are larger than 20 μm. Additionally, the flow conditions are assumed to be laminar, having a Reynolds number lower than 2300 ($\mathrm{Re} < 2300$).

In CFD analysis, the domain is divided into a finite number of elements. The following steps are used to solve a general CFD simulation problem:

1. Provide an outline of the geometrical domain.
2. Divide the domain into nonoverlapping grid cells or control volumes.
3. Assign the physical and chemical phenomena.
4. Define material (build in library or new assignment) and fluid properties.
5. Specify boundary conditions that occur within the geometrical domain.

Consequently, the physical and chemical phenomena are described by the equations of mass transport, momentum, heat, and concentration of chemical genera involved in the CVD process.

The mass conservation equation is referred to as the continuity equation in continuum mechanics, identifying the net accumulation of mass in any fluid element, and is equal to the net rate of flow of mass into the fluid element.

$$\frac{\partial_\rho}{\partial_t} = -\nabla \cdot (\rho \vec{v}) \tag{8.2}$$

where ρ is the average density of the fluid, $\vec{v}$ is the flow velocity vector field in the fluid, (x, y) are the 2DS spatial coordinates, and t is time.

The momentum conservation equation in an inertial (nonaccelerating) is given by:

$$\frac{\partial}{\partial t}(\rho \vec{v}) + \nabla \cdot (\rho v \vec{v}) = -\rho + \nabla \cdot (\bar{\bar{\tau}}) + \rho \vec{g} + F$$

$$\tau = \mu[(\nabla u + (\nabla u)^T - \frac{2}{3}\delta \nabla u], \tag{8.3}$$

where p is the density of the fluid, u is the speed, τ indicates the tensor forces associated with the rate of shear, μ is the dynamic viscosity, δ is the identity matrix, and T is the temperature.

The equations for thermal energy are given below.

$$\nabla \cdot (\rho \cdot u \cdot h) = \nabla \cdot (\lambda \cdot \nabla T) + \tau : \nabla u + S_{\mathrm{E}}$$

$$h = \sum_{i}^{Nc} Y_i \cdot h_i$$

$$h_i = h_{i,T\,\mathrm{ref}} + \int_{T\,\mathrm{ref}}^{T} c_{p,i}\,\mathrm{d}t$$

$$S_{\mathrm{E}} = \sum_{j=1}^{Nr} (-\Delta H_{\mathrm{R}})_{\mathrm{j}} \cdot R_{\mathrm{j}}, \tag{8.4}$$

where p is the density of the fluid, u is the speed, τ refers to the tensor forces associated with the rate of shear, μ is the dynamic viscosity, δ is the identity matrix, T is the temperature, h is the thermodynamic enthalpy, S_{E} is the term of power generation, R_{j} is the reaction rate for the component j, λ is the thermal conductivity, $c_{p,i}$ is the specific heat at constant pressure, $\Delta\ H_{\mathrm{R}}$ is the change in enthalpy of the reaction, P is the pressure, and R is the universal gas constant.

If the fluid is incompressible ($p = 0$) and inviscid ($\mu = 0$), then the Navier–Stokes equations become the Euler equations, as seen below.

$$\nabla \cdot v = 0$$
$$\frac{\partial v}{\partial t} + (\nabla \cdot v)\, v = -\frac{1}{\rho}\nabla \rho + f. \tag{8.5}$$

And by inserting the vector identity, it becomes

$$\sum (v \cdot \nabla)\, v = \frac{1}{2}\nabla |v|^2 + (v \cdot \nabla)\, v. \tag{8.6}$$

Assuming the external force field has a potential

$$f = -\nabla V, \tag{8.7}$$

on combining the Eqs. (8.5–8.7), we will get

$$\frac{\partial v}{\partial t} + \frac{1}{2}\nabla |v|^2 + (v \cdot \nabla)\, v. \tag{8.8}$$

The solution of the above equations gives results such as the distribution of velocity and temperature profiles of the elements for the geometry of the components. While using these equations, the domain of the part to be simulated is discretized or meshed into control volumes or areas (3D or 2D shapes) of micro- or nanodimensions. It is obvious that these volumes or areas represent the actual or scaled (up or down) part, which will be used under the same conditions as those of the experimental procedure.

The fluid boundaries in the simulations considered are the walls of each part, the bottom, and the free surface of the fluid. On the walls, the boundary condition is that the velocity of the flow at the wall is zero in the normal direction of the wall.

8.6.2 Thermodynamics of an APCVD Reactor

An APCVD reaction is governed by both kinetics and thermodynamics. Kinetics defines the chemical reactions with respect to the reaction rates, effect of various variables, rearrangement of atoms, formation of intermediate species, etc. On the other hand, thermodynamics is the driving force that shows the direction the reaction is going to continue. Chemical thermodynamics is concerned with the interrelation of various forms of energy and the transfer of energy from one chemical system to another under the first and second laws of thermodynamics. In the case of APCVD, this transfer occurs when gaseous compounds are introduced into the reaction chamber.

The Gibbs free energy of a reaction (ΔG_r) can be used to determine the initial feasibility of the reaction for a given pressure and temperature. For the calculation of the ΔG_r, the Gibbs free energy of formation contributions (ΔG_f) for both product and reactant species need to be determined. The Gibbs free energy for a deposition is expressed in the following differential equation, where T is the temperature in kelvin and ΔH_f^0 is the standard enthalpy of formation at 298 K, S^0 is the entropy at 298 K, and C^p is the heat capacity.

$$\Delta G_f(T) = \Delta H_f^0(298) + \int_{298}^{T} C_P \mathrm{d}T - TS^0(298) - \int_{298}^{T} \frac{C_P}{T} \mathrm{d}T \quad (8.9)$$

$$\Delta G_r = \sum \Delta G_f(\text{products}) - \sum \Delta G_f(\text{reactants}) \quad (8.10)$$

A positive ΔG_r means the reaction will lead to a thermodynamically adverse outcome. A negative ΔG_r means the reaction will lead to a thermodynamically favorable outcome.

In the event where many different reaction routes are considered possible by the thermodynamic laws, the one giving rise to the most negative ΔG_r will dominate. The equilibrium constant K for any such reaction mechanism can be found by

$$K = \exp\left(-\frac{\Delta G_r}{\mathrm{RT}}\right). \quad (8.11)$$

Once K is found for the various allowed reaction mechanisms, the partial pressure or activity of the gaseous precursor species can be calculated by the law of mass action, as follows:

$$K_T = \frac{\prod_{i=1}^{n} \mathrm{P}_i(\text{products})}{\prod_{i=1}^{n} \mathrm{R}_i(\text{reactants})} \quad (8.12)$$

All possible reaction routes, such as vapor pressure or aerosol saturation, homogeneous or heterogeneous reaction, precursor adsorption, desorption/removal of by-products produced, and surface diffusion, need to be considered. The film growth rate is estimated by the slowest—the rate determining—step of deposition.

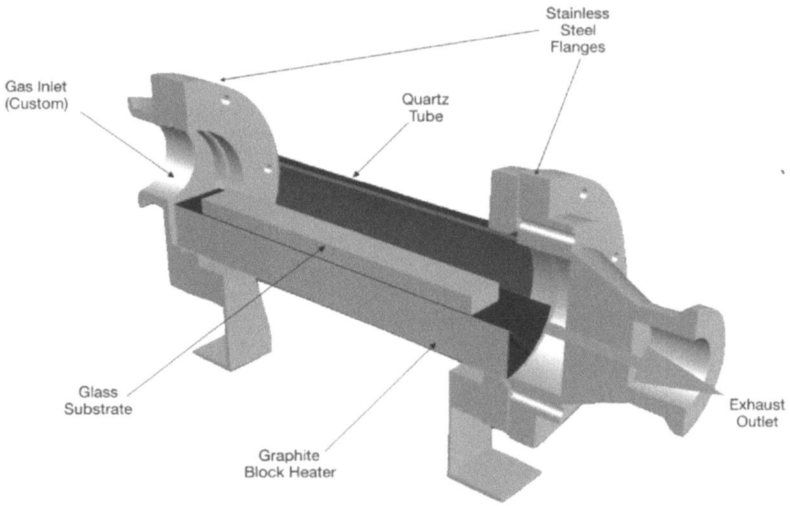

Figure 8.11 Symmetry plane of a laminar flow APCVD reactor (LFR-APCVD).

8.6.3 CVD Reactor Geometry

A detailed design of an APCVD reactor is depicted in Fig. 8.11. The reactor consists of metallic flanges (stainless steel 316L), a graphite heater, a quartz tube, and of course the substrate to be deposited, which in this case is a conductive glass (FTO). The quartz tube (cylindric geometry) is held between the two flanges. One flange is connected to a custom-made inlet, and the second flange is connected to a conical outlet section, leading to the exhaust. The inlet is made from an aluminum-bronze alloy for better corrosion resistance while it is used within various precursors, so that it does not face functional complications. The inlet part holds a specially designed baffle to spread the flow entering from a 1/4″ pipe over the width of the reactor. The graphite block placed below the glass substrate has three heating elements capable of setting up and supporting high temperatures.

Generally, in every CFD model, the target is to avoid spending a great amount of computer time and power to model the completely detailed part, in our case study the reactor, as described in the opening paragraph of this subchapter. The most crucial parts of the

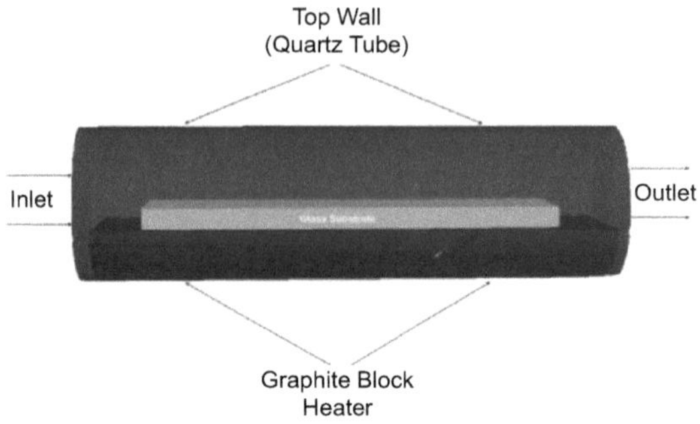

Figure 8.12 Components used in the computational fluid dynamics simulation.

reactor are selected as the domain for CFD simulation, specifically, the inlet part inside the quartz tube, the glass substrate that is heated by the graphite block, and the wall surroundings of the quartz tube, as seen in Fig. 8.12. Hence, due to symmetry, only half of the needs will be simulated. Additionally, some important assumptions are taken into consideration while simulating the model:

- There are no gas leaks in any part of the APCVD reactor. In the experimental procedures that took place in the laboratories of Delta Nano-Engineering Solutions Ltd. (DNES Ltd.), when ink was used to visualize the flow patterns inside the APCVD reactor, only a negligible amount of the ink leaked through.
- The temperature of the walls of the quartz tube will be set according to the experimental parameters that were completed before and after the CFD simulation.

The inlet flow is assumed to be uniform due to the small geometry of the latter.

8.6.4 Methodologies of CFD Simulation Steps Analysis

To support a close relationship between the experimental setup and the CFD model, it is important to highlight how and which important part of the system in question would be examined. This is highly significant since computer power and time need to be saved to ensure time- and cost-efficient results. Apart from evaluating

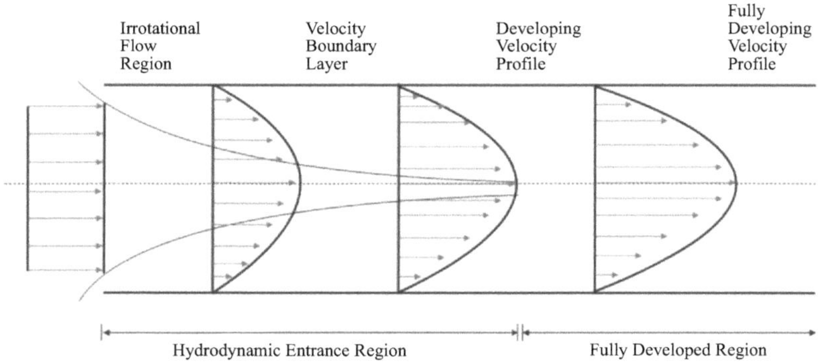

Figure 8.13 Velocity profile of the fluid in a quartz tube (circular) of the APCVD reactor (laminar flow distribution).

a simulation model, it is important to envisage how it will really work as a physical system in order to "interpret" the working order in terms of mathematical expressions calculations. This is imperative because since the fluid flow in the APCVD reactors is principally three dimensional, the mathematical expression of the transport equations should be constructed properly. For the requirements of this chapter, a steady-state simulation was considered (Poiseuille flow), as seen in Fig. 8.13, which stood for the conditions after the flow field has developed inside the APCVD reactor, long enough for it to reach equilibrium. Moreover, the physics of the system to be simulated during CFD modeling must be determined. Critical factors that influence the effect of the outcomes can be turbulence flow, heat transfer, chemical species transport, and radiation phenomena. Fluid physics, such as thermal sources, velocity profiles, and chemical sources, are essential details that need to be modeled appropriately for the simulation to run and produce results efficiently.

Reaching meticulous and descriptive results within the prerequisites of the research hypothesis is essential in order to get a full and clear picture of what is really going on inside and outside, occasionally, the reactor's system boundaries. Consequently, the aims of transport phenomena in chemical processes are three closely related topics: fluid dynamics, heat transfer, and mass transfer. Fundamentally, fluid dynamics involves the transport of momentum, heat transfer involves the transport of energy, and mass transfer is concerned with the transport of mass of various chemical species. The CFD code, as a numerical simulation of fluid flow, is used to describe the complex behavior of fluids by numerically solving the laws that govern the movement of fluids in or around a part of the system or subsystem.

The important CFD simulation steps can be summarized as follows:

1. Creation of the geometrical model (1D, 2D, or 3D)
2. Designation of boundary conditions on volumes and/or surfaces that may occur
3. Mesh generation of the model
4. Conduction of quality checks before the first simulation
5. Performance of the simulation process
6. Outcome evaluation of refinement of the grid; discretization of the model into a larger number of discrete elements
7. Modification of the simulation and reoccupation as needed
8. Postprocess analysis of the simulation results and conclusions

8.7 Simulation Results

In the mesh, the relevance center was set to *fine* and the smoothing was set to *high*. To obtain more accurate results within the simulation, *refinement* was used. The refinement level could be found in ANSYS Fluent and was increased to Level 3, obtaining the results as seen in Fig. 8.14. The boundary conditions that were used in the CFD simulation were the following: fully developed velocity profile of the N_2 flow rate at 0.1 L/min and the reactor temperature at 600°C, including the substrate and the walls of the quartz tube. In the outlet, there was zero pressure during an outlet outflow. At the internal walls, we assumed a no-slip condition for the velocity components and a fixed temperature based on experimental data and prior experiments. At the reactor outlet, zero axial gradients of velocity, outflow, and temperature were used.

The measured chemical mechanism has been used to simulate the growth profile in the laminar flow reactor. Figure 8.15 shows the

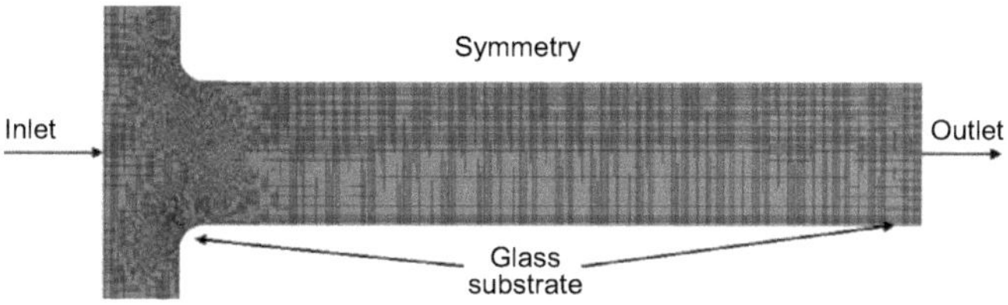

Figure 8.14 Representation of the quartz tube model after the discretization process.

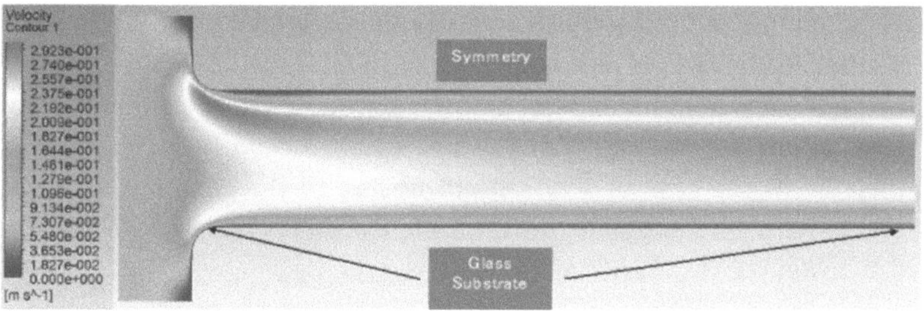

Figure 8.15 Total calculated velocity contour of DBDAS on an FTO glass, placed in the APCVD reactor at 600°C.

velocity vector magnitude of nitrogen flow on the symmetry axis of the bottom of the reactor, including the inlet, the glass substrate, and the top plate (symmetry).

Qualitatively, the growth rate profile on the substrate corresponds with the measurements done at DNES Ltd., using dibutoxydiacetoxysilane (DBDAS) and N_2 as the carrier gas. Additionally, it was found that the layer was the thickest at the middle and upward, to the end of the glass substrate.

Quantitatively, the velocity magnitude along with the rate distribution predictions made by the geometrical model were in good agreement with the measurements completed at DNES Ltd. laboratories. The predictions made by the simulation model were obtained purely on basis of the geometry of the reactor, the temperature profiles measured by the DNES Ltd., and the reaction parameters obtained from the experiments done in the LF-APCVD reactor.

Nanoparticles coming from the hot region have a greater velocity due to volumetric expansion as a result of decreased density than those coming from the cold region. That said, the highest growth rate was expected at the middle and upwards, to the end of the reactor, because the fluid has reached the highest temperature while the velocity magnitude is fully developed, leading to a high quality material layer deposited on the top of glass substrate.

Both quantitative and qualitative experimental results revealed the validity of the chemical model used, where the rate-limiting step was proposed to be a gas-phase reaction. In the very case where a diffusion-limited step at the surface of the glass substrate takes place, the highest growth rate along with the fully developed velocity magnitude would be expected at the entrance (inlet) of the reactor, where the concentration of the precursor would be at its highest. Instead, in the case of a rate-limiting step in the gas phase, the gas had to be preheated in order to prove the intermediate film forming precursor.

References

1. Choy, K. L. (2003). Chemical vapour deposition of coatings, *Prog. Mater. Sci.*, **48**(2), pp. 57–170.
2. Drosos, C. and Vernardou, D. (2015). Perspectives of energy materials grown by APCVD, *Sol. Energy Mater. Sol. Cells*, **140**, pp. 1–8.
3. Vernardou, D., Paterakis, P., Drosos, H., Spanakis, E., Povey, I. M., Pemble, M. E., Koudoumas, E. and Katsarakis, N. (2011). A study of the electrochemical performance of vanadium oxide thin films grown by atmospheric pressure chemical vapour deposition, *Sol. Energy Mater. Sol. Cells*, **95**(10), pp. 2842–2847.
4. Dobkin, D. M. and Zuraw, M. K. (2003). *Principles of Chemical Vapor Deposition* [electronic resource] (Springer Netherlands, Dordrecht).
5. Jones, A. C. and Hitchman, M. L. (2009). *Chemical Vapour Deposition: Precursors, Processes and Applications* [electronic resource] (Royal Society of Chemistry, Cambridge, UK).
6. Vernardou, D., Louloudakis, D., Spanakis, E., Katsarakis, N. and Koudoumas, E. (2015). Functional properties of APCVD VO_2 layers, *Int. J. Thin Films Sci. Technol.*, **4**, pp. 187–191.

7. Peters, E. S., Carmalt, C. J. and Parkin, I. P. (2004). Dual-source chemical vapour deposition of titanium sulfide thin films from tetrakisdimethylamidotitanium and sulfur precursors, *J. Mater. Chem.*, **14**(23), pp. 3474–3477.
8. Vernardou, D., Pemble, M. E. and Sheel, D. W. (2004). Vanadium oxides prepared by liquid injection MOCVD using vanadyl acetylacetonate, *Surf. Coat. Technol.*, **188–189**, pp. 250–254.
9. Pierson, H. O. (1999). 4.5 MOCVD reactions for the deposition of metals, in *Handbook of Chemical Vapor Deposition (CVD): Principles, Technology and Applications*, 2nd ed. (William Andrew Publishing/Noyes, USA).
10. Dobkin, D. M. and Zuraw, M. K. (2003). Gas discharge plasmas for CVD, in *Principles of Chemical Vapor Deposition* (Springer Netherlands, Dordrecht), pp. 149–194.
11. Warwick, M. E. A. and Binions, R. (2014). Chemical vapour deposition of thermochromic vanadium dioxide thin films for energy efficient glazing, *J. Solid State Chem.*, **214**, pp. 53–66.
12. Ritala, M. and Niinistö, J. (2009). Atomic layer deposition, in *Chemical Vapour Deposition: Precursors, Processes and Applications* (The Royal Society of Chemistry), pp. 158–206.
13. MacChesney, J. B., Potter, J. F. and Guggenheim, H. J. (1968). Preparation and properties of vanadium dioxide films, *J. Electrochem. Soc.*, **115**(1), pp. 52–55.
14. Vernardou, D., Pemble, M. E. and Sheel, D. W. (2006). The growth of thermochromic VO_2 films on glass by atmospheric-pressure CVD: a comparative study of precursors, CVD methodology, and substrates, *Chem. Vap. Deposition*, **12**(5), pp. 263–274.
15. Gaskell, J. M., Afzaal, M., Sheel, D. W., Yates, H. M., Delfanazari, K. and Muskens, O. L. (2016). Optimised atmospheric pressure CVD of monoclinic VO_2 thin films with picosecond phase transition, *Surf. Coat. Technol.*, **287**, pp. 160–165.
16. Manning, T. D. and Parkin, I. P. (2004). Vanadium(IV) oxide thin films on glass and silicon from the atmospheric pressure chemical vapour deposition reaction of $VOCl_3$ and water, *Polyhedron*, **23**(18), pp. 3087–3095.
17. Manning, T. D., Parkin, I. P., Clark, R. J. H., Sheel, D., Pemble, M. E. and Vernadou, D. (2002). Intelligent window coatings: atmospheric pressure chemical vapour deposition of vanadium oxides, *J. Mater. Chem.*, **12**(10), pp. 2936–2939.

18. Wilkinson, M., Kafizas, A., Bawaked, S. M., Obaid, A. Y., Al-Thabaiti, S. A., Basahel, S. N., Carmalt, C. J. and Parkin, I. P. (2013). Combinatorial atmospheric pressure chemical vapor deposition of graded TiO_2–VO_2 mixed-phase composites and their dual functional property as self-cleaning and photochromic window coatings, *ACS Comb. Sci.*, **15**(6), pp. 309–319.
19. Manning, T. D., Parkin, I. P., Pemble, M. E., Sheel, D. and Vernardou, D. (2004). Intelligent window coatings: atmospheric pressure chemical vapor deposition of tungsten-doped vnadium dioxide, *Chem. Mater.*, **16**(4), pp. 744–749.
20. Kritikos, L., Zambelis, L., Papadimitropoulos, G. and Davazoglou, D. (2007). Structure and electrical properties of selectively chemically vapor deposited vanadium oxide films from vanadium tri-i-propoxy oxide vapors, *Surf. Coat. Technol.*, **201**(22–23), pp. 9334–9339.
21. Louloudakis, D., Vernardou, D., Spanakis, E., Katsarakis, N. and Koudoumas, E. (2013). Thermochromic vanadium oxide coatings grown by APCVD at low temperatures, *Phys. Procedia*, **46**, pp. 137–141.
22. Papadimitropoulos, G., Kostis, I., Trantalidis, S., Tsiatouras, A., Vasilopoulou, M. and Davazoglou, D. (2015). Investigation of structural, morphological and electrical properties of APCVD vanadium oxide thin films, *Phys. Status Solidi C*, **12**(7), pp. 964–968.
23. Vernardou, D., Louloudakis, D., Spanakis, E., Katsarakis, N. and Koudoumas, E. (2014). Thermochromic amorphous VO_2 coatings grown by APCVD using a single-precursor, *Sol. Energy Mater. Sol. Cells*, **128**, pp. 36–40.
24. Maruyama, T. and Ikuta, Y. (1993). Vanadium dioxide thin films prepared by chemical vapour deposition from vanadium(III) acetylacetonate, *J. Mater. Sci.*, **28**(18), pp. 5073–5078.
25. Spanò, S. F., Toro, R. G., Condorelli, G. G., Messina, G. M. L., Marletta, G. and Malandrino, G. (2015). Phase-selective route to V-O film formation: a systematic MOCVD study into the effects of deposition temperature on structure and morphology, *Chem. Vap. Deposition*, **21**(10-11-12), pp. 319–326.
26. Barreca, D., Depero, L. E., Franzato, E., Rizzi, G. A., Sangaletti, L., Tondello, E. and Vettori, U. (1999). Vanadyl precursors psed to modify the properties of vanadium oxide thin films obtained by chemical vapor deposition, *J. Electrochem. Soc.*, **146**(2), pp. 551–558.
27. Zhang, J. G., Liu, P., Turner, J. A., Tracy, C. E. and Benson, D. K. (1998). Highly stable vanadium oxide cathodes prepared by plasma-enhanced

chemical vapor deposition, *J. Electrochem. Soc.*, ***145***(6), pp. 1889–1892.

28. Wang, S., Liu, G., Hu, P., Zhou, Y., Ke, Y., Li, C., Chen, J., Cao, T. and Long, Y. (2018). Largely lowered transition temperature of a VO_2/carbon hybrid phase change material with high thermal emissivity switching ability and near infrared regulations, *Adv. Mater. Interfaces*, **5**(21), p. 1801063.
29. Naik, A. J. T., Bowman, C., Panjwani, N., Warwick, M. E. A. and Binions, R. (2013). Electric field assisted aerosol assisted chemical vapour deposition of nanostructured metal oxide thin films, *Thin Solid Films*, **544**, pp. 452–456.
30. Piccirillo, C., Binions, R. and Parkin, I. P. (2007). Synthesis and functional properties of vanadium oxides: V_2O_3, VO_2, and V_2O_5 deposited on glass by aerosol-assisted CVD, *Chem. Vap. Deposition*, **13**(4), pp. 145–151.
31. Warwick, M. E. A., Dunnill, C. W., Goodall, J., Darr, J. A. and Binions, R. (2011). Hybrid chemical vapour and nanoceramic aerosol assisted deposition for multifunctional nanocomposite thin films, *Thin Solid Films*, **519**(18), pp. 5942–5948.
32. Blanquart, T., Niinisto, J., Gavagnin, M., Longo, V., Heikkila, M., Puukilainen, E., Pallem, V. R., Dussarrat, C., Ritala, M. and Leskela, M. (2013). Atomic layer deposition and characterization of vanadium oxide thin films, *RSC Adv.*, **3**(4), pp. 1179–1185.
33. Peter, A. P., Martens, K., Rampelberg, G., Toeller, M., Ablett, J. M., Meersschaut, J., Cuypers, D., Franquet, A., Detavernier, C., Rueff, J.-P., Schaekers, M., Van Elshocht, S., Jurczak, M., Adelmann, C. and Radu, I. P. (2015). Metal-insulator transition in ALD VO_2 ultrathin films and nanoparticles: morphological control, *Adv. Funct. Mater.*, **25**(5), pp. 679–686.
34. Premkumar, P. A., Toeller, M., Radu, I. P., Adelmann, C., Schaekers, M., Meersschaut, J., Conard, T. and Elshocht, S. V. (2012). Process study and characterization of VO_2 thin films synthesized by ALD using TEMAV and O_3 precursors, *ECS J. Solid State Sci. Technol.*, **1**(4), pp. P169–P174.
35. Tangirala, M., Zhang, K., Nminibapiel, D., Pallem, V., Dussarrat, C., Cao, W., Adam, T. N., Johnson, C. S., Elsayed-Ali, H. E. and Baumgart, H. (2014). Physical analysis of VO_2 films grown by atomic layer deposition and RF magnetron sputtering, *ECS J. Solid State Sci. Technol.*, **3**(6), pp. N89–N94.
36. Rampelberg, G., Schaekers, M., Martens, K., Xie, Q., Deduytsche, D., Schutter, B. D., Blasco, N., Kittl, J. and Detavernier, C. (2011). Semiconductor-metal transition in thin VO_2 films grown by ozone based atomic layer deposition, *Appl. Phys. Lett.*, **98**(16), p. 162902.

37. Kim, C.-Y., Kim, S. H., Kim, S. J. and An, K.-S. (2014). VO_2(1 1 0) film formation on TiO_2(1 1 0) through post-reduction of ALD grown vanadium oxide. *Appl. Surf. Sci.*, **313**, pp. 368–371.
38. Dagur, P., Mane, A. U. and Shivashankar, S. A. (2005). Thin films of VO_2 on glass by atomic layer deposition: microstructure and electrical properties, *J. Cryst. Growth*, **275**(1–2), pp. e1223–e1228.
39. Chen, X., Pomerantseva, E., Banerjee, P., Gregorczyk, K., Ghodssi, R. and Rubloff, G. (2012). Ozone-based atomic layer deposition of crystalline V_2O_5 films for high performance electrochemical energy storage, *Chem. Mater.*, **24**(7), pp. 1255–1261.
40. Musschoot, J., Deduytsche, D., Van Meirhaeghe, R. L. and Detavernier, C. (2009). ALD of vanadium oxide, *ECS Trans.*, **25**(4), pp. 29–37.
41. Matero, R., Rahtu, A., Ritala, M., Leskelä, M. and Sajavaara, T. (2000). Effect of water dose on the atomic layer deposition rate of oxide thin films, *Thin Solid Films*, **368**(1), pp. 1–7.
42. Suntola, T. (1989). Atomic layer epitaxy, *Mater. Sci. Rep.*, **4**(5), pp. 261–312.
43. Wang, S., Owusu, K. A., Mai, L., Ke, Y., Zhou, Y., Hu, P., Magdassi, S. and Long, Y. (2018). Vanadium dioxide for energy conservation and energy storage applications: synthesis and performance improvement, *Appl. Energy*, **211**, pp. 200–217.

Chapter 9

Physical Vapor Deposition and Its Application in Vanadium Dioxide Synthesis

Tuan Duc Vu,[a,*] Ping Jin,[b] Zhang Chen,[c,*] Yanfeng Gao,[c] and Yi Long[a]

[a]*School of Materials Science and Engineering, Nanyang Technological University, 50 Nanyang Avenue, Singapore 639798, Singapore*
[b]*State Key Laboratory of High Performance Ceramics and Superfine Microstructure, Shanghai Institute of Ceramics, Chinese Academy of Sciences, Dingxi Road 1295, Changning, Shanghai 200050, China*
[c]*School of Materials Science and Engineering, Shanghai University, Shanghai 200444, China*
p-jin@mail.sic.ac.cn, yfgao@mail.sic.ac.cn, longyi@ntu.edu.sg

Physical vapor deposition (PVD) refers to a variety of vacuum deposition methods used to deposit thin films by the condensation of a vaporized form of the desired film material onto various workpiece surfaces. The coating methods involve purely physical processes, such as high-temperature vacuum evaporation with subsequent condensation and plasma sputter bombardment. When preparing VO_2(M/R) thin films, pure vanadium, V_2O_3, VO_2, or V_2O_5 is commonly used as the target. The partial pressure of oxygen in the deposition chamber is accurately controlled to prevent the

*These authors contributed equally to this work.

Vanadium Dioxide-Based Thermochromic Smart Windows
Edited by Yi Long and Yanfeng Gao

ISBN 978-981-4877-06-0 (Hardcover), 978-1-003-17731-9 (eBook)
www.jennystanford.com

formation of other vanadium oxides with different valences, and the substrates are heated to relatively high temperatures to promote the crystallization of VO_2(M/R). Specifically, PVD includes pulsed laser deposition (PLD), sputtering, and evaporation decomposition (ED). The following sections provide a brief introduction to these methods of VO_2 thin film preparation, including aspects of film quality, the difficulty of process control, target requirements, and whether the methods are suitable for doping.

9.1 Reactive Pulse Laser Deposition

Reactive pulse laser deposition (RPLD) is a widely used method for VO_2 thin films preparation. In this method, a high-power pulsed laser beam is used to strike the deposited target in vacuum (Fig. 9.1). Although there are relatively few requirements in terms of suitable targets (available targets include metals, metallic oxides powder, and ceramics), the deposition processing is complex. The laser pulse is absorbed by the target and converted into other kinds of energies, like thermal, chemical, and mechanical energy. During this process, the target could be evaporated, ablated, ionized, and even exfoliated. Moreover, the forms of ejected species that are produced by laser include electrons, ions, atoms, molecules, clusters, particulates, and

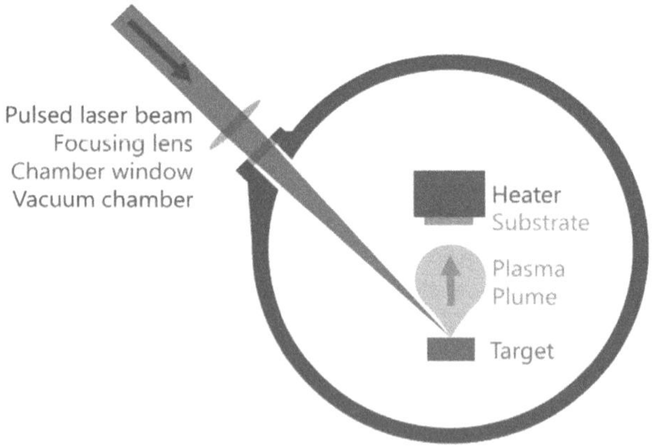

Figure 9.1 A typical experimental arrangement for PLD (https://en.wikipedia.org/wiki/Pulsed_laser_deposition).

molten globules [1]. RPLD is a powerful technique that enables the fabrication of high-quality VO_2 thin films by controlling their deposition parameters. And VO_2 films can be easily codoped by ablating dual targets.

Borek et al. were the first researchers to prepare VO_2(M/R) thin films using pulsed laser deposition (PLD) [2]. A pulsed excimer laser (at a wavelength of $\lambda = 248$ nm and a pulse duration of $\tau = 15$ ns) was adopted to ablate a metallic vanadium target in an ultra-high-vacuum deposition chamber. The total gas (Ar and O_2) pressure in the system was maintained between 100 and 200 mTorr (about 13–26 Pa). An oxygen concentration of approximately 10% was found to be optimal for the fabrication of pure VO_2 films. Thin films were deposited on r-cut sapphire substrates at temperatures ranging from 500°C to 525°C. The as-deposited thin films were all stored in the same atmosphere and temperature for approximately 1 h to obtain VO_2(M/R) polymorphs.

Various PLD processes have been reported ever since then. As shown in Table 9.1, the targets used include metallic vanadium, V_2O_3, VO_2, and V_2O_5. Besides the mostly commonly used KrF excimer laser (λ 248 nm), ArF (λ 193 nm), XeCl (λ 308 nm), and Nd-YAG (λ 532 nm) excimer lasers have also been adopted as the light source. During the thin film deposition, the oxygen partial pressure is usually controlled between 1.3 and 6.5 Pa by using pure O_2 or a mixed gas of O_2 and Ar [2, 3]. In most works, the substrate was heated to 400°C–630°C during the deposition, while in some works, a postannealing process was required to obtain high-quality VO_2 thin films [4, 5].

9.1.1 Oxygen Partial Pressure

Oxygen partial pressure is one of the key parameters during VO_2 thin film preparation. An improper oxygen partial pressure could not only lead to the formation of other vanadium oxides, such as V_2O_3 [6], V_6O_{13} [7], and V_2O_5 [8], but also influence the morphology and phase transition behavior of the VO_2 films (Fig. 9.2a–e) [9, 10]. Fan et al. [11] found that the obtained phases of VO_2 could be controlled by adjusting the oxygen partial pressure during the deposition. That is, a higher oxygen pressure favored the formation

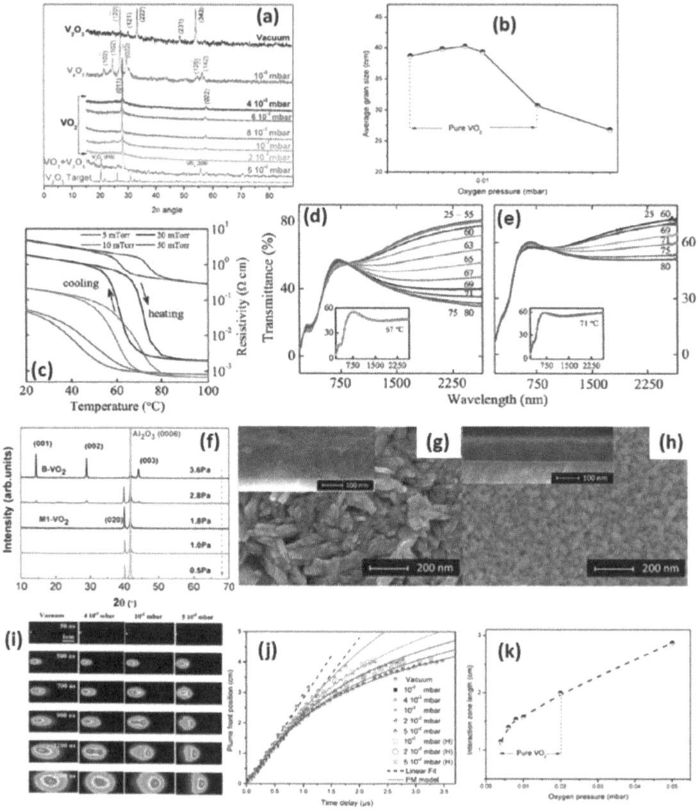

Figure 9.2 (a) XRD patterns of the V_2O_5 target and the films and (b) the calculated average grain size deposited at oxygen pressures ranging from 0 to 5 × 10^{-2} mbar (0 to 5 Pa). Reprinted by permission from Springer Nature: Ref. [10], Copyright 2012. (c) Temperature dependence of electrical resistivity for VO_2 films grown at various oxygen pressures. (d–e) Transmittance spectra for VO_2 films grown at 20 and 50 mTorr (2.67 and 6.67 Pa), respectively; (inset) the experimental (dotted lines) and best-fitted (solid lines) transmittance spectra of VO_2 films. Adapted from Ref. [12] with permission from The Royal Society of Chemistry. (f) XRD patterns of VO_2 films deposited on sapphire substrates at different oxygen pressure (0.5–3.6 Pa) with a VO_2 target. (g, h) The surface morphologies of films deposited at 3.6 Pa (B-VO_2) and 1.8 Pa (M1-VO_2) and (inset) the related cross sections. Reprinted from Ref. [11], Copyright 2012, with permission from Elsevier. (i) The temporal evolution of the visible plume under vacuum and at oxygen pressures of 4 × 10^{-3}, 10^{-2}, and 5 × 10^{-2} mbar represents the images of the plume obtained during deposition in the presence of a substrate heater. The position of the ablating surface is at the limit of the left side of the images. (j) Plume front position as a function of the time delay at vacuum and at different oxygen pressures. (k) The distance over which the plume-oxygen gas interaction occurs versus the oxygen pressure. Reprinted by permission from Springer Nature: Ref. [10], Copyright 2012.

of a VO_2(B) structure, while a lower oxygen pressure favored VO_2(M1/R) epitaxial growth (Fig. 9.2f–g). Lafane's work showed that the oxygen pressure influences the process of molecular oxygen dissociation and results in the deposition of films with different stoichiometries and morphologies [10]. There are two important steps: (i) species from the target hit the substrate, which leads to resputtering, and (ii) a shocked layer forms at the plume-oxygen gas boundary (Fig. 9.2i–k), which produces atomic oxygen that could be directly incorporated into the growing film. At lower oxygen partial pressure, a thin shocked layer forms, while with species of high velocities hitting the substrate, the strong resputtering energies produce homogeneous films but also favor the escape of light elements (O), forming V_2O_3, V_4O_7, etc. At higher oxygen partial pressure, a thick shocked layer forms, offering plenty of atomic oxygen, and the resputtering process is weak, so an inhomogeneous film with a stoichiometry close to the target is formed. However, the other parameters influence the appearance and performance of the deposited films. Different researchers working at the same oxygen partial pressure have reported different results (Figs. 9.2a and 9.2f).

9.1.2 Substrate Temperature

The substrate temperature is another key parameter during RPLD. It is well known that a higher substrate temperature would favor the formation of high-quality VO_2(M/R) films. Kim [13] prepared VO_2 films at a substrate temperature as high as 630°C; the range of electrical resistance changes (ERCs) reached 4×10^5, comparable to that of a single-crystal VO_2 of 10^5 [14]. Besides high temperatures, Nag et al. showed that postannealing after deposition would also help in the formation of high-quality samples [4]. Samples deposited at room temperature followed up by a postannealing process at 450°C showed obvious advantages in light regulation across the metal-insulator transition (MIT) compared to samples deposited at 500°C without a postannealing process (9.1.3). It should be mentioned that the cooling speed would also influence the performance of VO_2 thin films [15].

It is desirable to prepare VO_2 thin films at very low substrate temperatures. Soltani et al. [17] studied the process of depositing

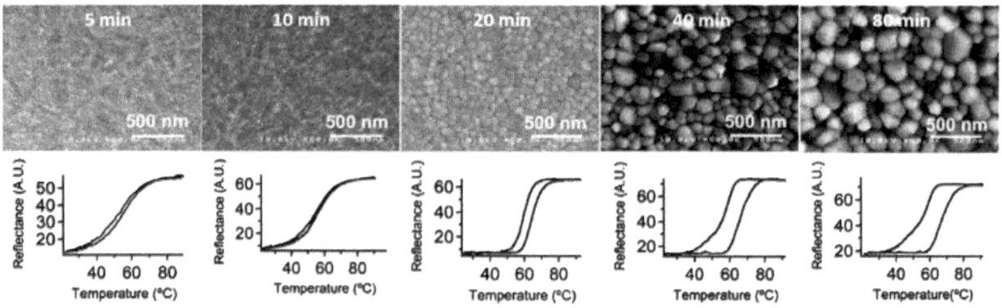

Figure 9.3 SEM and relative IR ($\lambda = 980$ nm) switching curves through the structural phase transition time for a 100 nm VO_2 thin film fabricated at 450°C annealing for 5–80 min under 250 mTorr (~33.3 Pa) of O_2. As the annealing time increased, the hysteresis enhanced, and the sharpness of the transition and grain size improved. Reprinted from [16], Copyright © 2004 with permission from AIP Publishing.

VO_2 films on Kapton substrates using a vanadium target. Although no optical switching was observed in the prepared VO_2/Kapton samples, it was discovered that the VO_2/Si(100) exhibited the well-known MIT with a transition temperature of about 68°C at deposition temperatures as low as 300°C. Maaza et al. [18] prepared high-stoichiometric VO_2 films using a VO_2 target without any substrate heating or atmosphere control. These films could be deposited on silicon, quartz, and sapphire substrates. They all showed obvious MIT at ~68°C with an ERC of ~10^2. Liu et al. [19] and Suh et al. [16] also reported VO_2 films with obvious MIT prepared with a vanadium target on MgO, quartz, and n-type silicon substrates. By improving the annealing time, Suh improved the crystalline quality of VO_2 (9.1.3d). However, all these films showed a weak performance in MIT processing. In recent years, few works have focused on lower-temperature deposition. Most studies have adopted higher deposition temperatures (500°C–550°C) (Table 9.1).

9.1.3 Epitaxial Growth

Epitaxial growth is one route to obtaining high-quality VO_2. Species produced by pulsed laser are high-energy species, which can assist in the in situ fabrication of oriented, epitaxial thin films, leading to PLD becoming a more widely used method for the epitaxial growth of VO_2 films. The first work on VO_2 films prepared by PLD was

Table 9.1 Main parameters of some work and the MIT performances of the obtained VO_2 films

Group	Target	Light source	Laser energy	Substrate	Substrate temperature	Background vacuum	Atmosphere	Post-annealing	Phase change amplitude
Borek et al. [2]	Metallic vanadium	KrF excimer laser (λ 248 nm), pulse duration 15 ns	—	*R*-cut sapphire substrate	500°C–525°C	Total gas pressure 100–200 mTorr	Ar and O_2: 100–200 mTorr O_2, partial pressure 10%	The same condition of the deposition for 1 h	$\text{Log}(R_s/R_m) =$ 4.5–4.6 (no thickness data)
Kim and Kwok [13]	Pressed V_2O_3 powder	ArF excimer laser (λ 193 nm)	2–3 J/cm^2	(0001) and (1010) of sapphire	630°C	$<10^{-5}$ torr	O_2: 20–30 mTorr	None	$\text{Log}(R_s/R_m)$: On (0001) 5.6, On (1010) 5 (no thickness data; the same method in other work involves 150 nm)
Chaker et al. [3]	Metallic vanadium	XeCl excimer laser (λ 308 nm), pulse duration 12 ns	—	Quartz	520°C	—	Ar and O_2: 90 mTorr O_2, partial pressure 5%	None	$\text{Log}(R_s/R_m) =$ 2.13 ($d = 300$ nm)

(*Contd.*)

Table 9.1 *(Continued)*

Group	Target	Light source	Laser energy	Substrate	Substrate temperature	Background vacuum	Atmosphere	Post-annealing	Phase change amplitude
Suh et al. [16]	Metallic vanadium	KrF excimer laser (λ 248 nm), 25 Hz pulse rate	4 mJ/cm^2	N-type silicon	No heating	3×10^{-6} torr	O_2: 5 mTorr [16] O_2: 10 mTorr [4]		$(T_s/T_m)/T_s \approx$ 70% ($\lambda = 980$ nm, $d = 80$ nm) [4]
Nag et al. [4]	Metallic vanadium			*C*-cut sapphire	(i) No heating (ii) 500°C	—	(i) O_2: 10 mTorr (ii) O_2: 50 mTorr	(i) 450°C and 250 mTorr O_2 (ii) None	(i) $(T_s/T_m)/T_s \approx$ 70% (ii) $(T_s/T_m)/T_s \approx$ 40% Both: white light, $d = 80$ nm
Narayan and Bhosle [43]	—	KrF excimer laser (λ 248 nm), pulse duration 25 ns	3–4 J/cm^2	Sapphire (0001)	500°C	—	—	None	$\text{Log}(R_s/R_m) \approx 5$ ($d = 65$ nm)

Yang et al. [44]	VO_2 ceramic	KrF excimer laser (λ 248 nm), pulse duration 25 ns	3–4 J/cm^2	Sapphire (0001)	600°C		O_2: 10 mTorr	None	—
Gupta et al. [45]	VO_2 ceramic	KrF excimer laser (λ 248 nm), pulse duration 25 ns	2–3 J $/cm^2$	Sapphire (0001)	500°C	110^{-6} torr	O_2: 10 mTorr	None	$\text{Log}(R_s/R_m) = 4.43$ ($d = 200$ nm)
Liu et al. (Fernandez) [19]	Metallic vanadium	KrF excimer laser (λ 248 nm), repetition rate 10 pps	14 J/cm^2	MgO and quartz	No heating	10^{-6} torr	O_2 flow: 0.3–1.5 sccm	None	MgO: $(T_s/T_m)/T_s \approx$ 88% ($\lambda = 1250$ nm) Quartz: $\text{Log}(R_s/R_m) =$ 2.82 ($d = 100$–150 nm)
Kim et al. [46]	Metallic vanadium	KrF excimer laser (λ 248 nm), pulse energy 300 mJ, repetition rate 10 Hz	Repetition rate of 300 mJ	Sapphire (0001)	600°C	—	30 mTorr	None	($d \approx 100$ nm)

(*Contd.*)

Table 9.1 (*Continued*)

Group	Target	Light source	Laser energy	Substrate	Substrate temperature	Background vacuum	Atmosphere	Post-annealing	Phase change amplitude
Kim et al. [5]	VO_2 ceramic	KrF excimer laser (λ 248 nm), pulse duration 30 ns	1.5 J/cm^2	Single-crystal *c*-plane sapphire	550°C	10^{-5} torr	O_2: 1–100 mTorr (optimized 10 mTorr)	The same condition of the deposition for 1 h	$\text{Log}(R_s/R_m) =$ 4.6 (160 nm), 1.3 (120 nm)
Lappalainen et al. [25]	V_2O_5 ceramic	XeCl excimer laser (λ 308 nm), pulse duration 25 ns, pulse repetition rate 5 Hz	3 J/cm^2	Magnesium oxide (100); *a*-, *c*-, and *r*-plane cut sapphires	400°C	10^{-5} mbar	2.6×0^{-5} mbar		$\text{Log}(R_s/R_m) =$ 3–4 (d = 100–200 nm) The sample in which the internal stress had been relaxed showed the sharpest MIT and also had the largest change in optical transmission.

Lafane et al. [10]	V_2O_5	KrF excimer laser (λ 248 nm), pulse duration 25 ns, repetition rate 5 Hz	2 J/cm^2	Glass	500°C	—	~37.6 mTorr	None	$(T_s/T_m)/T_s =$ 50%–65% ($\lambda =$ 1100 nm)
Maaza et al. [18]	VO_2	KrF excimer laser (λ 248 nm), repetition rate 10 Hz	0.5 ± 2.7 J/cm^2	Silicon, quartz, and sapphire	No heating	0.75–1.88 $\times 10^{-9}$ torr	None	None	$\text{Log}(R_s/R_m) \approx 2$ (150 nm)
Wu et al. [47]	Vanadium metal	—	1 W	Sapphire (0001)	500°C	1×10^{-5} torr	50 mTorr	None	$\text{Log}(R_s/R_m) \approx 4.3$ ($d = 480$ nm)

(*Contd.*)

Table 9.1 (*Continued*)

Group	Target	Light source	Laser energy	Substrate	Substrate temperature	Background vacuum	Atmosphere	Post-annealing	Phase change amplitude
Zhu et al. [48]	Vanadium metal	Nd-YAG laser (λ 532 nm), pulse duration 10 ns, repetition rate 10 Hz		Sapphire (0001)	450°C	1×10^{-6} torr	~45 mTorr	None	$\text{Log}(R_s/R_m) =$ 4–4.3 ($d = 500$ nm)
Jian et al. [49]	V_2O_5	KrF excimer laser (λ 248 nm)	3 J/cm^2	Amorphous glass, single-crystal *c*-cut sapphire, and Si (111)	Glass: 500°C *C*-cut sapphire and Si (111): 550°C	—	10 mTorr	None	$\text{Log}(R_s/R_m)$: Glass 3.5, *c*-cut sapphire 3.8, Si(111) 3
Borek et al. [2]	Metallic vanadium	KrF excimer laser (λ 248 nm), pulse duration 15 ns	—	*R*-cut sapphire	500°C–525°C	Total gas pressure 100–200 mTorr	Ar and O_2: 100–200 mTorr O_2, partial pressure 10%	The same condition of the deposition for 1 h	$\text{Log}(R_s/R_m) =$ 4.5–4.6 (no thickness data)

Kim and Kwok [13]	Pressed V_2O_3 powder	ArF excimer laser (λ 193 nm)	2–3 J/cm^2	(0001) and (1010) of sapphire	630°C	$<10^{-5}$ torr	O_2: 20–30 mTorr	None	Log(R_s/R_m): On (0001) 5.6, On (1010) 5 (no thickness data)
Chaker et al. [3]	Metallic vanadium	XeCl excimer laser (λ 308 nm), pulse duration 12 ns	—	Quartz	520°C	—	Ar and O_2: 90 mTorr O_2, partial pressure 5%	None	Log(R_s/R_m) = 2.13 ($d = 300$ nm)
Suh et al. [16]	Metallic vanadium	KrF excimer laser (λ 248 nm), pulse rate 25 Hz	4 mJ/cm^2	N-type silicon	No heating	3×10^{-6} torr	O_2: 5 mTorr [16] O_2: 10 mTorr [4]		$(T_s/T_m)/T_s \approx$ 70% ($\lambda = 980$ nm, $d = 80$ nm) [4]

(Contd.)

Table 9.1 (*Continued*)

Group	Target	Light source	Laser energy	Substrate	Substrate temperature	Background vacuum	Atmosphere	Post-annealing	Phase change amplitude
Nag et al. [4]	Metallic vanadium			*C*-cut sapphire	(i) No heating (ii) 500°C	—	(i) O_2: 10 mTorr (ii) O_2: 50 mTorr	(i) 450°C and 250 mTorr O_2 (ii) None	(i) $(T_s/T_m)/T_s \approx$ 70% (ii) $(T_s/T_m)/T_s \approx$ 40% Both: white light, $d = 80$ nm)
Narayan [43]	—	KrF excimer laser (λ 248 nm), pulse duration 25 ns	3–4 J/cm^2	Sapphire (0001)	500°C	—	—	None	$\mathrm{Log}(R_s/R_m) \approx 5$ ($d = 65$ nm)
Yang et al. [44]	VO_2 ceramic	KrF excimer laser (λ 248 nm), pulse duration 25 ns	3–4 J/cm^2	Sapphire (0001)	600°C	—	O_2: 10 mTorr	None	—

Gupta et al. [45]	VO_2 ceramic	KrF excimer laser (λ 248 nm), pulse duration 25 ns	2–3 J /cm^2	Sapphire (0001)	500°C	10^{-6} torr	O_2: 10 mTorr	None	$\text{Log}(R_s/R_m) = 4.43$ ($d = 200$ nm)
Liu et al. (Fernandez) [19]	Metallic vanadium	KrF excimer laser (λ 248 nm), repetition rate 10 pps	14 J/cm^2	MgO and quartz	No heating	10^{-6} torr	O_2 flow: 0.3–1.5 sccm	None	MgO: $(T_s/T_m)/T_s \approx 88\%$ ($\lambda = 1250$ nm) Quartz: $\text{Log}(R_s/R_m) = 2.82$ ($d = 100$–150 nm)
Kim et al. [46]	Metallic vanadium	KrF excimer laser (λ 248 nm), pulse energy 300 mJ, repetition rate 10 Hz	Repetition rate of 300 mJ	Sapphire (0001)	600°C	—	30 mTorr	None	($d \approx 100$ nm)

(*Contd.*)

Table 9.1 (*Continued*)

Group	Target	Light source	Laser energy	Substrate	Substrate temperature	Background vacuum	Atmosphere	Post-annealing	Phase change amplitude
Kim et al. [5]	VO_2 ceramic	KrF excimer laser (λ 248 nm), pulse duration 30 ns	1.5 J/cm^2	Single-crystal *c*-plane sapphire	550°C	10^{-5} torr	O_2: 1–100 mTorr (optimized 10 mTorr)	The same condition of the deposition for 1 h	$\mathrm{Log}(R_s/R_m) =$ 4.6 (160 nm), 1.3 (120 nm)
Lappalainen et al. [25]	V_2O_5 ceramic	XeCl excimer laser (λ 308 nm), pulse duration 25 ns, pulse repetition rate 5 Hz	3 J/cm^2	Magnesium oxide (100); *a*-, *c*-, and *r*-plane cut sapphires	400°C	10^{-5} mbar	2.6×10^{-5} mbar		$\mathrm{Log}(R_s/R_m) =$ 3–4 ($d = 100$–200 nm) The sample in which the internal stress had been relaxed showed the sharpest MIT and also had the largest change of optical transmission.

Lafane et al. [10]	V_2O_5	KrF excimer laser (λ 248 nm), pulse duration 25 ns, repetition rate 5 Hz	2 J/cm^2	Glass	500°C	—	~37.6 mTorr	None	$(T_s/T_m)/T_s =$ 50%–65% ($\lambda =$ 1100 nm)
Maaza et al. [18]	VO_2	KrF excimer laser (λ 248 nm), repetition rate 10 Hz	0.5 ± 2.7 J/cm^2	Silicon, quartz, and sapphire	No heating	0.75–1.88 $\times 10^{-9}$ torr	None	None	$\text{Log}(R_s/R_m) \approx 2$ (150 nm)
Wu et al. [47]	Vanadium metal	—	1 W	Sapphire (0001)	500°C	1×10^{-5} torr	50 mTorr	None	$\text{Log}(R_s/R_m) \approx 4.3$ ($d = 480$ nm)

(*Contd.*)

Table 9.1 (*Continued*)

Group	Target	Light source	Laser energy	Substrate	Substrate temperature	Background vacuum	Atmosphere	Post-annealing	Phase change amplitude
Zhu et al. [48]	Vanadium metal	Nd-YAG laser (λ 532 nm), pulse duration 10 ns, repetition rate 10 Hz		Sapphire (0001)	450°C	1×10^{-6} torr	~45 mTorr	None	$\text{Log}(R_s/R_m) =$ 4–4.3 ($d = 500$ nm)
Jian et al. [49]	V_2O_5	KrF excimer laser (λ 248 nm)	3 J/cm^2	Glass, single-crystal *c*-cut sapphire, and Si (111)	Glass: 500°C *c*-cut sapphire Si (111): 550°C	—	10 mTorr	None	$\text{Log}(R_s/R_m)$: Glass 3.5, *c*-cut sapphire 3.8, Si(111) 3

epitaxial [2]. Details of epitaxial growth of VO_2 thin films can be found in the review by Nag et al. [1]. Here, we will just focus on the phase transition properties of epitaxial VO_2 films prepared by PLD across MIT.

Phase-transition properties refer not only to the phase-transition temperature (τ_C), but also to the phase-transition width (ΔH), speed (ΔT), and amplitude (ΔA). DH is the difference in the phase transition temperature from the VO_2(M) phase to the VO_2(R) phase and from the VO_2(R) phase to the VO_2(M) phase (τ_{C}:$_{M\text{-}R}$ – τ_{C}:$_{R\text{-}M}$), ΔT is the temperature required for a complete phase transition from the (M) to the (R) phase or from the (R) to the (M) phase, and ΔA is the difference in the resistance of transmittance of VO_2 across the MIT.

For single crystals, the transition widths (ΔH) are less than 0.15 K and the ERC across MIT (ΔA) reaches values as large as 5 orders of magnitude, while for unoriented films, ΔH ranges from 5 to 40 K [20] and ΔA is usually less than 4 orders of magnitude. For most of reported epitaxial VO_2 films prepared by PLD, ΔH could be reduced to a value as narrow as 2 K and ΔA is usually larger than 4 orders of magnitude. These indicate the high quality of the oriented VO_2 films. Different from bulk materials, the MIT performance of VO_2 epitaxial films is diverse, mainly caused by the stress in the films. This stress is induced by lattice and expansion efficiency mismatch between substrates and films. It has been found that the tensile stress along the *c* axis of a tetragonal unit cell (metal phase) would reduce the MIT temperature while the compressive stress along this direction would increase the MIT temperature [21, 22], and the thinner the deposited epitaxial VO_2 films, the larger the difference in the MIT temperature [23, 24]. The MIT temperature of epitaxial VO_2 films could be modified over a wide temperature range ($>40°C$) [21].

It was thought that the smaller the mismatch between the substrate and the deposited films, the less the influence this would have on the MIT performance of VO_2 films. To confirm this, oriented VO_2 on different substrates was prepared by PLD, including Al_2O_3 with *r*-, *m*-, and *c*-cuts [7, 25, 26]; MgO(100) [7, 25] and (111) [27]; MgF_2(001) [28, 29]; and TiO_2(001) [30–32], (110) [30], (111) [33], (101) [33], and (100) [33]. Different MIT performances were obtained due to lattice mismatch (Fig. 9.4a–d). To modify the stress, a buffer of epitaxial layers of other materials was deposited prior to VO_2, such as TiO_2 [21, 34, 35], AlN [36], SnO [37], Cr_2O_3 [38], and YSZ/NiO [39]. However, small misfits require a large distance to relax the misfit strain (critical thickness) [40], but as the resistance of the dislocation glide in VO_2 is very large, this means the misfits are not able to relax totally [21]. Although $TiO_2(001)$ had the lowest lattice mismatch with $VO_2(R)$ ($<1\%$), the MIT temperature of $VO_2/TiO_2(001)/Al_2O_3$ was much lower than that of bulk VO_2($\sim$25°C) [21, 34, 35] (Fig. 9.4e).

It is possible to achieve high-quality epitaxial VO_2 films with suitable conditions, including substrate materials, orientation, and pre- and postdeposition processes. However, Jian et al.'s work showed that (Fig. 9.4f–j) due to lattice constant change and thermal expansions, during thermal cycles (17°C–97°C), large strain would accumulate around the domain boundaries in the VO_2 films, resulting in degradation of ΔT and ΔH [22]. It is thus still a challenge to prepare high-quality VO_2 films with long-term stability for practical usage.

PLD offers advantages in terms of the concise control of the chemical composition, crystallinity, and epitaxy growth. The thinnest VO_2 films with MIT prepared by PLD are about 2–3 nm [23, 41], which means that the wafer-level fabrication of VO_2 is possible for device applications. However, the PLD method cannot be adopted for the growth of VO_2 films of a larger size (up to 1–2 inches) if considering film uniformity and surface defects, which greatly hinder the practical applications based on the phase-transition property of VO_2 materials [42].

9.2 Ion Plating/Ion Implantation

Ion plating is a method for producing a new phase in a host material by injecting impurities and following it up with nucleation within the host. Strictly speaking, ion plating is not a method for thin film preparation, since the nanoparticles (NPs) prepared by this method are separated by the host materials. However, these NPs are usually dispersed up to a very limited depth due to the implanted ion energy (several hundred nanometers); they act as a thin film, especially in

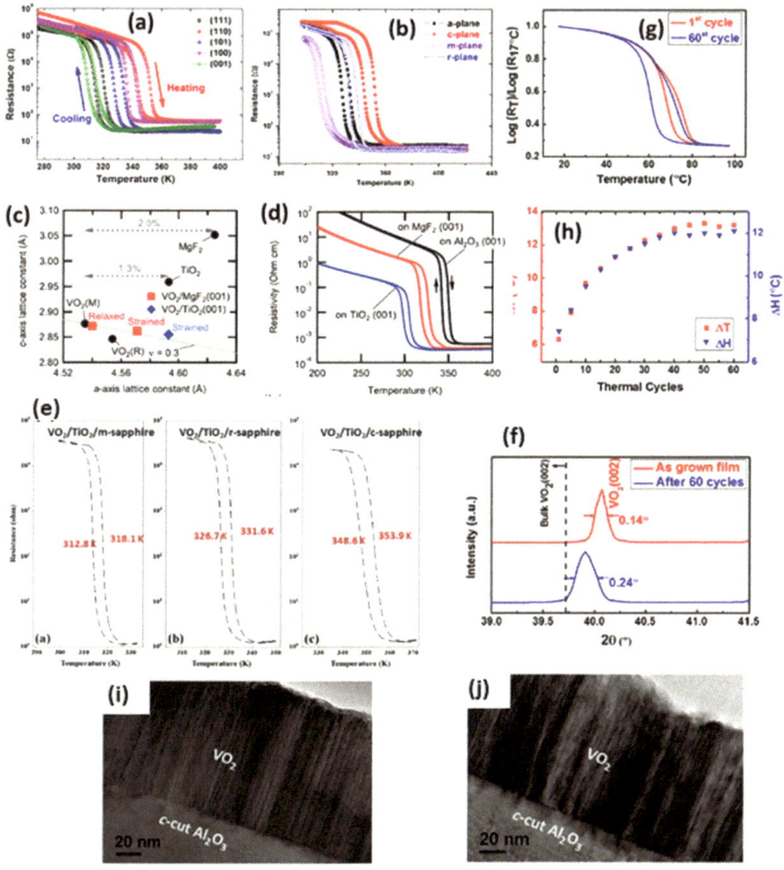

Figure 9.4 (a) Temperature-dependence resistance of VO_2 films grown on different orientation TiO_2 substrates. Reprinted from Ref. [33], Copyright (2014), with permission from Elsevier. (b) Temperature-dependence resistance of VO_2/TiO_2 bilayers grown on sapphire with different orientations of *c*-plane, *m*-plane, *r*-plane, and *a*-plane. Reprinted from Ref. [34], Copyright (2010), with permission from Elsevier. (c) Temperature-dependence resistivity of VO_2 thin films deposited on different substrates (TiO_2, MgF_2, or Al_2O_3) and (d) lattice constants of VO_2 films on TiO_2 and MgF_2 substrates; lattice constants of the R- and M-phases of VO_2 are also plotted; the dotted green line shows the strained VO_2(M) lattice constants. Reprinted from [28], Copyright © 2013 with permission from AIP Publishing. (e) Temperature-dependence resistance of $VO_2/TiO_2/m$-sapphire, $VO_2/TiO_2/r$-sapphire, and $VO_2/TiO_2/c$-sapphire heterostructures. Reprinted from Ref. [21], Copyright (2013), with permission from Elsevier. (f–j) Jian et al.'s work; (g) temperature-dependence resistance, (h) resistance changing rate, and (f) XRD of VO_2 with 1 cycle and 60 cycles of thermal treatment; (i–j) TEM cross-section images of 1 cycle and 60 cycles of thermal treatment samples, respectively. Reprinted from [22], Copyright © 2017 with permission from AIP Publishing.

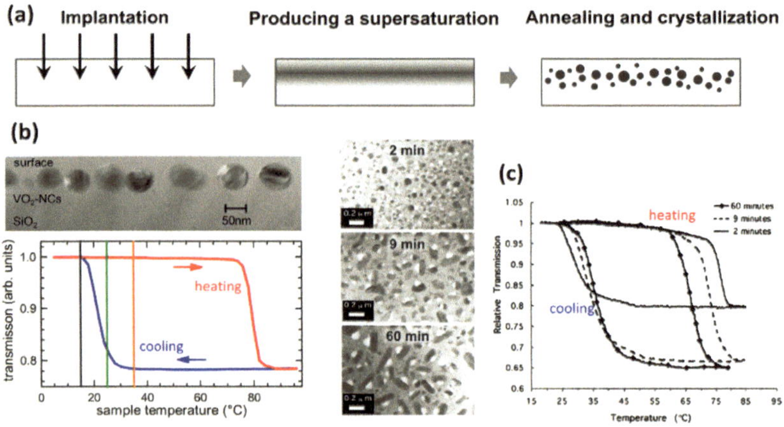

Figure 9.5 (a) Schematics of ion-beam-induced nucleation and the ion-implantation process. Reprinted by permission from Springer Nature: Ref. [53], Copyright 2009. (b) Cross-section TEM image (top) and thermal hysteresis loop of ion plating–synthesized VO_2 nanocrystals (NCs) in a fused silica matrix; the thermal hysteresis loop is obtained by tracking the transmission at 1500 nm; red and blue curves represent the heating and cooling processes, respectively. Reprinted from [54], Copyright © 2014 with permission from AIP Publishing. (c) TEM images (left) and the thermal hysteresis loop (right) of VO_2 NCs in a fused silica matrix formed at 1000°C in flowing argon with different annealing times; VO_2 NCs had sizes ranging from dozens to hundreds of nanometers and size-dependent ΔH as high as 40°C. Reprinted from [55], Copyright © 2002 with permission from AIP Publishing.

terms of their optical properties. These are called "nanocomposite" films/coatings in smart window studies. Thin films fabricated by this method perform better than pure VO_2 thin films in terms of luminous transmittance and their solar energy modulation ability [50–52].

A typical ion plating process for the preparation of VO_2 involves two steps (Fig. 9.5a): (i) at room temperature, stoichiometric vanadium and oxygen are coimplanted into a substrate, such as SiO_2 and Al_2O_3 [56], at an energy of about 150–300 [57] (100 [58]) keV for vanadium and 56–120 [57] (36 [58]) keV for oxygen, and (ii) the implanted substrate is annealed at a temperature, ranging from 700°C to 1000°C [53, 59], in high-purity flowing argon.

The energy of the ions determines the dispersion depth of the obtained NPs. In Lopez's work, VO_2 NPs of a sample prepared at a vanadium ion power of 300 keV were found within 500 nm of the top surface of the SiO_2 substrate or within 300 nm when the vanadium ions were at 150 keV. On the other hand, in Karl' work, the use of a 100 keV vanadium ion source resulted in a particle depth of about 150 nm. Particles 30–90 nm in size could be obtained by changing the annealing time (Fig. 9.5c). Doping could be easily realized in this method by coimplanting W and Mo [58]. Moreover, ion implanting could be a general method to perform nonequilibrium doping of VO_2 films predeposited with different methods [60, 61].

VO_2 NPs prepared by ion implanting are well separated by the matrix and have good crystallinity due to the high annealing temperature and a relatively stable surface (Fig. 9.5b). These attributes make them ideal samples to study the phase-transition process of VO_2 NPs [54, 62, 63], even in the case of single-NP studies [64]. A particle size–dependent loop width of VO_2 NPs during phase transition was found and thoroughly analyzed through thermodynamics and kinetics studies by Lopez. And the phase transition was found to possibly be induced by infrared light pulses of less than 200 fs duration [65].

The transmittance spectrum of these samples are rarely reported. Instead, the surface plasmon resonance phenomenon of VO_2 NPs in the metal phase is better studied [65, 66]. The most obvious difference caused by surface plasmon resonance is a strong absorption in the near-infrared region (mainly at 800–1800 nm). This greatly increases the switching efficiency of VO_2 and is later seen to be one of the greatest advantages of nanocomposite VO_2–based smart coatings. This helps to promote the usage of VO_2 for smart window application [51, 67].

However, ion implanting is not a widely used method to prepare VO_2. Compared to other methods, the limitation in substrate selection and the high annealing temperature increase the manufacturing cost and complexity of the process. At the same time, the composite structure of the VO_2 and the substrate does not show any advantages over nanocomposite coatings prepared using high-quality VO_2 NPs

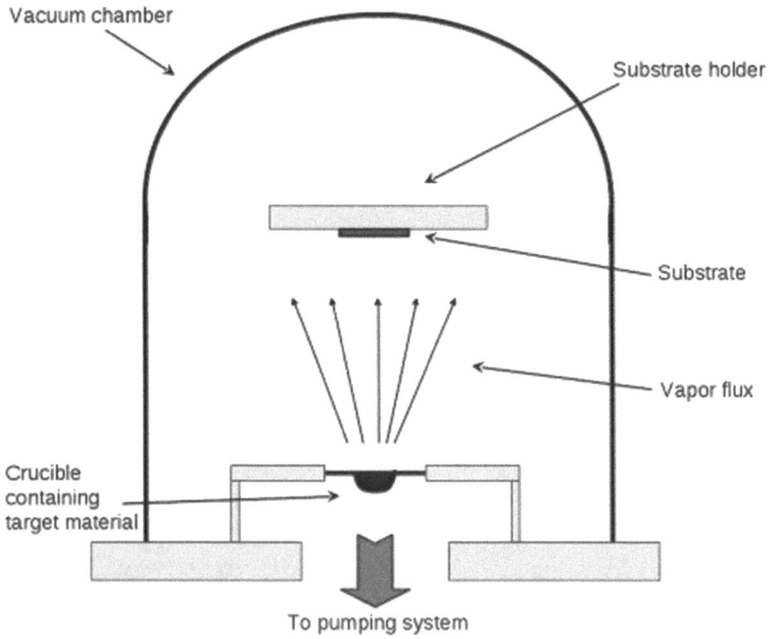

Figure 9.6 Diagram of the thermal evaporation process. Reprinted from Ref. [68], Copyright (2010), with permission from Elsevier.

[51]. On the contrary, this structure could limit VO_2 application in areas where good electrical properties are required.

9.3 Thermal Evaporation

Thermal evaporation (TE) is also called vacuum evaporation. In this method, the target is thermally evaporated in a high-vacuum chamber, where the vapor condenses on the substrate, forming a thin film. Since there is no interaction between the thermal source (such as laser, plasma, and electron beam) and the target, precursors, or substrates, the deposition process is simpler compared to other methods (Fig. 9.6).

A typical process for depositing VO_2 films by TE usually involves two steps [69, 70]: (i) A vanadium-rich metallic film [69, 70] or a VO_x film [71] is prepared. The target (typically metallic vanadium

[70, 72], VO_2 [70], or V_2O_5 [71]) is placed in a heated crucible made of Mo or W, located under the substrate in a high-vacuum chamber (2.7–6.7 $\times$ 10^{-3} Pa). (ii) The target is resistively heated to evaporate to the substrate, which is preheated to 250°C–300°C. The as-deposited films require postannealing at about 400°C–550°C in an Ar/O_2 mixed gas [70] or air (10.7–33.3 Pa) [69] to fully convert into VO_2 films.

Aside from oxidizing the metallic thin film, Ke et al. [73] reported a method to first grow a V_2O_5 nanowire film under atmospheric pressure and then reduce it to VO_2(R). Other phases, including VO_2(B) and V_2O_3, could also be obtained by changing the reduction atmosphere.

Jiang et al. [69] prepared VO_2 films of ~120 nm using TE with a metallic vanadium target. Their films showed good infrared control ability, which was comparable to that of films prepared by sputtering. The MIT temperatures of these films were less than 60°C. They also found that the MIT performance of the films was related to the postannealing time. The films showed a resistance change of 1 order of magnitude after 2 h of annealing and reached 2 orders of magnitude after 3 h of annealing but would then reduce for longer annealing times. Although TE is simple and promising, the reported works using this method only centered around preparing V_2O_3 [74, 75] or VO_x [73, 76]. And few works could be found on VO_2 film preparation after Jiang's work, where it was found that the MIT performance of VO_2 films prepared by this method was weak compared to that of films prepared by other methods [70, 71].

9.4 Electron Beam Deposition

Electron beam deposition (EBD) is a kind of vacuum evaporation coating method. Different from TE, EBD uses an electron beam to vaporize the evaporation material (Fig. 9.7). Any material can be evaporated by electron beam evaporation. EBD can be used to prepare multicomponent and high-purity films, benefiting from the fact that an electron beam can be accurately controlled and can directly affect the material to be evaporated. EBD equipment is also less expensive than that needed for other physical vapor-

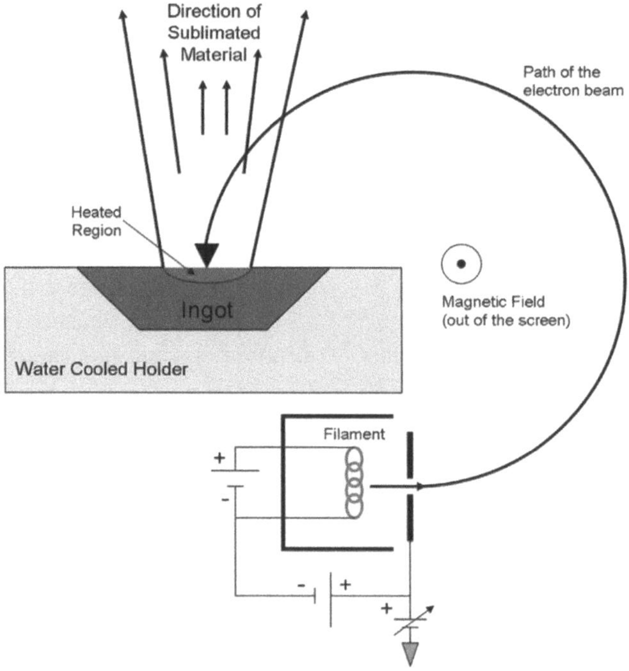

Figure 9.7 Diagram of the electron beam deposition process (https://en.wikipedia.org/wiki/Electron-beam_physical_vapor_deposition).

deposition techniques and further it shows the potential to produce very smooth films of large sizes, making it an excellent low-thermal budget deposition method for VO_2 films for optical and electronic applications [77].

The EBD of VO_2 requires a basic high-vacuum chamber (below 10^{-4} Pa). A vanadium target is adopted as the evaporation source [78–81]. The target is put in a crucible under the substrate (about 250 mm). An incident electron beam (accelerating voltage of 10 kV) is used to evaporate the vanadium under a pure oxygen atmosphere (total pressure 0.06–0.1 Pa). The deposition rate of the vanadium oxide films is controlled by a quartz oscillator at the nominal rate of 0.03–0.07 nm s^{-1}. Doping can be done using doped targets.

VO_2 powder can also be used as an evaporation source, either in powder form [77] or compressed into a sheet [82]. Different from using the vanadium target, here oxygen is unnecessary during

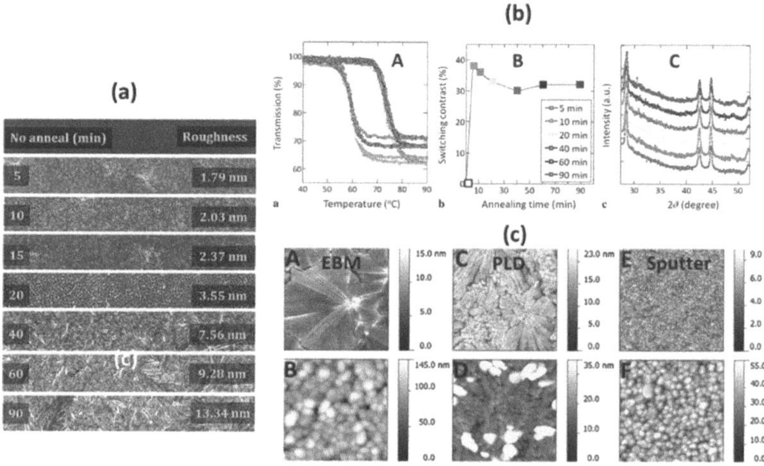

Figure 9.8 (a) Surface morphology of as-grown VO_2 film and VO_2 film exposed to annealing (for 5–60 min) prepared by EBD and their performance. (b-A) Temperature-dependence transmittance, (b-B) switching contrast vs. annealing time, and (b-C) XRD. Reprinted by permission from Springer Nature: Ref. [77], Copyright 2012. (c) AFM of VO_2 films prepared by EBD (c-A, f-B), PLD(c-C, f-D), and sputter (c-E, c-F). Reprinted from Ref. [83], Copyright (2015), with permission from Elsevier.

the evaporation, but a pressure of $\sim 0.5 \times 10^{-4}$ Pa is required for pretreating the VO_2 powder or sheet. The system pressure is maintained at $\sim 1.33 \times 10^{-2}$ Pa during evaporation. The deposition rate of this method is controlled at about 0.3 nm s^{-1} by a quartz crystal microbalance.

It seems that substrate heating during deposition is not the key parameter for obtaining high-quality VO_2 films with a thermchormic performance but a postdeposition oxygen annealing step (450°C–550°C) or an applied radio frequency (RF, incident power 10–50 W) is. In Marvel et al.'s work [77], a postannealing of as short as 5 min at 450°C could generate a pure VO_2(M) phase from the as-deposited film (Fig. 9.8a,b). Leroy et al. [79] used RF applied to the substrate to increase the adhesion and decrease the porosity of the forming films during deposition.

Théry et al. [81] comprehensively studied films deposited without any treatment (as-grown), those given a 50 W RF treatment (discharge), and those given postdeposition annealing at 550°C for 15 min (annealed). In this work, the as-grown films did not show thermochromic performance even with the substrate heated to 500°C. These films exhibited a poor structural quality (large deviation from stoichiometry, large mosaicity, and large strains) and a purely metallic behavior. Meanwhile, the annealed films exhibited excellent thermochromic performance, with resistivity changes of more than 4 orders of magnitude across the MIT. These films also exhibited excellent structural properties (a low level of heterogeneous strain, a mosaicity confined to narrow regions close to the interface, and a level of strain solely due to the film/substrate thermal expansion mismatch). They found that films with similar electrical properties could also be obtained by only RF discharge treatment during deposition. Compared to annealed films, these discharge films only suffered slightly degraded structural properties. It is crucial to add RF during the EBD of VO_2.

VO_2 films prepared by EBD could reach a performance comparable to those of films prepared by other physical vapor-deposition techniques. A resistance change of as high as 5 orders of magnitude during MIT indicates the fine crystallizability of the obtained VO_2 thin films [81]. Furthermore, uniform films with a size of 100×100 mm^2 were prepared by EBD, which evinced the potential of this method for large-scale preparation [81]. However, Marvel's work showed that the density of films prepared by EBD is slightly lower than that prepared by PLD and by sputtering with similar annealing processing [83]. A similar phenomenon could be inferred from the reflective spectrum of a sample prepared by other researchers [78], which was seen to be clearly lower than those of the samples prepared by sputtering. The maximum reflectance values were 38% and 43%, respectively, as shown in Figs. 9.5h and 9.5j [84]. The denser the films, the higher is the reflection. In Marvel's work, the EBD films also show lower high-temperature stability, which may result from the lower density [83]. A much lower energy of the ejected target material (0.2 eV vs. 50 eV and 10–40 eV of PLD and sputter, respectively, [83]) may be one of the reasons for the lower density. That is why an applied RF could decrease the porosity of the formed films in Bessaudou et al.'s work [79]. It is clear that the morphological evolution is dominated by the strain

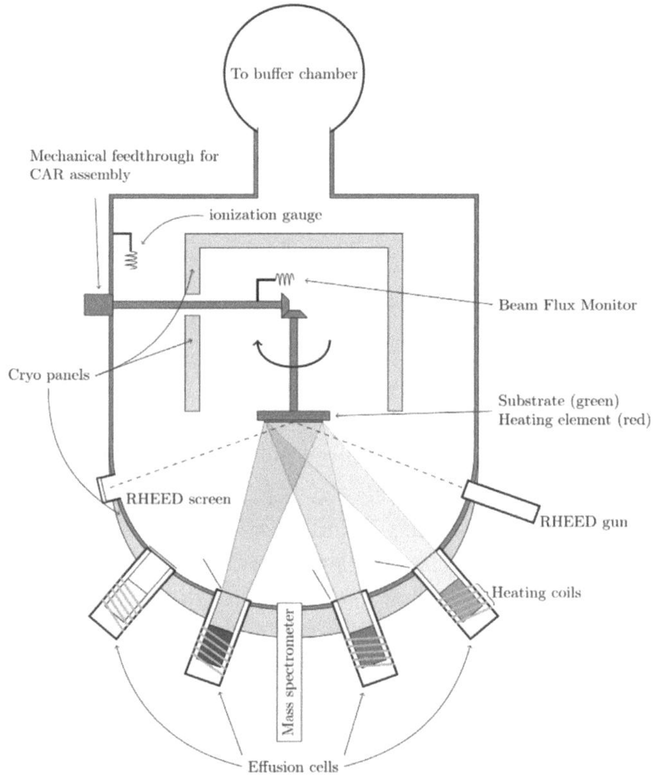

Figure 9.9 A simple sketch of a molecular beam epitaxy system (https://en.wikipedia.org/wiki/Molecular-beam_epitaxy).

imparted during the sputtering process. For films deposited by PLD and EBD, the substrate plays a more important role, with substrate dewetting and epitaxy determining the film structure during solid-phase crystallization [83].

9.5 Molecular Beam Epitaxy

Molecular beam epitaxy (MBE) (Fig. 9.9) is a widely used method to prepare high-quality and homogeneous epitaxial crystal layers. It is highly reproducible in terms of both film thickness and composition control. Although various methods to prepare VO_2 thin films have been studied, a process to prepare VO_2 thin films with a perfect V–O stoichiometry and near single-crystal structure is still very attractive. Studies on the preparation of VO_2 thin films using MBE and their MIT performance have been conducted only in recent years.

There are mainly two methods for VO_2 thin film preparation by MBE. The first one, developed by Sambi et al. [85, 86], is called the "periodic annealing" method. Here, before deposition, a $TiO_2(110)$ substrate is cleaned with cycles of Ar-ion sputtering, followed by annealing at 573 K in O_2 (1×10^{-6} mbar), to produce sufficient bulk oxygen vacancies to avoid the charging problem. Then, 0.2–0.5 monolayers (MLs), corresponding to 5.2×10^{14} vanadium atoms/cm^2, of amorphous V metal are deposited in vacuum (5×10^{-11} mbar) at room temperature by electron beam evaporation. A postannealing process (at 423 K) in oxygen atmosphere (7.5×10^{-7}–1.5×10^{-6} mbar) is followed to obtain an epitaxial VO_2 layer. The sample is then cooled to room temperature, and the next 0.2–0.5 ML of amorphous V metal is deposited. This cycle is repeated until a 3–5 ML thick epitaxial VO_2 layer is obtained [87]. The films prepared by Sambi [87] showed ordered VO_2 phase grown epitaxially on $TiO_2(110)$ with a rutile-type structure, but no MIT was observed in these films.

Tashman et al. [87] followed Sambi's work and deposited VO_2 on $TiO_2(001)$. The resulting films did not show MIT either. They developed the following process and alterations:

- They used distilled ozone instead of O_2.
- They increased the temperature from room temperature to 395 K and from 423 K to 473 K for deposition and annealing, respectively.
- They added a rapid postannealing process at 673 K (3K s^{-1}, 1×10^{-6} mbar of distilled ozone).

With these alterations, obvious MIT was observed in films even as thin as 2.3 nm, with a resistance change of 1.4 orders of magnitude ($\log[\Delta R/R]$). Tashman's work also showed that the transition width decreases monotonically with film thickness, while the hysteresis increases monotonically with film thickness [87].

Paik et al. [88] adopted this process for the study VO_2 thin films, but with an even higher deposition and annealing temperature (523 K) and a postannealing process up to 623 K. This annealing step would enhance the coalescence of (001)-oriented VO_2 islands and improve the abruptness of the MIT [88]. They also developed a process for TiO_2 substrate pretreatment [88].

The other method is RF-plasma-assisted oxide MBE, reported by Fan et al. [42]. In this method, a standard RF-plasma source was used to provide reactive oxygen radicals. Pure metallic vanadium powder was used as the target for e-beam evaporation. An Al_2O_3 crystal slice or TiO_2 could be used as the substrates, which were degassed and annealed at 823 K in vacuum (4×10^{-9} mbar). The chamber pressure was maintained at 1.3–4 $\times 10^{-5}$ mbar during the film preparation. The evaporation rate of vanadium was controlled at 0.1 A s^{-1}. An optimized flow rate of 1.8 sccm for pure VO_2 film depositions was used as a lower oxygen gas flow would cause oxygen deficiency and could result in a V_2O_3 film being formed instead.

Extremely thin VO_2 films could be prepared by MBE, the thinnest one with MIT reported being 1.5 nm, by Paik [88]. It seems that the thinner films have lower MIT temperatures [87–89]. Fan et al. studied this phenomenon using interfacial strain dynamics and theoretical calculations and claimed that the electronic orbital occupancy is strongly affected by the interfacial strain, which also changes the electron-electron correlation and controls the phase-transition temperature [89, 90]. Up to now, the lowest MIT temperature of pure VO_2 thin films reported by MBE was 280.5 K, in which a 3.3 nm thick film has a resistance change of 2.3 orders of magnitude. The thinnest film (1.6 nm) did not show the lowest MIT temperature, probably due to diffusion of the Ti substrate into the thin film layers [88]. However, compared with the epitaxial VO_2 films prepared by PLD on the similar TiO_2(001) substrate with a similar thickness (from 3–15 nm) [32], these thin epitaxial VO_2 films show lower MIT temperatures and smaller resistance changes during MIT. This may indicate higher stresses in these epitaxial VO_2 thin films. More work could still be done in the MIT investigation of MBE epitaxial VO_2.

Besides the advantages in preparing extremely thin films, MBE also shows advantages in the preparation of large-sized thin films, accurate stoichiometry, and especially low-temperature deposition (350°C [87]). Low-temperature deposition for VO_2 means a reduced lattice mismatch due to the thermal expansion and diffusion of substrate elements and higher crystal quality, which can be used in basic theory research [89, 90] and in the development of precision components [91, 92]. It could also enable the application of VO_2 in electronic devices, such as in uncooled infrared focal plane arrays. However, MBE requires a crystallized substrate with very little lattice mismatch with the deposition materials and expensive equipment, which limit its wide usage.

9.6 Sputtering

9.6.1 Introduction

Sputtering is a physical vapor deposition (PVD) process in which surface particles are physically knocked off the source material and condense on the substrate. The depositing source, called a "target," is bombarded by energetic gaseous ions in a "plasma." The kinetic energy from the incoming ions is transferred to the particles on the surface of the target, breaking the bonds between them and the bulk of the target. These surface particles are changed from solid to gas phase through the mechanical forces of the bombarding ions. Figure 9.10 is a schematic of a basic sputtering system.

The first crucial factor of this system is the sputtering gas, which normally consists of ultrapure argon for nonreactive deposition or a mixture of argon and another reactive gas (O_2, N_2, CH_4, H_2S, etc.) for reactive deposition. The sputtering gas mixture is ionized and turned into plasma through electrical discharge, which happens with the target as the cathode and the substrate as the anode. At this stage, the kinetic energy of the ions in the plasma depends on the power of the electricity applied to the system. Subsequently, this also controls the kinetic energy of the sputtered particles. The operating pressure of the sputtering gas mixture is also important as it controls the flux of gaseous ions and thus the deposition rate. After being sputtered out of the target surface, the ejected particles are deposited on the substrate, forming a thin film of the sputtered

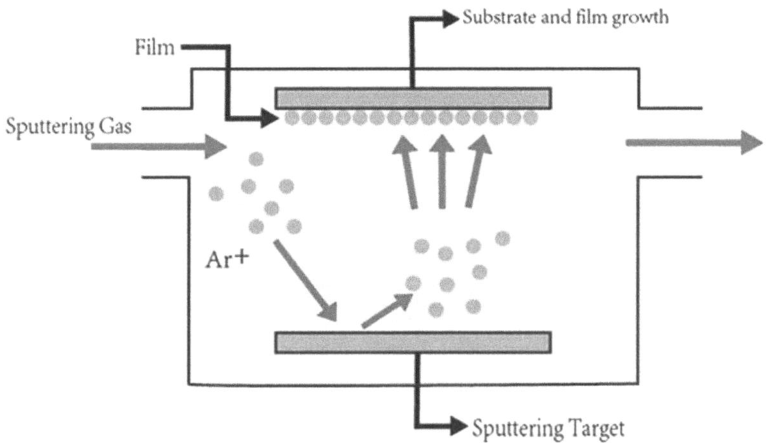

Figure 9.10 Basic setup of a sputtering system (reprinted with permission from Stanford Advanced Materials (SAM) https://samaterials.wordpress.com/2015/10/28/what-is-sputtering/).

target, or of the product in the case of reactive sputtering. The properties of the thin film are greatly dependent on the structure and temperature of the substrate. Postannealing can also be done after deposition to improve or change the quality of the deposited thin film. In summary, there are four main parameters that define a basic sputtering setup: (i) the power supplied, (ii) the operating pressure, (iii) the substrate type, and (iv) the substrate temperature. For the same sputtering system, changes made to these parameters are the key to changing and improving the thin film quality.

Different techniques of sputtering are direct current magnetron sputtering (DCMS), RF or medium frequency (MF) magnetron sputtering, ion beam sputtering, and high-power impulse magnetron sputtering (HiPIMS). It can also be classified into reactive sputtering and nonreactive sputtering, depending on the type of sputtering gases.

Sputtering is characterized by a suitable particle impinge energy to provide sufficient adhesion to the substrate, rich and convenient ways to control the microstructure of the deposited film, high deposition rates when using metal or conductive ceramic targets, and high homogeneity for large areas. The technique has been

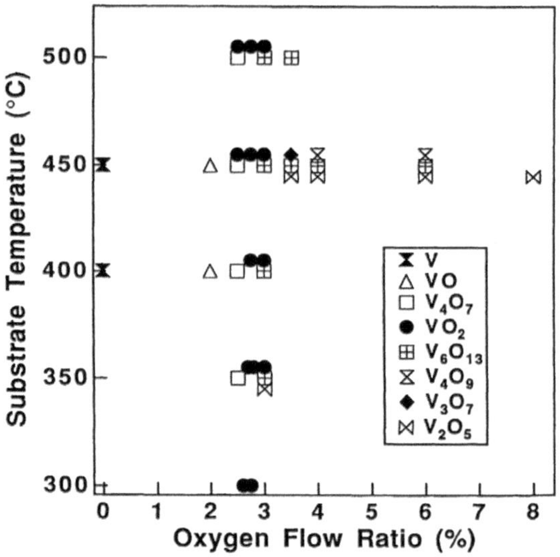

Figure 9.11 Phase formation in sputtered vanadium oxides under various substrate temperatures and oxygen flow ratios. Reprinted from Ref. [93]. Copyright (1994) The Physical Society of Japan and The Japan Society of Applied Physics.

matured and used in glass coating industry with off-line continuous coating systems to produce silver-based low-emissivity (low-E) glass.

Sputtering provides a convenient path to preparing vanadium dioxide films not only for small research samples but also for large-scale production, as has been demonstrated in the glass coating industry.

The difference and difficulties in the sputter-deposition of vanadium dioxide thin films compared with other oxides are the harsh conditions for the formation of a single-phase crystalline VO_2 due to the complexity of the vanadium-oxygen system. A comparatively high substrate temperature, normally in excess of 400°C; careful control of other parameters, such as the oxygen partial pressure and the total pressure; and proper substrate bias are required. Studies have shown that the deposition parameters of the oxygen flow ratio and the substrate temperature are the most important factors for VO_2 formation [93]. As shown in Fig. 9.11, at least eight crystalline phases (V, VO, V_4O_7, V_6O_{13}, V_4O_9, V_3O_7, V_2O_5, and VO_2) can be detected when the substrate temperature and the oxygen flow ratio are changed. The single phase of VO_2 is formed within a very narrow range of oxygen flow ratio, and its formation window depends on the substrate temperature—the higher the temperature, the wider the window. Precise control of the deposition parameters reduces the substrate temperature for VO_2. The predeposition of a vanadium seed layer and the application of a suitable substrate bias can further decrease the substrate temperature to under 300°C.

9.6.2 Reactive Direct Current Magnetron Sputtering

Reactive DCMS is characterized by three main features:

- The sputtering gas is a combination of argon and a reactive gas.
- DC power is used as the electrical source for ionization during sputtering.
- Magnetron sputtering is an add-on to the traditional system. It uses a magnetic field to trap and restrain electrons in the plasma for a longer amount of time, creating more gaseous ions in the plasma, subsequently increasing the deposition rate of the system.

Using a DC power source ensures that this is the lowest-cost and simplest technique for metal deposition. However, this is also a limitation because only conductive materials can be used as the target.

In the case of VO_2 deposition, vanadium is a suitable target as a transition metal and a mixture of argon and oxygen can be used as the reactive sputtering gas. Vanadium is known for its various oxidation states (II to V), so it is important to achieve the correct oxidation state for VO_2. According to Kang [95], the composition of a VO_x system can be influenced by the temperature during deposition and the mole fraction of V and O. Despite knowing the required mole fraction from literature, it is hard to achieve perfect VO_2 deposition in practice due to the lack of control between the ratio of sputtered vanadium and oxygen in the gas mixture. Thus,

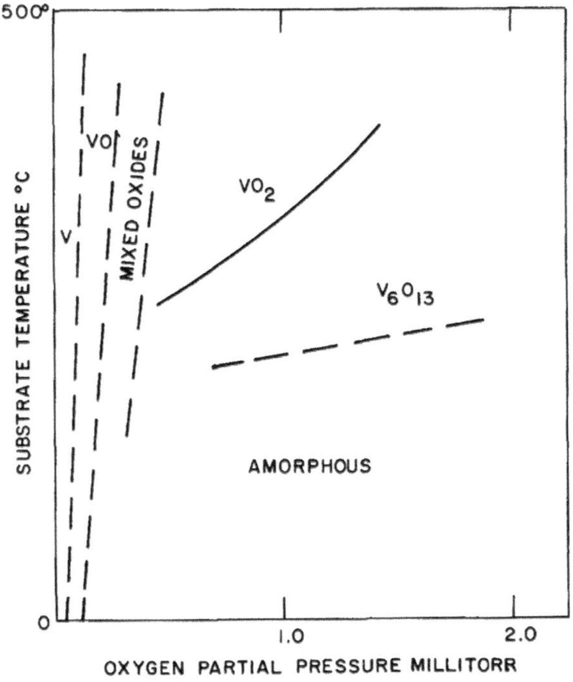

Figure 9.12 Phase diagram of VO_x. Reprinted from [94], Copyright © 1974 with permission from AIP Publishing.

factors such as the O_2/Ar ratio, overall operation pressure, and the substrate temperature and structure are tuned so that the V/O mole fraction can be controlled. Figure 9.12 is the phase diagram of VO_x as a function of substrate temperature and oxygen partial pressure reported by Griffiths and Eastwood in 1974 [94]. This gives a general idea of how each stated parameter would affect the phases in sputtered VO_x film. The detailed effects of each parameter are discussed in the following sections.

9.6.2.1 Without a postannealing process

Without any postannealing process, the effects of factors such as reactive gas ratio and substrate temperature can be observed more clearly. Yuce et al. fabricated a VO_2 thin film on a *c*-cut sapphire (Al_2O_3(0001)) substrate by optimizing the reactive oxygen flow ratio (Fig. 9.13b) while keeping the substrate temperature constant at 550°C, 50 W power, and 0.85 Pa operational pressure [96]. Raman spectra of thin films in Fig. 9.13a show a direct correlation between the presence of V_2O_5 in the deposited film and the oxygen flow ratio, with the former increasing with an increase in the latter. This is consistent with how a higher ratio of reactive oxygen flow (2.5%–3%) leads to a higher oxygen mole fraction, which then results in a higher oxidation state of vanadium after reaction. Vanadium(V) was formed instead of vanadium(IV). The optical properties (Fig. 9.13c and 9.13d) of the deposited VO_2 also confirmed the abundance of VO_2 at 2.25% O_2 as the MIT was clear at this level but was not significant at a higher oxygen level. It can be concluded that the flow ratio of O_2 directly affects the O_2 mole fraction and, thus, the phase formation of VO_x.

Zhao et al. conducted an experiment in which seven samples of VO_2 films of 120 nm thickness were produced on a *c*-sapphire substrate at different substrate temperatures, from 550°C to 700°C, while other parameters were kept constant (0.4 Pa, 11% O_2/Ar flow ratio) [97]. In the SEM images (Fig. 9.13e–h), the samples are divided into three represented groups: G1, G2, and G3. G1 (S550, S575, and S600) has a coarse, rough surface with a relatively larger grain size. From cross-section SEM images, it was reported that G1 has a porous structure. G2 (S625, and S650) has a smooth surface and a compact morphology with fine grains. On the other hand, G3 (675, and S700) has a rougher surface than G2 but less than G1. According to the report, no porous structure was detected in G3. These groupings in terms of the microstructure also translate into differences seen in the X-ray diffraction (XRD) spectra. From Fig. 9.13i–l, the author concluded that the double peak in XRD results could be contributed to the transition of vanadium oxide from V^{5+} to V^{4+} to V^{3+} from a low temperature to a high temperature. The presence of intermediate phases during the transition from V_2O_5 to VO_2 and from VO_2 to V_2O_3 is the main reason there were double peaks in the XRD spectra. This result shows the influence of the substrate temperature on the stoichiometry of the deposited vanadium oxide film.

A crucial aspect of any sputtering deposition setup is the choice of substrate. A *c*-cut sapphire was used by both Zhao et al. [97] and

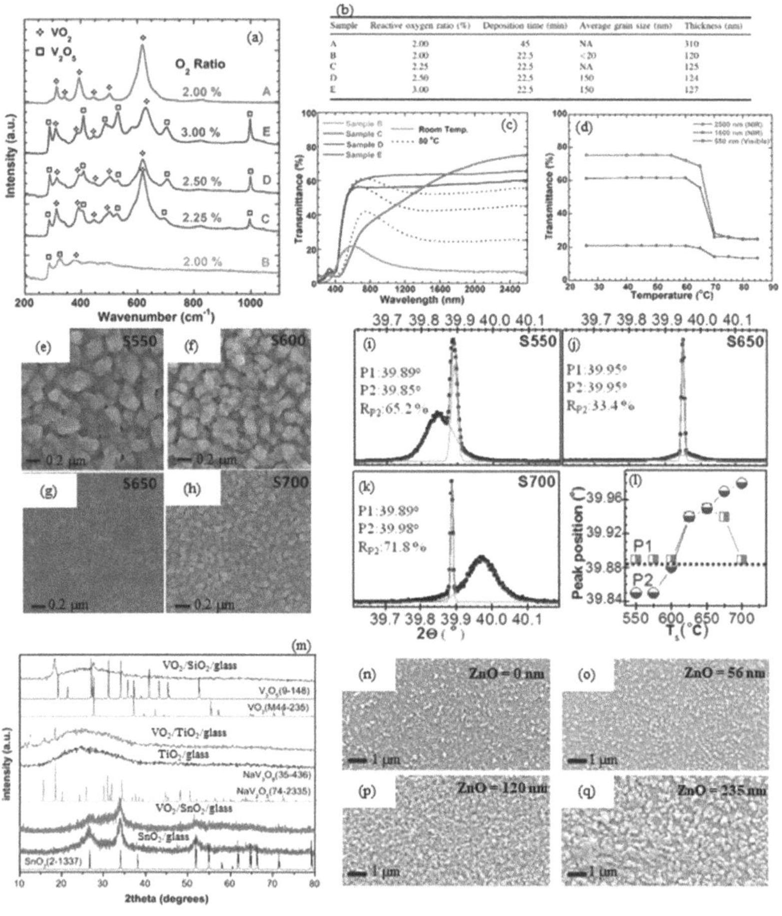

Figure 9.13 Characterizations of VO_2 deposited by reactive DCMS without annealing. (a, b) Raman spectra of thin films produced at varying O_2/Ar flow ratios. (c, d) Optical properties of samples in (a). Reprinted from Ref. [96]. © IOP Publishing. Reproduced with permission. All rights reserved. (e–h) SEM images of VO_2 deposited using different substrate temperatures. (i–l) XRD spectra of samples in (e–h). Reprinted from [97], Copyright © 2013 with permission from AIP Publishing. (m) XRD spectra of VO_2 on different substrates. Reprinted from Ref. [98], Copyright (2016), with permission from Elsevier. (n–q) SEM images of VO_2 on ZnO buffers of different thicknesses. Reprinted from Ref. [99], Copyright (2018), with permission from Elsevier.

Yuce et al. [98] due to its ability to operate at high temperatures, which was higher than 550°C for both reports. In practice, soda-lime glass is a better substrate for testing because of its potential real-life application. However, soda-lime glass at a high temperature leads to the diffusion of sodium (Na) into the thin film, diminishing the transformation ability of VO_2.

Zhu et al. [98] deposited VO_2 on a soda-lime glass substrate at the low temperature of 300°C with three different buffer layers: TiO_2, SnO_2, and SiO_2. Other parameters did not deviate much from the other reported reactive DCMS setup: 80 W power, 0.55 Pa operation pressure, and 3.25% O_2/Ar flow ratio. The XRD patterns in Fig. 9.13m show that TiO_2 is not a suitable buffer layer because it was not able to hinder the diffusion of Na into VO_2, thus causing the detection of NaV_3O_8 and NaV_2O_5. The SiO_2-buffered sample showed the presence of (M) phase VO_2 with its characteristic peak. However, peaks from the V_3O_5 phase were also present. This means that the SiO_2 buffer layer did not manage to help create a pure VO_2 thin film. Only the SnO_2 buffered sample was relatively pure. Zhu et al. [99] proceeded further into this area with the deposition of VO_2 on ZnO-coated soda-lime glass of different thicknesses in order to investigate the effect of buffer layer thickness on deposited VO_2. A low temperature of 320°C was used, with other typical parameters being constant: 120 W power, 0.5 Pa operational pressure, and 4.5% O_2/Ar flow ratio. It was found that a thicker buffer layer promoted crystallinity of VO_2 at a low substrate temperature. This could be observed by comparing the SEM image of VO_2 with and without a ZnO buffer of different thicknesses in Fig. 9.13n–q.

9.6.2.2 With a postannealing process

Postannealing in reactive DCMS is typically an extra step to guarantee the quality of the deposited thin film. This is done to crystallize an amorphous film, continue the reaction in the reactive gas without further sputtering, or purge the excessive reactive gas from the deposited film. Yu et al. deposited VO_2 film on a polished silicon substrate at room temperature using varying O_2/Ar flow ratios, 0.06 Pa operation pressure, and 90 W power [100]. The low temperature of the substrate undoubtedly resulted in amorphous

VO_2, which required recrystallization at 400°C for 1 h in a nitrogen atmosphere. Nitrogen was used as an inert gas so it would not affect the established stoichiometry of the amorphous VO_2 film. The atmospheric or ambience pressure also contributed to retaining the stoichiometry.

Not only is postannealing used to remedy low-temperature depositions, it is also used to further improve the quality of the VO_2 deposited at a high temperature. Dou et al. reported comprehensive findings on this by conducting a two-step annealing process on deposited VO_2 (*m*-/*a*-/*r*-plane sapphire, 2% O_2/Ar flow ratio, 1 Pa, 550°C) [101]. After deposition, the VO_2 film was crystallized at 500°C for 400 s in 220 Pa atmosphere, followed by air oxidizing in atmospheric pressure at 450°C for 200 s, 400 s, 600 s, and 1000 s. XRD patterns (Fig. 9.14a) showed a significant decrease in the VO_2 presence with an increase in air oxidizing time. As expected, the V_2O_5 presence also increased due to further oxidation of vanadium to V^{5+} state. The oxidation of vanadium in air or in an oxygen-rich atmosphere is covered in a later section as part of the nonreactive sputtering processes. In contrast, Huang et al. [102] conducted a postannealing process in an O_2-lacking environment to transform the mixed-phase vanadium oxide obtained from deposition to a purer stoichiometry. In this report, a vanadium oxide film was deposited in the following conditions: 12.5% O_2/Ar flow ratio, 0.11 Pa operation pressure, 200 W power, and 300°C substrate temperature. XRD patterns of the deposited film revealed the presence of mixed VO_2 and V_2O_5 phase. After being annealed in an Ar atmosphere, the V_2O_5 was eliminated, as seen in the XRD patterns in Fig. 9.14d.

VO_2 deposition on *m*-, *a*-, and *r*-plane sapphire was also reported by Dou et al. [101]. Before annealing, different substrates yielded thin films with different microstructures (Fig. 9.14d). After 1000 s of air oxidation, all three samples had similar microstructures. This exhibits the inability of different substrates to affect the structure of the VO_2 film after prolonged annealing.

While the effect of having different substrates may not last after annealing, the effect of different O_2/Ar flow ratios is reported to be conserved, as reported by Xu et al. In their report, a VO_2 thin film was deposited on a Si (100) substrate at room temperature and under

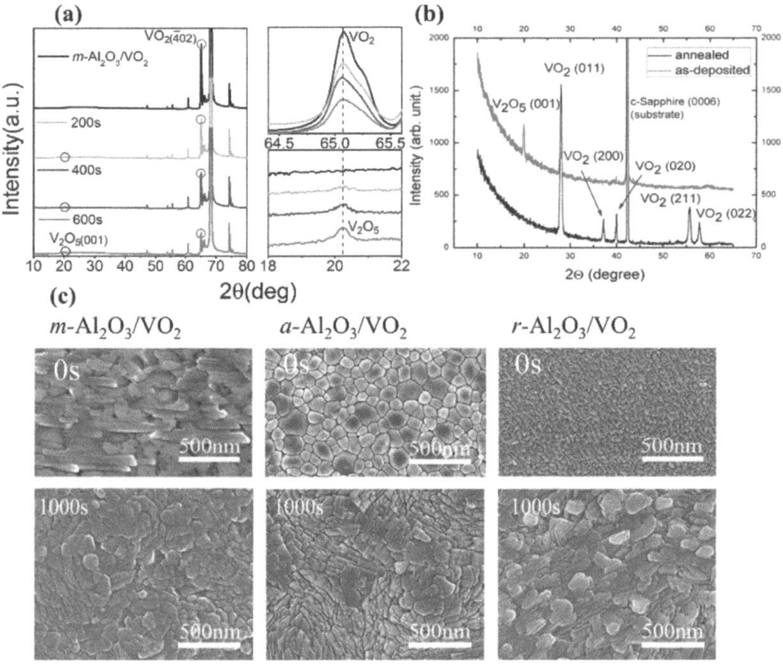

Figure 9.14 Characterization of VO_2 thin film deposited by reactive DCMS with annealing. (a) XRD spectra of oxidized VO_2 on an m-plane sapphire. Reprinted from Ref. [101], Copyright (2018), with permission from Elsevier. (b) XRD patterns of a vanadium oxide film before and after annealing. Reprinted from [102], Copyright © 2012 with permission from AIP Publishing. (c) SEM images of VO_2 samples on different substrates before and after 1000 s of annealing. Reprinted from [101], Copyright © 2015 with permission from Elsevier.

the following conditions: 200 W power; 0.63 Pa; and O_2/Ar ratios of 25%, 32.5%, 42.5%, and 50%. The deposited film was annealed in vacuum for 2 h at 450°C [103]. According to XPS results in the report, it was observed that the vanadium oxide undergoes oxidation from V^{3+} to V^{4+} to V^{5+} with increasing amounts of O_2. This is consistent with what was reported by Yuce et al. in the previous section. It can be concluded that if the annealing time does not transform VO_2 to V_2O_5, it has no influence on the effect of different reactive gas flow ratios.

9.6.3 Reactive RF-Magnetron Sputtering

The traditional sputtering system using a DC power source is only capable of using conductive materials as the target. The need to sputter a wider range of materials thus led to the development of RF sputtering, in which an alternate current (AC) is used as the power source instead. This AC with a standard frequency of 13.56 MHz is used to create an alternate bias between the substrate and the target. By constantly change the target to act as the cathode and the anode, the process prevent non-conductive targets from accumulating a positive charge over the course of sputtering. Thus, RF sputtering can be used for conducting, insulating, and semiconducting materials. This means that metallic VO_x materials can be used as a sputtering target. This is further discussed in the nonreactive sputtering section.

Reactive RF-magnetron sputtering (RFMS) is very similar to its DC counterpart and so the traditional parameters, such as O_2/Ar flow ratio, substrate heating, and substrate microstructure, still play a crucial role in deciding the quality of the deposited films. However, using an AC power source generally allows the use of higher sputtering power (from 100 W [104] to as high as 450 W [105]). This increased sputtering power also eliminates the annealing step in most reported papers except when room-temperature deposition is attempted [106].

Despite the similarity of RFMS to its DC counterpart, the effects of conventional parameters have been investigated for VO_2 thin films deposited by reactive RFMS. Jiang et al. reported investigations on a series of VO_2 samples on quartz glass at 450°C with 200 W RF power under 1 Pa of varying O_2 flow [107, 108]. Instead of the traditional stoichiometry change with varying O_2/Ar flow ratios, the report gives a fresh viewpoint on the effect of the O_2/Ar flow ratio on a single-phase VO_2 thin film. While XRD patterns (Fig. 9.15a) show no signs of other vanadium oxide phases in the deposited thin film, atomic force microscope (AFM) images (Fig. 9.15d) show an increasing crystal size as the O_2/Ar flow ratio increases from 2% to 5%. Variations in the optical properties and transition temperatures were also reported with an exceptional transition temperature of 46°C for the 2% sample.

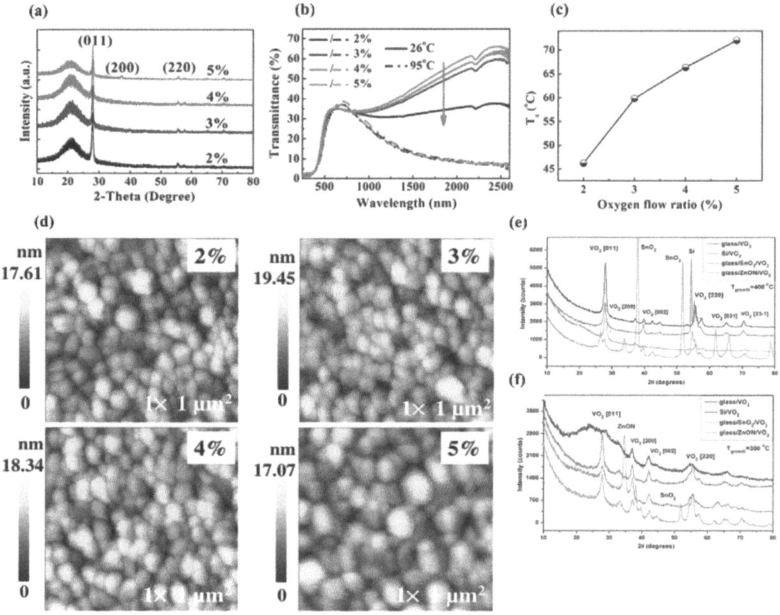

Figure 9.15 Characterization of VO_2 deposited by RFMS: (a) XRD patterns, (b) transmission spectra, (c) transition temperature for each sample, and (d) AFM images of each sample. Reprinted from Ref. [107], Copyright (2014), with permission from Elsevier. (e, f) XRD of VO_2 deposited on different substrates at different temperatures. Reprinted from Ref. [109], Copyright (2015), with permission from Elsevier.

Some effects of different substrates and substrate temperatures were also reported. Panagopoulou et al. deposited VO_2 at 400°C and 300°C on various substrate materials: glass/ZnON, glass/SnO_2, commercial glass, and silicon [109]. XRD analysis (Fig. 9.15e,f) produced nonsurprising results indicating a better-formed VO_2 using the higher substrate temperature of 400°C. Broader peaks from 300°C samples provided evidence of secondary phases, whether different VO_2 phases or different VO_x stoichiometries, formed during deposition.

Aside from the traditional parameters, different methods to modify the sputtering process are also possible under RFMS. One of these is to create a secondary bias acting on the substrate concurrently with the RF acting on the target. Azhan et al. conducted a series of experiments using reactive RFMS and RF substrate bias to deposit VO_2 on a single-crystal sapphire substrate [110, 111]. On using RF substrate bias during deposition, the reported transition temperature of VO_2 film is 36°C for 40 W biasing power. This is an extremely good value compared to the conventional transition temperature of 68°C. However, films deposited with a higher biasing power showed a significant decrease in the MIT ability. A balance between the transition temperature and MIT properties is thus needed so that VO_2 films deposited by this method can be considered for further application.

9.6.4 Reactive Pulsed DC Magnetron Sputtering

Pulsed direct current magnetron sputtering (pDCMS) is the result of combining the DC power source from conventional DCMS and the concept of alternating bias to prevent target charging from RFMS. Instead of using an AC power source with a frequency in the range of megahertz, pDCMS still uses a DC power source with a pulsing capability in the range of kilohertz. This prevents charge accumulation on the insulator target surface, allowing a wider range of materials to be used with a DC power source. Additional systems such as a substrate bias can also be used in this case though it is traditionally redundant in conventional DCMS [112].

9.6.5 Reactive High-Power Impulse Magnetron Sputtering

HiPIMS is a recently developed technology that combines magnetron sputtering and pulsed power. Similar to how DC and RF sputtering are characterized by their method of create plasma from a sputtering gas, HiPIMS is characterized by the use of high voltage and a short-duration energy burst (pulse) to generate plasma. Because of this plasma generation mechanism, high-density plasma can be formed on the surface of target materials, causing a higher degree of ionization of the sputtered particles as well as a higher rate of molecular gas dissociation compared to other sputtering methods. In other words, an HiPIMS-deposited thin film often has a higher density than thin films deposited by other methods.

Using HiPIMS for VO_2 deposition is still a relatively new research area. Due to the theoretically higher film density, it is expected that high-quality VO_2 could be deposited at a much lower temperature (<400°C) compared to its DC and RF counterparts without using any postannealing process. This allows for selecting from a wider range of substrates; even a polymeric substrate has been attempted [113]. Aside from conventional parameters, like in other previously mentioned reactive sputtering methods, HiPIMS systems are also identified by the frequency and duty cycle of their pulses. Unfortunately, there has been no report yet on the effect of these parameters on similarly prepared VO_2 samples. HiPIMS systems are compatible with additional substrate bias. Results both with [105, 112–114] and without [115] substrate bias have been reported.

HiPIMS vs. RFMS: Loquai et al. [105] conducted a comparison between VO_2 deposited on B270 glass using HiPIMS and that by using RFMS. The processing parameters utilized, as seen in Fig. 9.16a, are comparable between both deposition processes, except for different O_2/Ar flow ratios. From the transmission spectra (Fig. 9.16b,c), the optical properties of VO_2 films by both deposition processes are also similar despite the difference in O_2 flow. This means that HiPIMS could produce VO_2 of a similar grade with half the amount of O_2 as compared to RFMS.

HiPIMS vs. pDCMS: Aijaz et al. conducted an extensive study on the effect of substrate temperature, O_2/Ar flow ratio, and substrate bias on the quality of VO_2 thin films deposited by both HiPIMS and pDCMS [112]. An average power of 600 W was used for both HiPIMS (500 Hz, pulse on-time of 100 μs) and pDCMS (100 kHz, pulse off-time of 1.6 μs). By using a measured discharge voltage vs. O_2 fraction plot (Fig. 9.16d), the point of transition between metallic film and insulating film can be singled out. Thus, the stoichiometry of the VO_2 film could be identified. The thermochromic performance of VO_2 by HiPIMS at different O_2 fractions, temperatures, and substrate biases is shown in Fig. 9.16e–l. As concluded in previous sections, an increase in the O_2 fraction leads to a transition from V^{3+} to V^{4+} to V^{5+}. This was reaffirmed in this report. A comparison with VO_2 by pDCMS showed that HiPIMS is able to produce functional thermochromic VO_2 at a much lower substrate temperature.

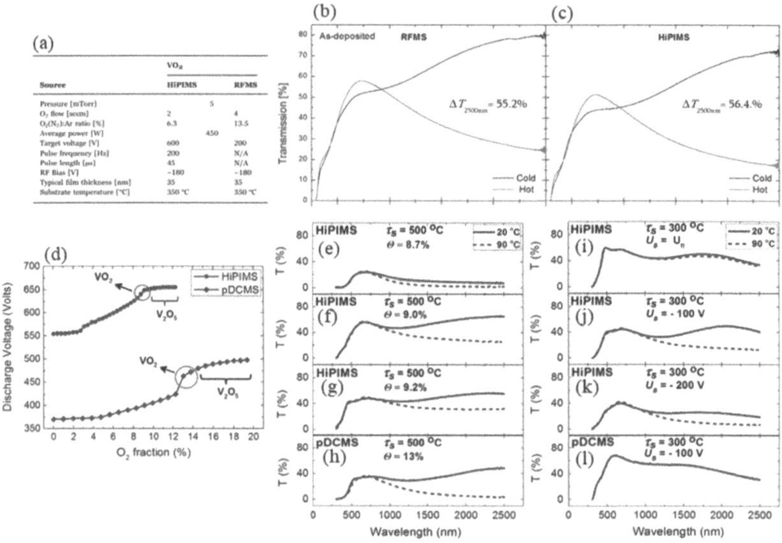

Figure 9.16 (a) Deposition parameter of a VO_2 thin film by RFMS (b) and by HiPIMS (c). Reprinted from Ref. [105], Copyright (2017), with permission from Elsevier. (d) Measured discharge voltage during deposition vs. O_2 fraction. Transmission spectra of a VO_2 thin film with no bias (e–h) and a varying bias and a constant O_2 fraction of 8.7% (i–l). Reprinted from Ref. [112], Copyright (2016), with permission from Elsevier.

9.6.6 Inductively Coupled Plasma-Assisted Sputtering

Inductively coupled plasma-assisted sputtering (ICPS) is an unconventional modification to conventional sputtering methods. In addition to the magnetron plasma produced by a magnetic field near the target surface, conductive coils are installed in the space between the target and the substrate (Fig. 9.17a). This configuration can further enhance the plasma density during sputtering, ensuring it reaches a density comparable to that produced by HiPIMS but without the use of extremely high energy pulses.

Reports on VO_2 deposition using ICPS are limited to a couple of research groups due to the niche setup [116–120]. Mian and Okimura compared conventional RFMS and ICPS using low-temperature (250°C) deposition and deposition using traditional temperature (400°C) [116]. An XRD spectra comparison

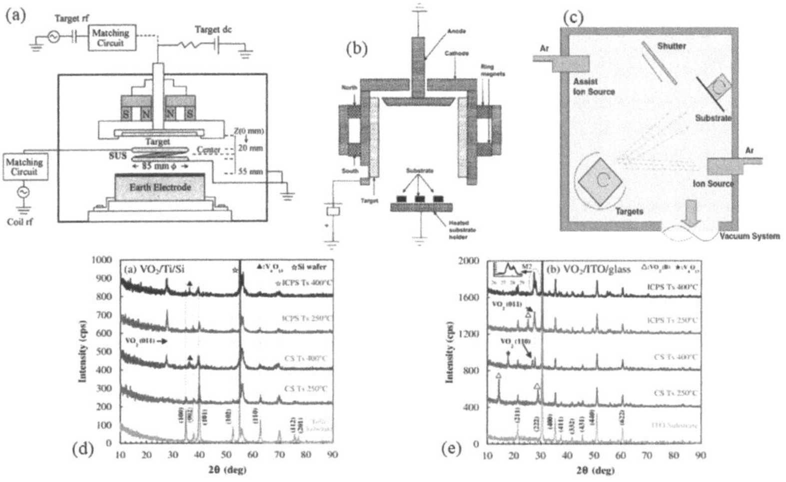

Figure 9.17 (a) Schematic of an ICPS system. Reprinted from Ref. [118], Copyright (2004), with permission from Elsevier. (b) Schematic of an ICMS system. Reprinted from Ref. [121], Copyright (2008), with permission from Elsevier. (c) Schematic of an ion-beam-assisted sputtering system. XRD spectra of VO_2 on a Ti/Si substrate (d) and an ITO/glass substrate (e). Reprinted from Ref. [116]. Copyright (2014) The Japan Society of Applied Physics.

(Fig. 9.17d,e) showed a VO_2 thin film of an exceptional quality deposited at 250°C by ICPS. XRD peaks from the 250°C samples from both ITO and Si substrates were comparable, even sharper than the 400°C peaks by conventional RFMS.

9.6.7 Inverted Cylindrical Magnetron Sputtering

Inverted cylindrical magnetron sputtering (ICMS) is another modification of the conventional RFMS setup. Instead of additional magnetic coils, like in ICPS, a setup involving a perpendicular cathode and anode is used (Fig. 9.17b). This setup is useful for deposition on a substrate with a complex geometry. Aside from this niche usage, other parameters for ICPS do not deviate too much from those for conventional RFMS. Research on VO_2 deposition by ICPS is rare, but the reported results to date are comparable to those achieved with a conventional system [121, 122].

9.6.8 Ion-Beam-Assisted Sputtering

Ion-beam-assisted sputtering (IBS) is different from all the previously discussed sputtering methods. Instead of generating bombarding ions inside a plasma of sputtering gas mixture, ions are directly propagated toward target materials at an angle so that the ejected particles are deposited on the substrate surface (Fig. 9.17c). Because ions are injected into the chamber using ion guns, the operating pressure in an IBS system is much lower than that in conventional plasma sputtering. IBS has the advantage of controlling the stoichiometry of the deposited film by tuning the power of the ion guns. This is especially useful for hard-to-control stoichiometry compounds, such as VO_x. However, IBS is still mainly used for the nonreactive sputtering of vanadium metal thin films instead [123, 124]. Reactive IBS requires dual ion gun configuration to control O_2 and Ar separately [125].

9.6.9 Nonreactive Sputtering Methods

A nonreactive VO_2 sputtering system either uses a VO_x ceramic target as the source material or consists of two steps: (i) deposition of a vanadium metal thin film on a substrate and (ii) oxidation of the said film into VO_2.

9.6.9.1 Ceramic target system

As mentioned in the previous section, the molar ratio of vanadium and oxygen during sputtering is one of the most crucial factors in achieving stoichiometric VO_2. A method of bypassing this issue is to use a VO_2 ceramic target as the source material during sputtering. Theoretically, the vanadium-to-oxygen ratio in the produced film should be the same as in the source. Yu et al. reported a typical process to produce stoichiometric VO_2 using this method [126]. Nonreactive RFMS was used to sputter VO_2 ceramic onto Eagle XG glass with a pure Ar atmosphere. Postannealing was done under vacuum to enhance the thin film's crystallinity. As expected, RFMS was used because it is the most common system for ceramic sputtering. This is because most ceramics are insulators and are not suitable for sputtering using DC power. However, VO_2 is unique because it exhibits metallic properties at room temperature. Thus, DCMS was also attempted and shown to be successful by Sun et al. [127]. VO_2 is not the only ceramic target that can be used as a thin film source. A V_2O_5 target has also been tested [128]. Because it has a different stoichiometry, an additional annealing process in an O_2-rich environment is required to reach the final VO_2 thin film.

9.6.9.2 Vanadium target system with pure argon sputtering

A VO_2 thin film can also be achieved by oxidizing a sputtered vanadium film. DCMS is the most common method for producing a vanadium film as it is the cheapest and most accessible sputtering system [129–134]. There are also reports, however, on IBS being used instead [123, 124]. Vanadium films can be deposited on a wide range of substrates because it is common practice to conduct deposition at room temperature on a substrate. The crystallinity and morphology of the as-deposited vanadium film are not the emphasis of these studies. Most of them instead have been focused on the effect of the annealing and oxidation parameters on the final VO_2 film. Details are not discussed here as they are not the focus of this paper. However, it is important to note that this is one of the main synthesis routes of a VO_2 film through a sputtering system.

9.6.10 Modification to a Sputtering System for High-Performance VO_2 films

9.6.10.1 Low-temperature deposition of VO_2 films by sputtering

The deposition of VO_2 single-phase films at low temperatures has particular importance when considering the selectivity of the substrate and compatibility for large-scale production. It is known that the selection of certain single-crystal substrates, for example, sapphire, rutile, ZnO, or even mica, facilitates the formation of VO_2 crystalline film due to epitaxial or partially epitaxial growth of VO_2 on them. Therefore, in addition to the precise control of the deposition parameters stated earlier, the deposition of a seed layer on a polycrystalline or glass substrate before VO_2 growth in sputtering enables the formation of a single crystalline VO_2 film at a

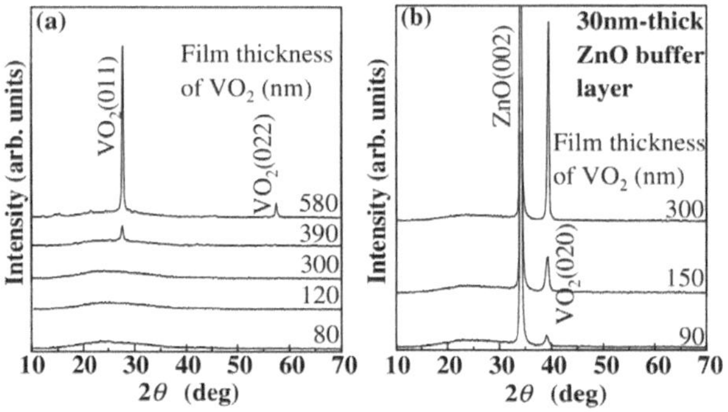

Figure 9.18 XRD patterns of (a) VO_2/glass with VO_2 thicknesses from 80 to 580 nm and (b) VO_2/ZnO/glass with VO_2 thicknesses of 90, 150, and 300 nm deposited at 400°C. Reprinted from Ref. [136]. Copyright (2003) The Japan Society of Applied Physics.

lower-than-typical substrate temperature. Such a seed layer can be vanadium metal (probably containing some oxygen) [93], but oxides having a crystal structure similar to that of VO_2 are most favorable. Because the formation of VO_2 films by sputtering is at a substrate temperature higher than its phase transition temperature (68°C), either by deliberate heating or by natural particle bombardment during the film growth, the deposited VO_2 film can be considered to have a rutile structure. Therefore, the typical seed materials should be those with the same rutile structure and small difference in lattice parameters compared to VO_2. Furthermore, not only the oxides with a rutile structure but also those with a similar atomic arrangement for the crystal lattice demonstrated good performance in reducing the substrate temperature required for sputtered VO_2. Such materials are ZnO, ITO, Al_2O_3, Ga_2O_3, Cr_2O_3, V_2O_3, Ti_2O_3, Fe_2O_3, etc. [135–139].

Figure 9.18 shows the effect of a ZnO seed layer on the deposition of VO_2 [136]. When VO_2 films were deposited on a bare soda-lime glass substrate by RFMS at 400°C, a film thickness of over 390 nm was required for the film to show the growth of a VO_2 crystalline film (Fig. 9.18a). Under the same conditions, the VO_2 crystal phase was observed at a thickness of 90 nm with the use of 30 nm thick ZnO as a buffer layer (seed layer). Although VO_2 (tetragonal at 400°C) and ZnO (hexagonal) belong to different crystal systems, they have very

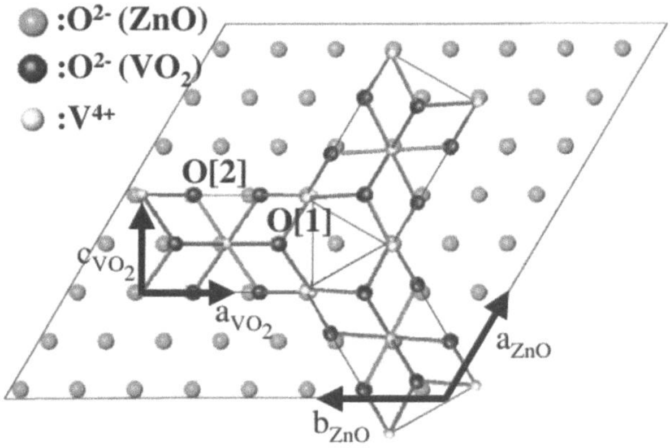

Figure 9.19 Heteroepitaxial relationships of $VO_2(010)$ on $ZnO(001)$. Reprinted from Ref. [136]. Copyright (2003) The Japan Society of Applied Physics.

similar oxygen networks (Fig. 9.19) for both crystal systems, so a heteroepitaxial growth of VO_2 on ZnO took place to promote the VO_2 film growth.

Similar heteroepitaxial growth of VO_2 on a seed layer or even on a bare substrate of a single crystal having a different crystalline symmetry (e.g., sapphire) was also reported. It was reported that a multilayer epitaxial film of W-doped VO_2/V_2O_3 was grown on a sapphire (110) substrate by reactive magnetron sputtering [137]. The films were deposited by reactively sputtering a V-W alloy target in an Ar-O_2 gas mixture at linearly increasing oxygen flow without interrupting the film growth. Multilayer epitaxy took place with a W-VO_2 top layer grown on W-V_2O_3 on a sapphire (110) substrate. For the film deposited in the oxygen flow from 10 to 26 sccm, a W-VO_2/W-V_2O_3 epitaxial multilayer growth was formed by XRD, as shown in Fig. 9.20. Detailed investigation by XRD pole-figure plots confirmed the epitaxial relationship being (001)f//(110)s, (110)f//(001)s for W-V_2O_3, and (010)f//(110)s, (100)f//(001)s for W-VO_2, where f and s denote the film and the substrate, respectively. The heteroepitaxial relationship is summarized in Fig. 9.21.

Recent results have given a more detailed demonstration of the effect of V_2O_3 for VO_2 growth on sapphire (001) [138]. The thermochromic films of VO_2 were deposited by magnetron

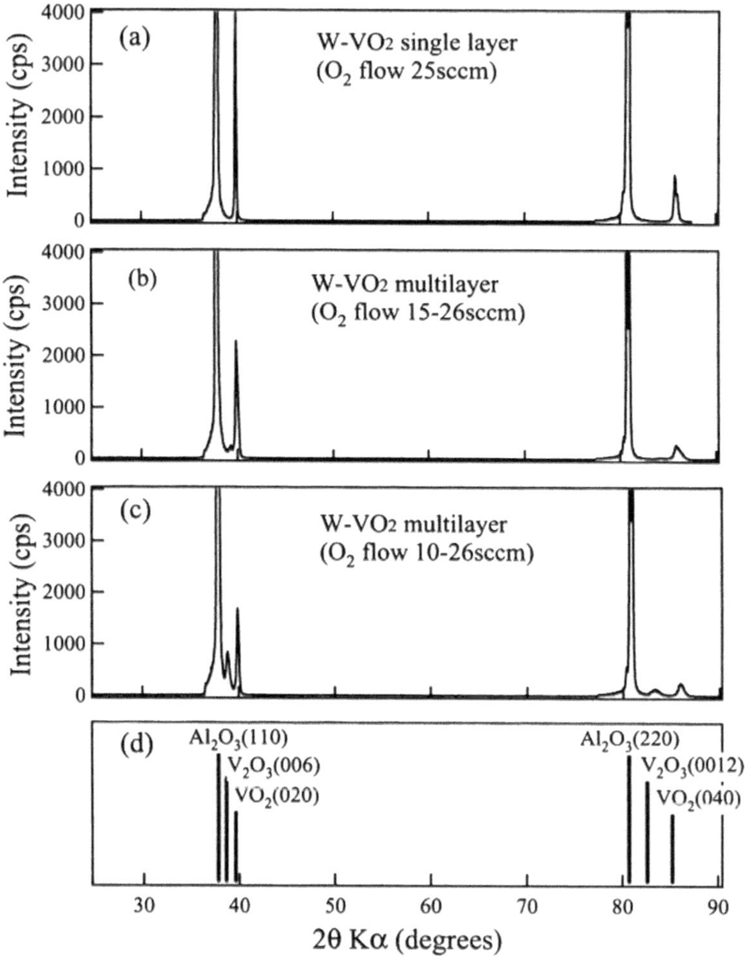

Figure 9.20 XRD patterns for a W-VO_2 film deposited at a constant oxygen flow of 25 sccm (a); the multilayers deposited at an oxygen flow of 15–26 sccm (b) and 10–26 sccm (c), in comparison to the JCPDS standards. Reprinted from Ref. [137], Copyright (2000), with permission from Elsevier.

sputtering at a fairly low temperature of 250°C without subsequent annealing by embedding a V_2O_3 interlayer. The V_2O_3 acts as a seed layer, leading to the heteroepitaxial growth of a VO_2 film at a low substrate temperature. The growth relationship was confirmed and is demonstrated in Fig. 9.22. The microstructure, using cross-

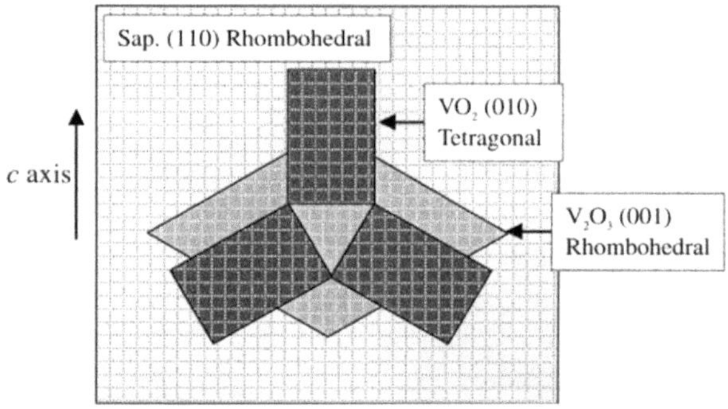

Figure 9.21 Epitaxial relationship for the VO_2/V_2O_3 multilayer on sapphire (110). Reprinted from Ref. [137], Copyright (2000), with permission from Elsevier.

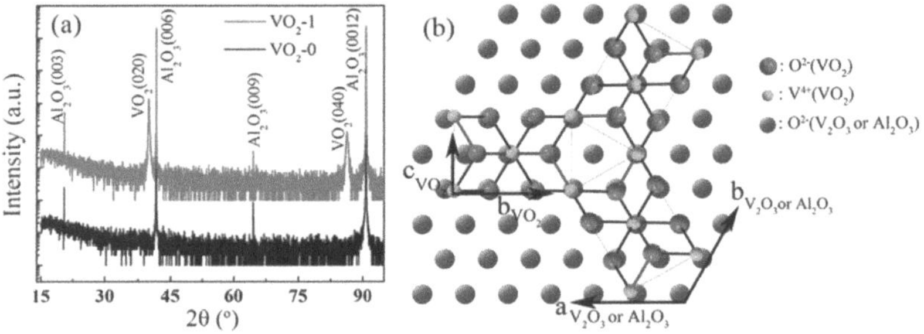

Figure 9.22 XRD patterns and epitaxial relationship of VO_2/V_2O_3 or VO_2/Al_2O_3. Reprinted from [138], Copyright © 2016 with permission from AIP Publishing.

sectional TEM, is shown in Fig. 9.23. The simplicity of target selection and a relatively low substrate temperature imply the possibility of use in large-scale off-line continuous glass coating.

A VO_2 smart window with a Cr_2O_3 seed layer was deposited on a glass substrate by magnetron sputtering at low temperatures from 250°C to 350°C, which is almost suitable for a large-scale off-line continuous glass coating process [139]. The Cr_2O_3 layer was designed to not only provide a heteroepitaxial template for low-temperature growth of VO_2 but also serve as an antireflection (AR) layer for improved luminous transmittance. The results demonstrated that the well-designed smart window shows quite a large

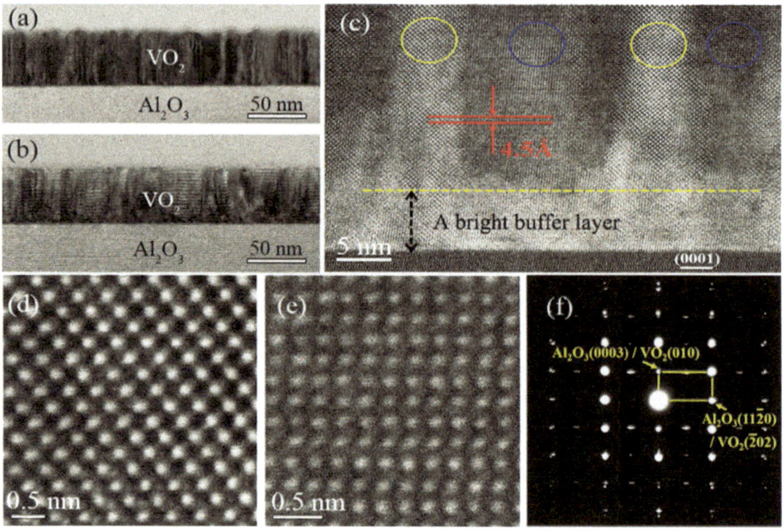

Figure 9.23 STEM images of VO_2 films without (a) and with (b) a V_2O_3 seed layer. (c) STEM-HAADF high-resolution image of (b); (d) STEM-HAADF high-resolution image of the bright region in (c) (yellow circles); (e) STEM-HAADF high-resolution image of the dark region in (c) (blue circles); and (f) SAED patterns of (b). Reprinted from [138], Copyright © 2016 with permission from AIP Publishing.

solar modulation ($\Delta T_{sol} = 12.2\%$) at a luminous transmittance of $T_{lum} = 46\%$, even without the use of a surface AR coating. Figure 9.24 gives a summary of a VO_2/Cr_2O_3/glass smart window. Figure 9.25 demonstrates the transmittance spectra with a comparison with values reported in literatures.

9.6.10.2 Transition temperature control of VO_2 by element doping

The phase transition of VO_2 crystals takes place when the temperature is increased to a value above 68°C. The transition temperature (τ_c) is too high for smart window applications. It is known that doping VO_2 with certain elements can reduce the τ_c to a more accommodating temperature range. There are a number of elements that can be used for doping VO_2, each with its unique effects on both the τ_c and the optical properties of the resulting film.

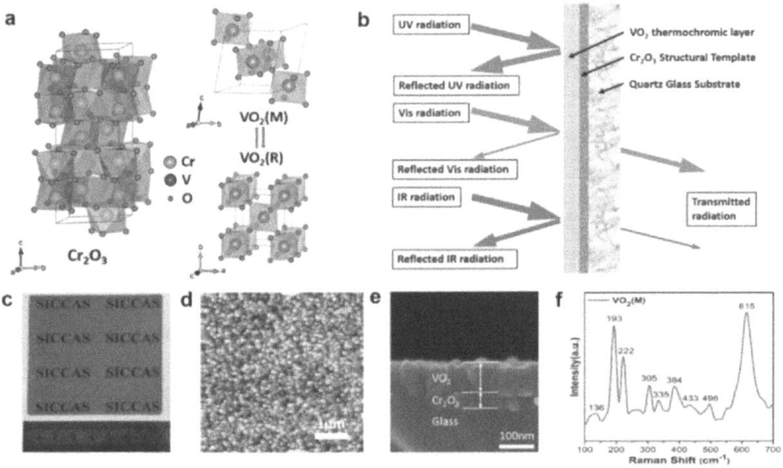

Figure 9.24 (a) The crystal structure of hexagonal Cr_2O_3 and monoclinic and rutile VO_2; (b) schematic illustration of a Cr_2O_3/VO_2 smart window; (c) photo of the smart window (75×75 mm^2); (d) AFM image of the coated surface; (e) cross-section SEM image of the window; (f) Raman spectra of VO_2/Cr_2O_3 on glass. Reprinted with permission from Ref. [139]. Copyright (2017) American Chemistry Society.

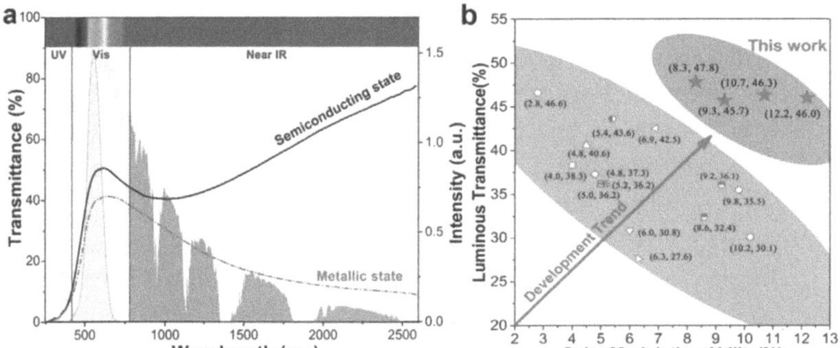

Figure 9.25 Transmittance (a) and a comparison in terms of property with literature (b). Reprinted with permission from Ref. [139]. Copyright (2017) American Chemistry Society.

Normally, the replacement of the V in VO_2 by metal elements of high valences, such as W, Mo, Nb, Ta, and Re, can reduce the τ_c, while that by metal elements of low valences, such as Cr, Ti, and Al, increases the τ_c. Doping VO_2 with the metal Mg and the nonmetal element F and codoping with a metal and a nonmetal, such as F + W and Mg + F, were also reported [139–141].

A convenient way of doping VO_2 with metal elements is cosputtering. Multiple targets are mounted in a sputtering chamber and sputtered at the same time. The amount of dopant can be controlled by changing the power ratio applied to the vanadium (or vanadium oxide) and the doping metal (or metal oxide). Thin films with different doping levels are formed under similar conditions, which hep to study the dopant-to-deposition rate, the film composition and structure, and the dopant's effect on the τ_c. This method is effective for the basic study of small samples with a wide selection of materials.

The effect of W as a dopant on the τ_c of VO_2 was investigated thoroughly [142]. Thin films of $V_{1-x}W_xO_2$ were deposited on a glass and Si substrate by dual-target cosputtering. The crystal phase formation was identified by thin film XRD, and the composition (particularly the W/V ratio) was measured by Rutherford backscattering spectrometry. Films of the crystalline single-phase $V_{1-x}W_xO_2$ with $x = 0 - 0.026$ were successfully formed and their τ_c was measured. Figure 9.26a shows the transmittance at 2000 nm versus the temperature for the $V_{1-x}W_xO_2$ films (a), and Fig. 9.26b shows the τ_c (the middle point of the transmittance-temperature curve on the temperature increasing portion) versus the W doping level (b). It was found that the W doping reduced the τ_c by 23°C/at.% of W, with a significant reduction in the hysteresis loop width and phase transition sharpness.

Similar research was done for Mo-doped VO_2, with a τ_c reduction efficiency of 12°C/at.% of Mo for $V_{1-x}W_xO_2$ films by sputtering [143]. The use of a V-M (M is the dopant metal) alloy target containing the desired dopant concentration is the way for obtaining a stable deposition process for metal-doped VO_2 films on a large scale. A ceramic target with vanadium oxides and a dopant metal or compound can also be used, although reports on vanadium oxide targets are still scarce.

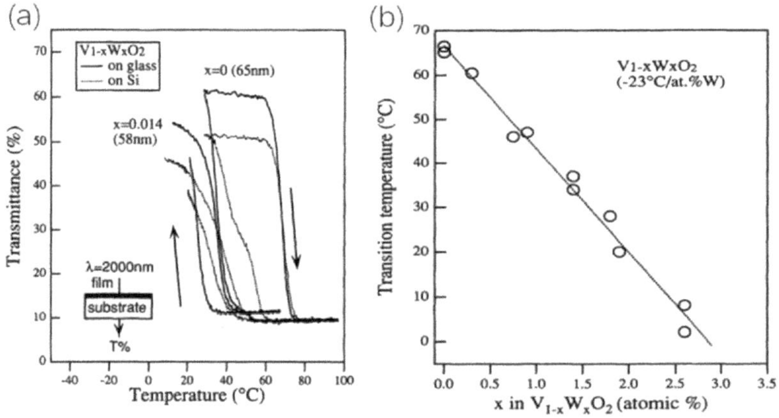

Figure 9.26 Transmittance at 2000 nm vs. temperature of $V_{1-x}W_xO_2$ films (a) and the transition temperature vs. the W doping level (b). Reprinted from Ref. [142]. Copyright (1995) The Physical Society of Japan and The Japan Society of Applied Physics.

VO_2 was doped with Mg by reactively cosputtering targets of V and Mg [141]. Thermochromic films of $V_{1-x}W_xO_2$ were made by reactive DCMS on heated glass. The MIT temperature decreased by 3K/at.% of Mg, while the optical transmittance increased concomitantly. Specifically, the transmittance of visible light and of solar radiation was enhanced by 10% when the Mg content was 7 at.%. Figure 9.27 shows the transmittance spectra (a) and the luminous transmittance increase with the doping level [141].

Nonmetal doping is usually done by introducing a reactive gas containing the dopant element during sputtering. For example, F-doped VO_2 was deposited by sputtering a V target in a gas mixture containing Ar, O_2, and CF_4 [140]. The F-VO_2 films demonstrated significantly enhanced transmittance with a marginally depressed τ_c.

VO_2 can be codoped with multiple metal elements by using cosputtering of multiple dopant targets or multielement alloys for large-scale deposition. The codoping of metal with nonmetal elements can be done with metal (or alloy) targets sputtered in a nonmetal containing a gaseous mixture, as shown by the F + W codoping sputtering process [144]. The results of codoping seem to

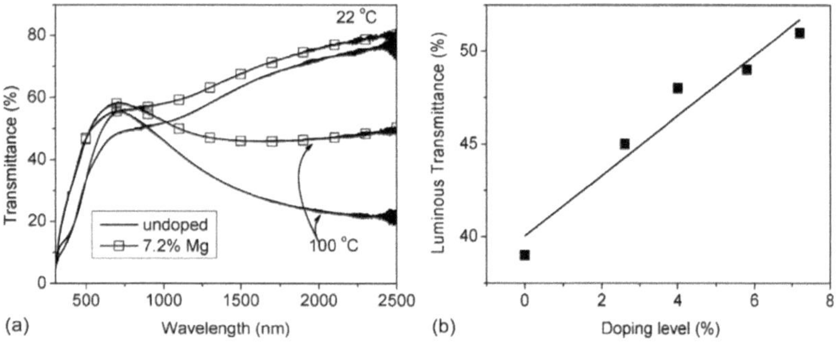

Figure 9.27 Transmittance spectra (a) and luminous transmittance increase with doping level (b) for sputter-deposited $Mg_xV_{1-x}O_2$ films. Reprinted from [141], Copyright © 2009 with permission from AIP Publishing.

affect the original properties of VO_2 independently with each doping element.

9.6.10.3 VO_2 multilayer structure for high performance and multifunction

Although the simplest structure of VO_2-based thermochromic glass can be a VO_2 thin film of dozens of nanometers on glass. A multilayer design is usually required for better performance in color, luminous transmittance, solar modulation, and lifetime. Theoretically, a 50 nm VO_2 film on glass demonstrate a luminous transmittance of 33% and a solar modulation of 5%. The performance of a VO_2 single layer is far from being satisfactory for practical application.

The use of suitable AR coatings is naturally onc of the most important solutions for improvement. The selection of suitable materials with optimized film stacks can not only improve the luminous transmittance (AR) but also increase the solar modulation and lifetime and even leads to multifunctional performance.

A detailed investigation was done for the selection of AR coating materials for VO_2 smart windows [145]. TiO_2 was considered as one of the best AR coating materials for VO_2 because of the high refractive index, matching that of VO_2, unlike other materials reported, such as SiO_2. Earlier publications [145–147] reported that the use of TiO_2 as an AR coating for 50 nm thick VO_2 improved the luminous transmittance from 30.9% to 57.6%, with an increase as high as 86%.

The effect of a TiO_2 AR coating for VO_2 was recently demonstrated by the fabrication of a 400 × 400 mm VO_2 smart window with a pilot off-line sputtering system [148]. A multifunctional $TiO_2(A)/VO_2(M)/TiO_2(R)$/glass multilayer film is designed and deposited by MF reactive magnetron sputtering. An exciting fact that the antifogging, self-cleaning, and energy-saving effects are integrated into the multilayer film could offer significant potential for wider applications of the smart window. First, the bottom layer, of rutile-phase TiO_2, is formed on glass, and it plays an important role in the formation of the monoclinic-phase VO_2 layer and serves as an AR layer as well. Then, the VO_2 layer with the monoclinic phase is formed and performs automatic solar/heat control for energy saving. Finally, the top TiO_2 layer, containing a dominant phase of anatase, is formed, which displays remarkable photocatalytic properties. The multilayer film shows satisfactory optical properties with excellent solar regulation efficiency (ΔT_{sol} = 10.2%) and an applicable luminous transmittance (T_{lum} = 30.1%) at a low temperature. In addition, the multilayer film implements

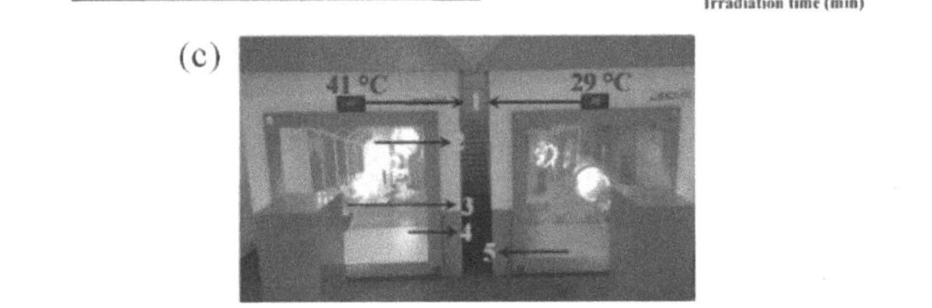

Figure 9.28 VO_2 smart glass (a), change in the water contact angle with UV irradiation (b), and energy performance with temperature change in a house model (c). Reprinted from Ref. [148], Copyright (2015), with permission from Elsevier.

photoinduced superhydrophilicity (water contact angle of 2.1°) through UV irradiation, resulting in an antifogging/self-cleaning effect. A high level of photocatalytic activity is detected on the surface of the multilayer film through degradation of stearic acid and rhodamine B. Figure 9.28 shows the appearance of the smart glass (a), the change of the water contact angle by UV irradiation of the glass surface (b), and the energy efficiency with an inside temperature change in a house model (c).

The multilayer structure in VO_2 glass can also be designed so the smart window performs other functions. An interesting thermochromic smart window was designed targeting improved skin comfort in addition to other energy performances [149]. In this design, the multilayer structure was optimized to have the highest modulation at the solar infrared wavelength range (1350–2500 nm), to which the human skin is the most sensitive. Basically, the thermochromic behavior of VO_2 was modulated by TiO_2 AR layers

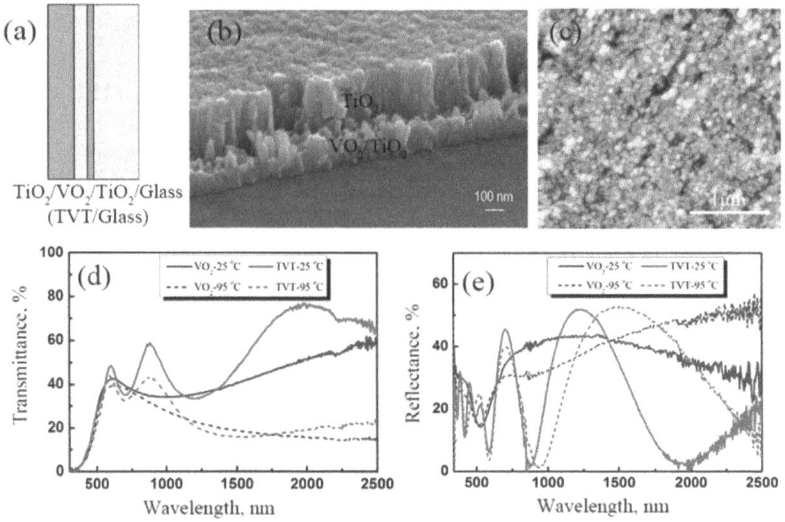

Figure 9.29 Structure and performance of $TiO_2/VO_2/TiO_2$/glass (TVT/glass): (a) glass structure; (b) cross-sectional SEM image; (c) AFM image of the glass surface; (d) transmittance and (e) reflectance spectra of the glass upon switching. Reprinted from Ref. [149], Copyright (2017), with permission from Elsevier.

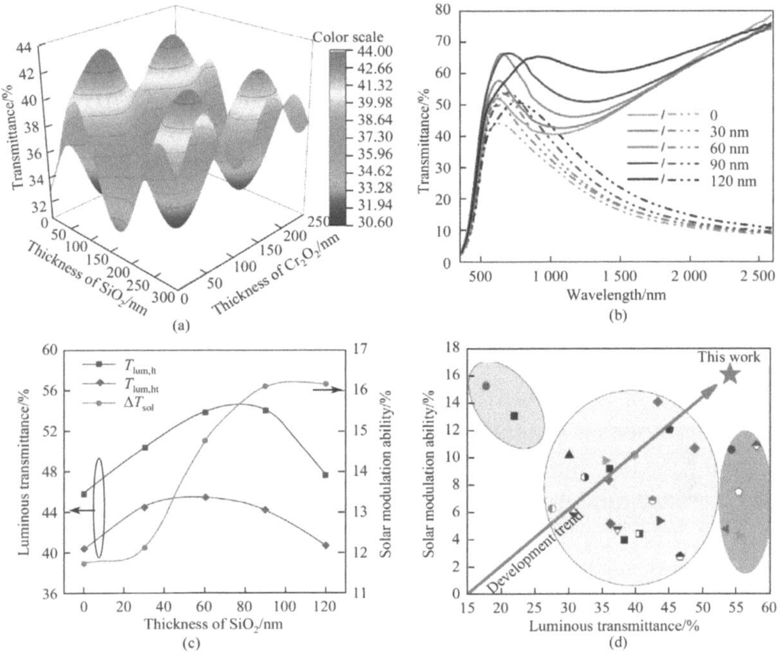

Figure 9.30 (a) Optimization design of the structure targeting the luminous transmittance ($T_{lum,lt}$) of the $SiO_2/VO_2/Cr_2O_3$/glass structure; (b) transmittance spectra of $SiO_2/VO_2/Cr_2O_3$/glass upon switching for a SiO_2 top layer of different thicknesses, (c) change in optical properties with a SiO_2 top layer of different thicknesses, (d) a comparison of the performance with literature. Reprinted from Ref. [145], Copyright (2004), with permission from Elsevier.

taking into account skin comfort for the first time. A $TiO_2/VO_2/TiO_2$ film, with thicknesses of 240 nm, 70 nm, and 10 nm, respectively, was designed and deposited on a glass substrate to show the highest modulation at 1350–2500 nm. The work provides a new insight into thermochromic smart window design that takes into consideration human comfort, which will improve the acceptance of such products by people. Figure 9.29 shows their results.

A carefully designed multilayer leads to thermochromic glass with the best optical performance and can be deposited by sputtering at a low substrate temperature in a relatively large area [145]. Novel $SiO_2/VO_2/Cr_2O_3$/glass structures were fabricated on

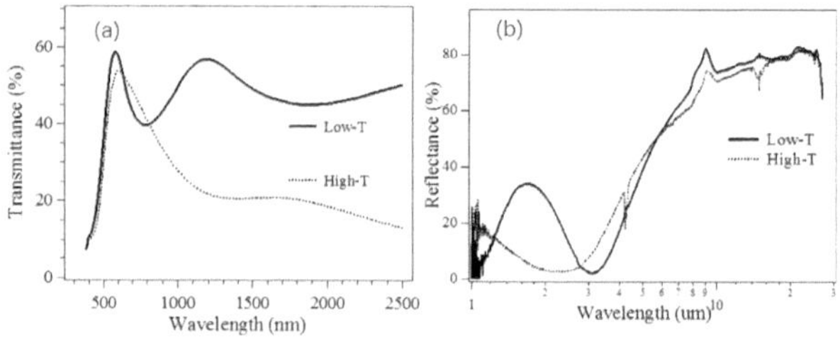

Figure 9.31 The transmittance spectra (a) and FTIR reflectance (b) for an ITO/VO_2/TiO_2/glass window at a low temperature and a high temperature, respectively [146].

the basis of an optimized multilayer design. The bottom (Cr_2O_3) layer provides a structural template for improving the crystallinity of VO_2 at a low substrate temperature and also serves as an AR layer to increase the luminous transmittance of the structure. As a result, a high solar modulation of $\Delta T_{sol} = 16.1\%$ was obtained at a fairly good luminous transmittance of $T_{lum} = 54\%$, which are the reported best data for the VO_2-based thermochromic smart window. Fig. 9.30 shows the results.

A high-insulation performance can be added to the VO_2-based thermochromic window by combining it with a transparent conducting oxide (TCO) layer. It was published that a TCO layer (ITO, doped ZnO, doped SnO, etc.) can be used to increase the insulation performance in addition to its AR effect [146]. An ITO/VO_2/TiO_2/glass structure was demonstrated to have high luminous transmittance, high solar modulation, and high insulation

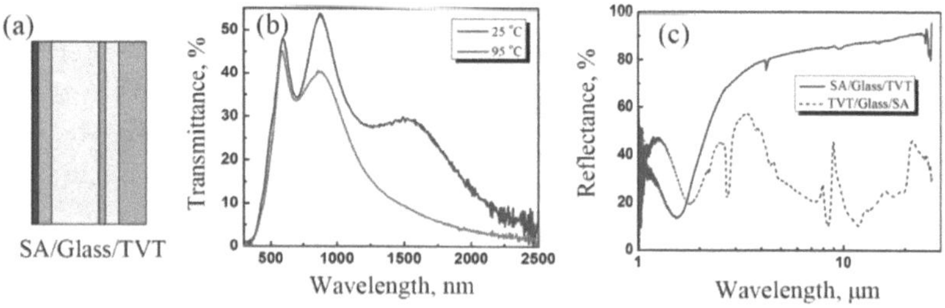

Figure 9.32 A novel design to realize automatic solar control and high insulation. Reprinted from Ref. [143], Copyright (1996), with permission from Elsevier.

(high reflection or low-E) in the far infrared wavelength, as shown in Fig. 9.31 [146].

Arranging the low-E glass and VO_2-based thermochromic glass to form an insulation glass unit is a very effective method to realize automatic solar control and high insulation at the same time. A recent publication demonstrates such a design (as shown in Fig. 9.32a) [143]. The film stacks are deposited on the different surfaces of a sheet glass, with SA indicating a SiO_2/AZO/glass multilayer structure and TVT a TiO_2/VO_2/TiO_2/glass multilayer structure. Large solar modulation was obtained as shown in Fig. 9.32b, and a low-E effect for insulation was demonstrated in Fig. 9.32c.

References

1. Nag, J. and Haglund, R. F., Jr. (2008). Synthesis of vanadium dioxide thin films and nanoparticles, *J. Phys.: Condens. Matter*, **20**(26), p. 264016.
2. Borek, M., Qian, F., Nagabushnam, V. and Singh, R. K. (1993). Pulsed laser deposition of oriented VO_2 thin films on *R*-cut sapphire substrates, *Appl. Phys. Lett.*, **63**(24), pp. 3288–3290.
3. Soltani, M., Chaker, M., Haddad, E., Kruzelecky, R. V. and Margot, J. (2004). Effects of Ti-W codoping on the optical and electrical switching of vanadium dioxide thin films grown by a reactive pulsed laser deposition (in English), *Appl. Phys. Lett.*, **85**(11), pp. 1958–1960.
4. Nag, J., Haglund, R. F., Jr. and Payzant, E. A. (2011). Effects of growth temperature on epitaxial thin films of vanadium dioxide grown by pulsed laser deposition, in *2011 Conference on Lasers and Electro-Optics*, pp. 1–2.
5. Kim, H., Charipar, N., Osofsky, M., Qadri, S. B. and Pique, A. (2014). Optimization of the semiconductor-metal transition in VO_2 epitaxial thin films as a function of oxygen growth pressure, *Appl. Phys. Lett.*, **104**(8), p. 081913.
6. Yang, T.-H., Wei, W., Jin, C. and Narayan, J. (2009). Controlled synthesis and structure-property correlations in vanadium based oxides, in *Materials and Devices for Smart Systems III*, 1129, Su, J., Wang, L. P., Furuya, Y., Trolier McKinstry, S. and Leng, J., eds. (Materials Research Society Symposium Proceedings), p. 325.

7. Lappalainen, J., Heinilehto, S., Jantunen, H. and Lantto, V. (2009). Electrical and optical properties of metal-insulator-transition VO_2 thin films, *J. Electroceram.*, **22**(1–3), pp. 73–77.
8. Masina, B. N., Lafane, S., Wu, L., Abdelli-Messaci, S., Kerdja, T. and Forbes, A. (2015). Optimizing the synthesis of vanadium-oxygen nanostructures by plasma plume dynamics using optical imaging, *Opt. Eng.*, **54**(3), p. 037106.
9. Guo, Y. X., et al. (2014). Oxygen pressure induced structure, morphology and phase-transition for VO_2/c-sapphire films by PLD, *Appl. Phys. A*, **115**(4), pp. 1245–1250.
10. Lafane, S., Kerdja, T., Abdelli-Messaci, S., Khereddine, Y., Kechouane, M. and Nemraoui, O. (2013). Correlation of plume dynamics and oxygen pressure with VO_2 stoichiometry during pulsed laser deposition (in English), *Appl. Phys. A*, **112**(1), pp. 159–164.
11. Fan, L. L., et al. (2012). Oxygen pressure dependent VO_2 crystal film preparation and the interfacial epitaxial growth study, *Thin Solid Films*, **520**(19), pp. 6124–6129.
12. Zhang, P., et al. (2015). Manipulations from oxygen partial pressure on the higher energy electronic transition and dielectric function of VO_2 films during a metal-insulator transition process (in English), *J. Mater. Chem. C*, **3**(19), pp. 5033–5040.
13. Kim, D. H. and Kwok, H. S. (1994). Pulsed laser deposition of VO_2 thin films, *Appl. Phys. Lett.*, **65**(25), pp. 3188–3190.
14. Morin, F. J. (1959). Oxides which show a metal-to insulator transition at the neel temperature, *Phys. Rev. Lett.*, **3**(1), pp. 34–36.
15. Ngom, B. D., et al. (2014). Competitive growth texture of pulsed laser deposited vanadium dioxide nanostructures on a glass substrate, *Acta Mater.*, **65**, pp. 32–41.
16. Suh, J. Y., Lopez, R., Feldman, L. C. and Haglund, R. F. (2004). Semiconductor to metal phase transition in the nucleation and growth of VO_2 nanoparticles and thin films, *J. Appl. Phys.*, **96**(2), pp. 1209–1213.
17. Soltani, M., Chaker, M., Haddad, E. and Kruzelesky, R. V. (2006). Thermochromic vanadium dioxide smart coatings grown on Kapton substrates by reactive pulsed laser deposition, *J. Vac. Sci. Technol., A*, **24**(3), pp. 612–617.
18. Maaza, M., et al. (2000). Direct production of thermochromic VO_2 thin film coatings by pulsed laser ablation (in English), *Opt. Mater.*, **15**(1), pp. 41–45.

19. Liu, H., Vasquez, O., Santiago, V. R., Diaz, L. and Fernandez, F. E. (2004). Semiconductor-to-metallic phase transition of VO_2 by laser excitation, *J. Electron. Mater.*, **33**(10), pp. 1171–1175.
20. Kang, L. T., et al. (2010). Effects of annealing parameters on optical properties of thermochromic VO_2 films prepared in aqueous solution, *J. Phys. Chem. C*, **114**(4), pp. 1901–1911.
21. Bayati, M. R., et al. (2013). Correlation between structure and semiconductor-to-metal transition characteristics of $VO_2/TiO_2/$ sapphire thin film heterostructures, *Acta Mater.*, **61**(20), pp. 7805–7815.
22. Jian, J., Chen, A., Chen, Y., Zhang, X. and Wang, H. (2017). Roles of strain and domain boundaries on the phase transition stability of VO_2 thin films, *Appl. Phys. Lett.*, **111**(15), p. 153102.
23. Zhi, B., et al. (2014). Thickness-dependent metal-to-insulator transition in epitaxial VO_2 films, *Mater. Res. Express*, **1**(4), p. 046402.
24. Molaei, R., Bayati, R., Wu, F. and Narayan, J. (2014). A microstructural approach toward the effect of thickness on semiconductor-to-metal transition characteristics of VO_2 epilayers, *J. Appl. Phys.*, **115**(16), p. 164311.
25. Lappalainen, J., Heinilehto, S., Saukko, S., Lantto, W. and Jantunen, H. (2008). Microstructure dependent switching properties of VO_2 thin films, *Sens. Actuators, A*, **142**(1), pp. 250–255.
26. Wu, Z. P. (2000). Epitaxial growth and defect microstructures of vanadium dioxide thin films on sapphire substrates, *Defects Diffus. Ceram.*, **186**(1), pp. 137–143.
27. Koo, H., et al. (2012). Effect of lattice misfit on the transition temperature of VO_2 thin film, *J. Mater. Sci.*, **47**(17), pp. 6397–6401.
28. Shibuya, K., Tsutsumi, J. Y., Hasegawa, T. and Sawa, A. (2013). Fabrication and Raman scattering study of epitaxial VO_2 films on MgF_2 (001) substrates, *Appl. Phys. Lett.*, **103**(2), p. 021604.
29. Shibuya, K. and Sawa, A. (2014). Epitaxial growth and structural transition of VO_2/MgF_2(001), *Jpn. J. Appl. Phys.*, **53**(5), p. 05ff03.
30. Muraoka, Y., Ueda, Y. and Hiroi, Z. (2002). Large modification of the metal-insulator transition temperature in strained VO_2 films grown on TiO_2 substrates, *J. Phys. Chem. Solids*, **63**(6–8), pp. 965–967.
31. Nagashima, K., Yanagida, T., Tanaka, H. and Kawai, T. (2006). Influence of ambient atmosphere on metal-insulator transition of strained vanadium dioxide ultrathin films, *J. Appl. Phys.*, **100**(6), p. 063714.

32. Nagashima, K., Yanagida, T., Tanaka, H. and Kawai, T. (2007). Interface effect on metal-insulator transition of strained vanadium dioxide ultrathin films, *J. Appl. Phys.*, **101**(2), p. 026103.
33. Li, J. and Dho, J. (2014). Characteristics of phase transition of VO_2 films grown on TiO_2 substrates with different crystal orientations, *J. Cryst. Growth*, **404**, pp. 84–88.
34. Li, J. A. and Dho, J. (2010). Controlling metal-insulator transition in the hetero-epitaxial VO_2/TiO_2 bilayer grown on Al_2O_3 (in English), *J. Cryst. Growth*, **312**(22), pp. 3287–3291.
35. Breckenfeld, E., et al. (2017). Strain effects in epitaxial VO_2 thin films on columnar buffer-layer TiO_2/Al_2O_3 virtual substrates, *ACS Appl. Mater. Interfaces*, **9**(2), pp. 1577–1584.
36. Slusar, T., Cho, J.-C., Kim, B.-J., Yun, S. J. and Kim, H.-T. (2016). Epitaxial growth of higher transition-temperature VO_2 films on AIN/Si, *APL Mater.*, **4**(2), p. 026101.
37. Kim, H., Bingham, N. S., Charipar, N. A. and Pique, A. (2017). Strain effect in epitaxial VO_2 thin films grown on sapphire substrates using SnO_2 buffer layers, *AIP Adv.*, **7**(10), p. 105116.
38. Yang, T.-H., Mal, S., Jin, C., Narayan, R. J. and Narayan, J. (2011). Epitaxial VO_2/Cr_2O_3/sapphire heterostructure for multifunctional applications, *Appl. Phys. Lett.*, **98**(2), p. 022105.
39. Molaei, R., Bayati, M. R. and Narayan, J. (2012). Thin film epitaxy and near bulk semiconductor to metal transition in VO_2/NiO/YSZ/Si(001) heterostructures, *J. Mater. Res.*, **27**(24), pp. 3103–3109.
40. Narayan, J. and Larson, B. C. (2003). Domain epitaxy: a unified paradigm for thin film growth (in English), *J. Appl. Phys.*, **93**(1), pp. 278–285.
41. Martens, K., Aetukuri, N., Jeong, J., Samant, M. G. and Parkin, S. S. P. (2014). Improved metal-insulator-transition characteristics of ultrathin VO_2 epitaxial films by optimized surface preparation of rutile TiO_2 substrates, *Appl. Phys. Lett.*, **104**(8), p. 081918.
42. Fan, L. L., et al. (2013). Growth and phase transition characteristics of pure M-phase VO_2 epitaxial film prepared by oxide molecular beam epitaxy (in English), *Appl. Phys. Lett.*, **103**(13), p. 131914.
43. Narayan, J. and Bhosle, V. M. (2006). Phase transition and critical issues in structure-property correlations of vanadium oxide, *J. Appl. Phys.*, **100**(10), p. 103524.
44. Yang, T.-H., Nori, S., Zhou, H. and Narayan, J. (2009). Defect-mediated room temperature ferromagnetism in vanadium dioxide thin films, *Appl. Phys. Lett.*, **95**(10), p. 102506.

45. Gupta, A., Narayan, J. and Dutta, T. (2010). Near bulk semiconductor to metal transition in epitaxial VO_2 thin films, *Appl. Phys. Lett.*, **97**(15), p. 151912.
46. Kim, B.-J., Seo, G., Choi, J., Kim, H.-T. and Lee, Y. W. (2012). Laser-assisted control of electrical oscillation in VO_2 thin films grown by pulsed laser deposition, *Jpn. J. Appl. Phys.*, **51**(10), p. 107302.
47. Wu, Z. P., et al. (1999). Molybdenum substitutional doping and its effects on phase transition properties in single crystalline vanadium dioxide thin film, *J. Appl. Phys.*, **86**(9), pp. 5311–5313.
48. Zhu, P. R., Yamamoto, S., Miyashita, A., Wu, Z. P., Narumi, K. and Naramoto, H. (1999). Semiconductor-metal phase transition in VO_2 films synthesized on alpha-Al_2O_3 by oxygen-reactive deposition using a neodymium-doped yttrium aluminium garnet laser, *Philos. Mag. Lett.*, **79**(8), pp. 603–608.
49. Jian, J., Chen, A., Zhang, W. and Wang, H. (2013). Sharp semiconductor-to-metal transition of VO_2 thin films on glass substrates, *J. Appl. Phys.*, **114**(24), p. 244301.
50. Gao, Y. F., et al. (2012). Enhanced chemical stability of VO_2 nanoparticles by the formation of SiO_2/VO_2 core/shell structures and the application to transparent and flexible VO_2-based composite foils with excellent thermochromic properties for solar heat control, *Energy Environ. Sci.*, **5**(3), pp. 6104–6110.
51. Chen, Z., Gao, Y. F., Kang, L. T., Cao, C. X., Chen, S. and Luo, H. J. (2014). Fine crystalline VO_2 nanoparticles: synthesis, abnormal phase transition temperatures and excellent optical properties of a derived VO_2 nanocomposite foil (in English), *J. Mater. Chem. A*, **2**(8), pp. 2718–2727.
52. Chen, Z., Cao, C. X., Luo, H. J., Gao, Y. F. (2016). VO_2-based thermochromic smart window: from energy savings to generation (in Chinese), *Chin. Sci. Bull.*, **61**, pp. 1661–1678.
53. Meldrum, A., Lopez, R., Magruder, R. H., Boatner, L. A. and White, C. W. (2010). Structure and properties of nanoparticles formed by ion implantation, in *Materials Science with Ion Beams*, 116, Bernas, H., ed., Topics in Applied Physics (Springer-Verlag Berlin, Berlin), pp. 255–285.
54. Jostmeier, T., Zimmer, J., Karl, H., Krenner, H. J. and Betz, M. (2014). Optically imprinted reconfigurable photonic elements in a VO_2 nanocomposite (in English), *Appl. Phys. Lett.*, **105**(7), p. 071107.

55. Lopez, R., Boatner, L. A., Haynes, T. E., Feldman, L. C. and Haglund, R. F. (2002). Synthesis and characterization of size-controlled vanadium dioxide nanocrystals in a fused silica matrix, *J. Appl. Phys.*, **92**(7), pp. 4031–4036.
56. Gea, L. A., Boatner, L. A., Evans, H. M. and Zuhr, R. (1997). Optically active surfaces formed by ion implantation and thermal treatment, *Nucl. Instrum. Methods Phys. Res., Sect. B*, **127**, pp. 553–556.
57. Lopez, R., Boatner, L. A., Haynes, T. E., Haglund, R. F. and Feldman, L. C. (2001). Enhanced hysteresis in the semiconductor-to-metal phase transition of VO_2 precipitates formed in SiO_2 by ion implantation, *Appl. Phys. Lett.*, **79**(19), pp. 3161–3163.
58. Karl, H., Dreher, J. and Stritzker, B. (2009). Semiconductor-metal phase transition in doped ion beam synthesized VO_2 nanoclusters, in *Functional Metal-Oxide Nanostructures*, 1174, Wu, J., Han, W. Q., Janotti, A. and Kim, H. C., eds. (Materials Research Society Symposium Proceedings), pp. 107–112.
59. Gea, L. A. and Boatner, L. A. (1996). Optical switching of coherent VO_2 precipitates formed in sapphire by ion implantation and annealing, *Appl. Phys. Lett.*, **68**(22), pp. 3081–3083.
60. Lim, H., Stavrias, N., Johnson, B. C., Marvel, R. E., Haglund, R. F. and McCallum, J. C. (2014). Optical switching and photoluminescence in erbium-implanted vanadium dioxide thin films, *J. Appl. Phys.*, **115**(9), p. 093107.
61. Jin, P., Nakao, S. and Tanemura, S. (1998). Tungsten doping into vanadium dioxide thermochromic films by high-energy ion implantation and thermal annealing, *Thin Solid Films*, **324**(1–2), pp. 151–158.
62. Lopez, R., Feldman, L. C. and Haglund, R. F. (2004). Size-dependent optical properties of VO_2 nanoparticle arrays (in English), *Phys. Rev. Lett.*, **93**(17), p. 177403.
63. Lopez, R., Haynes, T. E., Boatner, L. A., Feldman, L. C. and Haglund, R. F. (2002). Size effects in the structural phase transition of VO_2 nanoparticles (in English), *Phys. Rev. B*, **65**(22), p. 224113.
64. Donev, E. U., Lopez, R., Feldman, L. C. and Haglund, R. F. (2009). Confocal Raman microscopy across the metal-insulator transition of single vanadium dioxide nanoparticles (in English), *Nano Lett.*, **9**(2), pp. 702–706.
65. Rini, M., et al. (2005). Photoinduced phase transition in VO_2 nanocrystals: ultrafast control of surface-plasmon resonance, *Opt. Lett.*, **30**(5), pp. 558–560.

66. Jostmeier, T., et al. (2016). Thermochromic modulation of surface plasmon polaritons in vanadium dioxide nanocomposites (in English), *Opt. Express*, **24**(15), pp. 17321–17331.

67. Zhou, Y., Huang, A., Li, Y., Ji, S., Gao, Y. and Jin, P. (2013). Surface plasmon resonance induced excellent solar control for VO_2@SiO_2 nanorods-based thermochromic foils, *Nanoscale*, **5**(19), pp. 9208–9213.

68. Martín-Palma, R. J. and Lakhtakia, A. (2013). Chapter 15: Vapor-deposition techniques, in *Engineered Biomimicry*, Lakhtakia, A. and Martín-Palma, R. J., eds. (Elsevier, Boston), pp. 383–398.

69. Jiang, S. J., Ye, C. B., Khan, M. S. and Granqvist, C. G. (1991). Evolution of thermochromism during oxidation of evaporated vanadium films, *Appl. Opt.*, **30**(7), pp. 847–851.

70. Golan, G., Axelevitch, A., Sigalov, B. and Gorenstein, B. (2003). Metal-insulator phase transition in vanadium oxides films (in English), *Microelectron. J.*, **34**(4), pp. 255–258.

71. Manthrammel, M. A., Fatehmulla, A., Al-Dhafiri, A. M., Alshammari, A. S. and Khan, A. (2017). Temperature dependent surface and spectral modifications of nano V_2O_5 films, *Opt. Spectrosc.*, **122**(3), pp. 420–425.

72. Paone, A., et al. (2015). Influence of doping in thermochromic $V_{1-x}W_xO_2$ and $V_{1-x}Al_xO_2$ thin films: twice improved doping efficiency in $V_{1-x}W_xO_2$, *J. Alloys Compd.*, **621**, pp. 206–211.

73. Ke, J.-S., Weng, S.-F., Wu, M.-C. and Lee, C.-S. (2013). Controlled synthesis of VO_2(R), VO_2(B), and V_2O_3 vanadium-oxide nanowires, *J. Nanopart. Res.*, **15**(7), p. 1632.

74. Axelevitch, A., Gorenstein, B. and Golan, G. (2011). Investigation of the electrical transport mechanism in VOx thin films, *Microelectron. Reliab.*, **51**(12), pp. 2119–2123.

75. Guo, Q., Lee, S. and Goodman, D. W. (1999). Vanadium oxides thin films grown on rutile TiO_2(110)-(1 x 1) and (1 x 2) surfaces (in English), *Surf. Sci.*, **437**(1–2), pp. 38–48.

76. Ortega-Reyes, L. and Avila-Garcia, A. (2015). Thermally grown vanadium oxide films and their electrical properties, *Mater. Sci. Semicond. Process.*, **37**, pp. 123–128.

77. Marvel, R. E., Appavoo, K., Choi, B. K., Nag, J. and Haglund, R. F. (2013). Electron-beam deposition of vanadium dioxide thin films (in English), *Appl. Phys. A*, **111**(3), pp. 975–981.

78. Crunteanu, A., et al. (2015). Electric field-assisted metal insulator transition in vanadium dioxide (VO_2) thin films: optical switching behavior and anomalous far-infrared emissivity variation, in *Oxide-*

Based Materials and Devices VI, 9364, Teherani, F. H., Look, D. C. and Rogers, D. J., eds. (Proceedings of SPIE), p. 93640J1-J11.

79. Leroy, J., Bessaudou, A., Cosset, F. and Crunteanu, A. ( ). Structural, electrical and optical properties of thermochromic VO_2 thin films obtained by reactive electron beam evaporation (in English), *Thin Solid Films*, **520**(14), pp. 4823–4825.
80. Thery, V., et al. (2016). Role of thermal strain in the metal-insulator and structural phase transition of epitaxial VO_2 films, *Phys. Rev. B*, **93**(18), p. 184106.
81. Thery, V., et al. (2017). Structural and electrical properties of large area epitaxial VO_2 films grown by electron beam evaporation, *J. Appl. Phys.*, **121**(5), p. 055303.
82. Lee, M. H., Kim, M. G. and Song, H. K. (1996). Thermochromism of rapid thermal annealed VO_2 and Sn-doped VO_2 thin films (in English), *Thin Solid Films*, **290**, pp. 30–33.
83. Marvel, R. E., Harl, R. R., Craciun, V., Rogers, B. R. and Haglund, R. F., Jr. (2015). Influence of deposition process and substrate on the phase transition of vanadium dioxide thin films, *Acta Mater.*, **91**, pp. 217–226.
84. Jin, P., Xu, G., Tazawa, M. and Yoshimura, K. (2002). A VO_2-based multifunctional window with highly improved luminous transmittance (in English), *Jpn. J. Appl. Phys.*, **41**(3A, Part 2), pp. L278–L280.
85. Sambi, M., Della Negra, M., Granozzi, G., Li, Z. S., Jorgensen, J. H. and Moller, P. J. (1999). An ARPEFS study of the structure of an epitaxial VO_2 monolayer at the TiO_2(110) surface, *Appl. Surf. Sci.*, **142**(1–4), pp. 146–151.
86. Sambi, M., Sangiovanni, G., Granozzi, G. and Parmigiani, F. (1997). Growth and the structure of epitaxial VO_2 at the TiO_2(110) surface, *Phys. Rev. B*, **55**(12), pp. 7850–7858.
87. Tashman, J. W., et al. (2014). Epitaxial growth of VO_2 by periodic annealing (in English), *Appl. Phys. Lett.*, **104**(6), p. 063104.
88. Paik, H., et al. (2015). Transport properties of ultra-thin VO_2 films on (001) TiO_2 grown by reactive molecular-beam epitaxy, *Appl. Phys. Lett.*, **107**(16), p. 163101.
89. Fan, L. L., et al. (2014). Strain dynamics of ultrathin VO_2 film grown on TiO_2 (001) and the associated phase transition modulation (in English), *Nano Lett.*, **14**(7), pp. 4036–4043.
90. Fan, L. L., et al. (2016). The epitaxial growth and interfacial strain study of VO_2/MgF_2 (001) films by synchrotron based grazing incidence X-ray diffraction (in English), *J. Alloys Compd.*, **678**, pp. 312–316.

91. Lin, Y. C., et al. (2016). Photoluminescence of monolayer transition metal dichalcogenides integrated with VO_2 (in English), *J. Phys.: Condens. Matter*, **28**(50), p. 504001.

92. Fan, L. L., et al. (2016). Infrared response and optoelectronic memory device fabrication based on epitaxial VO_2 film (in English), *ACS Appl. Mater. Interfaces*, **8**(48), pp. 32971–32977.

93. Jin, P. and Tanemura, S. (1994). Formation and thermochromism of VO_2 films deposited by RF magnetron sputtering at low substrate temperature, *Jpn. J. Appl. Phys.*, **33**(Part 1, No. 3A), pp. 1478–1483.

94. Griffiths, C. H. and Eastwood, H. K. (1974). Influence of stoichiometry on the metal-semiconductor transition in vanadium dioxide, *J. Appl. Phys.*, **45**(5), pp. 2201–2206.

95. Kang, Y.-B. (2012). Critical evaluation and thermodynamic optimization of the VO-$VO_{2.5}$ system, *J. Eur. Ceram. Soc.*, **32**(12), pp. 3187–3198.

96. Yuce, H., Alaboz, H., Demirhan, Y., Ozdemir, M., Ozyuzer, L. and Aygun, G. (2017). Investigation of electron beam lithography effects on metal-insulator transition behavior of vanadium dioxide, *Phys. Scr.*, **92**(11), p. 114007.

97. Zhao, Y., et al. (2013). Tuning the properties of VO_2 thin films through growth temperature for infrared and terahertz modulation applications, *J. Appl. Phys.*, **114**(11), p. 113509.

98. Zhu, B., Tao, H. and Zhao, X. (2016). Effect of buffer layer on thermochromic performances of VO_2 films fabricated by magnetron sputtering, *Infrared Phys. Technol.*, **75**, pp. 22–25.

99. Zhu, M., Qi, H., Wang, B., Wang, H., Guan, T. and Zhang, D. (2018). Thermochromism of vanadium dioxide films controlled by the thickness of ZnO buffer layer under low substrate temperature, *J. Alloys Compd.*, **740**, pp. 844–851.

100. Yu, Q., et al. (2013). Oxygen pressure manipulations on the metal-insulator transition characteristics of highly (0?1?1)-oriented vanadium dioxide films grown by magnetron sputtering, *J. Phys. D: Appl. Phys.*, **46**(5), p. 055310.

101. Dou, Y.-K., et al. (2015). Oxidizing annealing effects on VO_2 films with different microstructures, *Appl. Surf. Sci.*, **345**, pp. 232–237.

102. Huang, Z., Chen, S., Lv, C., Huang, Y. and Lai, J. (2012). Infrared characteristics of VO_2 thin films for smart window and laser protection applications, *Appl. Phys. Lett.*, **101**(19), p. 191905.

103. Xu, H. Y., Huang, Y. H., Li, J. P., Ma, F. and Xu, K. W. (2015). Effect of porous morphology on phase transition in vanadium dioxide thin films, *J. Vac. Sci. Technol., A*, **33**(6), p. 061508.

104. Kakiuchida, H., Jin, P. and Tazawa, M. (2009). Control of optical performance in infrared region for vanadium dioxide films layered by amorphous silicon, *Int. J. Thermophys.*, **31**(10), pp. 1964–1971.

105. Loquai, S., Baloukas, B., Klemberg-Sapieha, J. E. and Martinu, L. (2017). HiPIMS-deposited thermochromic VO_2 films with high environmental stability, *Sol. Energy Mater. Sol. Cells*, **160**, pp. 217–224.

106. Ben-Messaoud, T., Landry, G., Gariépy, J. P., Ramamoorthy, B., Ashrit, P. V. and Haché, A. (2008). High contrast optical switching in vanadium dioxide thin films, *Opt. Commun.*, **281**(24), pp. 6024–6027.

107. Jiang, M., Cao, X., Bao, S., Zhou, H. and Jin, P. (2014). Regulation of the phase transition temperature of VO_2 thin films deposited by reactive magnetron sputtering without doping, *Thin Solid Films*, **562**, pp. 314–318.

108. Jiang, M., et al. (2014). Room temperature optical constants and band gap evolution of phase pure M1-VO_2 thin films deposited at different oxygen partial pressures by reactive magnetron sputtering, *J. Nanomater.*, **2014**, pp. 1–6.

109. Panagopoulou, M., et al. (2015). The effect of buffer layer on the thermochromic properties of undoped radio frequency sputtered VO_2 thin films, *Thin Solid Films*, **594**, pp. 310–315.

110. Azhan, N. H., Okimura, K., Ohtsubo, Y., Kimura, S.-i., Zaghrioui, M. and Sakai, J. (2016). Large modification in insulator-metal transition of VO_2 films grown on Al_2O_3 (001) by high energy ion irradiation in biased reactive sputtering, *J. Appl. Phys.*, **119**(5), p. 055308.

111. Azhan, N. H., Su, K., Okimura, K. and Sakai, J. (2015). Radio frequency substrate biasing effects on the insulator-metal transition behavior of reactively sputtered VO_2 films on sapphire (001), *J. Appl. Phys.*, **117**(18), p. 185307.

112. Aijaz, A., Ji, Y.-X., Montero, J., Niklasson, G. A., Granqvist, C. G. and Kubart, T. (2016). Low-temperature synthesis of thermochromic vanadium dioxide thin films by reactive high power impulse magnetron sputtering, *Sol. Energy Mater. Sol. Cells*, **149**, pp. 137–144.

113. Loquai, S., Baloukas, B., Zabeida, O., Klemberg-Sapieha, J. E. and Martinu, L. (2016). HiPIMS-deposited thermochromic VO_2 films on polymeric substrates, *Sol. Energy Mater. Sol. Cells*, **155**, pp. 60–69.

114. Fortier, J. P., Baloukas, B., Zabeida, O., Klemberg-Sapieha, J. E. and Martinu, L. (2014). Thermochromic VO_2 thin films deposited by HiPIMS, *Sol. Energy Mater. Sol. Cells*, **125**, pp. 291–296.

115. Houska, J., Kolenaty, D., Rezek, J. and Vlcek, J. (2017). Characterization of thermochromic VO_2 (prepared at 250°C) in a wide temperature range by spectroscopic ellipsometry, *Appl. Surf. Sci.*, **421**, pp. 529–534.

116. Mian, M. S. and Okimura, K. (2014). Effects of energetic substrate-incident ions on the growth of crystalline vanadium dioxide films in inductively coupled plasma-assisted sputtering, *Jpn. J. Appl. Phys.*, **53**(3), p. 035802.

117. Mian, M. S. and Okimura, K. (2014). Effect of conductive TiN buffer layer on the growth of stoichiometric VO_2 films and the out-of-plane insulator–metal transition properties, *J. Vac. Sci. Technol., A*, **32**(4), p. 041502.

118. Nakamura, T. and Okimura, K. (2004). Ti ion density in inductively coupled plasma enhanced dc magnetron sputtering, *Vacuum*, **74**(3–4), pp. 391–395.

119. Nihei, Y., Sasakawa, Y. and Okimura, K. (2008). Advantages of inductively coupled plasma-assisted sputtering for preparation of stoichiometric VO_2 films with metal-insulator transition, *Thin Solid Films*, **516**(11), pp. 3572–3576.

120. Watanabe, T., Okimura, K., Hajiri, T., Kimura, S.-i. and Sakai, J. (2013). Phase selective growth and characterization of vanadium dioxide films on silicon substrates, *J. Appl. Phys.*, **113**(16), p. 163503.

121. Kana, J. B. K., et al. (2008). Thermochromic VO_2 thin films synthesized by RF-inverted cylindrical magnetron sputtering, *Appl. Surf. Sci.*, **254**(13), pp. 3959–3963.

122. Kana, J. B. K., et al. (2010). High substrate temperature induced anomalous phase transition temperature shift in sputtered VO_2 thin films, *Opt. Mater.*, **32**(7), pp. 739–742.

123. Ba, C. O. F., Bah, S. T., D'Auteuil, M., Fortin, V., Ashrit, P. V. and Vallée, R. (2014). VO_2 thin films based active and passive thermochromic devices for energy management applications, *Curr. Appl. Phys.*, **14**(11), pp. 1531–1537.

124. Yi, X., et al. (2003). A new fabrication method for vanadium dioxide thin films deposited by ion beam sputtering, *Infrared Phys. Technol.*, **44**(2), pp. 137–141.

125. Liang, J., et al. (2011). Infrared transition properties of vanadium dioxide thin films across semiconductor-metal transition, *Rare Met.*, **30**(3), pp. 247–251.

126. Yu, J. H., Nam, S. H., Lee, J. W. and Boo, J. H. (2016). Enhanced visible transmittance of thermochromic VO_2 thin films by SiO_2 passivation layer and their optical characterization, *Materials (Basel)*, **9**(7), p. 556.

127. Sun, G., et al. (2017). Low-temperature deposition of VO_2 films with high crystalline degree by embedding multilayered structure, *Sol. Energy Mater. Sol. Cells*, **161**, pp. 70–76.

128. Kang, M. I., Kim, I. K., Oh, E. J., Kim, S. W., Ryu, J. W. and Park, H. Y. (2012). Dependence of optical properties of vanadium oxide films on crystallization and temperature, *Thin Solid Films*, **520**(6), pp. 2368–2371.

129. Chu, X., et al. (2014). Effect of annealing duration and substrates on structure and property of vanadium dioxide films, *J. Wuhan Univ. Technol.-Mater. Sci. Ed.*, **29**(6), pp. 1117–1123.

130. Liang, J.-R., et al. (2014). Fabrication of VO_2 thin film by rapid thermal annealing in oxygen atmosphere and its metal—insulator phase transition properties, *Chin. Phys. B*, **23**(7), p. 076801.

131. Liu, X., Wang, S.-W., Chen, F., Yu, L. and Chen, X. (2015). Tuning phase transition temperature of VO_2 thin films by annealing atmosphere, *J. Phys. D: Appl. Phys.*, **48**(26), p. 265104.

132. Luo, Z., Wu, Z., Xu, X., Wang, T. and Jiang, Y. (2010). Growth, electrical, and optical properties of nanocrystalline VO_2 (011) thin films prepared by thermal oxidation of magnetron sputtered vanadium films, *J. Vac. Sci. Technol., A*, **28**(4), pp. 595–599.

133. Xiao, H., et al. (2016). Microstructures and thermochromic characteristics of VO_2/AZO composite films, *Infrared Phys. Technol.*, **76**, pp. 580–586.

134. Zhu, N.-W., Hu, M., Xia, X.-X., Wei, X.-Y. and Liang, J.-R. (2014). Preparation and modification of VO_2 thin film on R-sapphire substrate by rapid thermal process, *Chin. Phys. B*, **23**(4), p. 048108.

135. Jin, P. (2006). A high-insulation thermochromic glass and preparation method, Japan Patent P2006-206398A.

136. Kato, K., Song, P. K., Odaka, H. and Shigesato, Y. (2003). Study on thermochromic VO_2 films grown on ZnO-coated glass substrates for "smart windows," *Jpn. J. Appl. Phys.*, **42**, pp. 6523–6531.

137. Jin, P., et al. (2000). Epitaxial growth of W-doped VO_2/V_2O_3 multilayer on α-$Al_2O_3(110)$ by reactive magnetron sputtering, *Thin Solid Films*, **375**(1–2), pp. 128–131.

138. Sun, G. Y., Cao, X., Gao, X., Long, S., Liang, M. and Jin, P. (2016). Structure and enhanced thermochromic performance of low-temperature fabricated VO_2/V_2O_3 thin film, *Appl. Phys. Lett.*, **109**, p. 143903.

139. Chang, T. C., et al. (2017). Facile and low-temperature fabrication of thermochromic Cr_2O_3/VO_2 smart coatings: enhanced solar modulation ability, highluminous transmittance and UV-shielding function, *ACS Appl. Mater. Interfaces*, **9**, pp. 26029–26037.

140. Khan, K. A. and Granqvist, C. G. (1989). Theremochromic sputter-deposited vanadium oxyfluoride coatings with low luminous absorptance, *Appl. Phys. Lett.*, **55**, pp. 4–6.

141. Mlyuka, N. R., Niklasson, G. A. and Granqvist, C. G. (2009). Mg doping of thermochromic VO_2 films enhances the optical transmittance and decreases the metal-insulator transition temperature, *Appl. Phys. Lett.*, **95**(17), p. 171909.

142. Jin, P. and Tanemura, S. (1995). Relationship between transition temperature and x in $V_{1-x}W_xO_2$ films deposited by dual-target magnetron sputtering, *Jpn. J. Appl. Phys.*, **34**(5R), pp. 2459–2460.

143. Jin, P. and Tanemura, S. (1996). $V_{1-x}Mo_xO_2$ thermochromic films deposited by reactive magnetron sputtering, *Thin Solid Films*, **281–282**, pp. 239–242.

144. Burkhardt, W., et al. (2002). Tungsten and fluorine co-doping of VO_2 films, *Thin Solid Films*, **402**(1–2), pp. 226–231.

145. Xu, G., Jin, P., Tazawa, M. and Yoshimura, K. (2004). Optimization of antireflection coating for VO_2-based energy ececient window, *Sol. Energy Mater. Sol. Cells*, **83**(1), pp. 29–37.

146. Jin, P. (2003). A high-performance automatic solar control glass coating materials, Japanese Patent P2003-94551A.

147. Jin, P., Xu, G., Tazawa, M. and Yoshimura, K. (2003). Design, formation and characterization of a novel multifunctional window with VO_2 and TiO_2 coatings, *Appl. Phys. A*, **77**(3–4), pp. 455–459.

148. Zheng, J., Bao, S. and Jin, P. (2015). $TiO_2(R)/VO_2(M)/TiO_2(A)$ multilayer film as smart window: combination of energy-saving, antifogging and self-cleaning functions, *Nano Energy*, **11**, pp. 136–145.

149. Sun, G., Cao, X., Zhou, H., Bao, S. and Jin, P. (2017). A novel multifunctional thermochromic structure with skin comfort design for smart window application, *Sol. Energy Mater. Sol. Cells*, **159**, pp. 553–559.

Chapter 10

Sol-Gel Synthesis of Thermochromic VO_2 Coatings

Haibo Jin and Jingbo Li
School of Materials Science and Engineering, Beijing Institute of Technology, Zhongguancun South Street No. 5, Haidian District, Beijing 100081, China
hbjin@bit.edu.cn

10.1 Introduction

Recently, thermochromic VO_2 coatings have induced great interest as a result of their promising application in smart windows because of their unique ability to regulate solar heat gain [1–4]. VO_2(M) undergoes a first-order reversible phase transition at a critical temperature τ_c (68°C for the bulk) from a low-temperature monoclinic phase ($P2_1/c$, $M1$) to a high-temperature rutile phase ($P4_2/mnm$, R), which is accompanied by a remarkable electric conductivity change from a semiconductor state to a metallic state and a unique optical switching from infrared (IR) transparency to IR reflectance [3, 5, 6]. The particular optical properties of VO_2 in response to temperature have made it promising for application in smart thermochromic windows for energy-efficient livable conditions [3, 7].

Vanadium Dioxide-Based Thermochromic Smart Windows
Edited by Yi Long and Yanfeng Gao

ISBN 978-981-4877-06-0 (Hardcover), 978-1-003-17731-9 (eBook)
www.jennystanford.com

There are many techniques to prepare thermochromic VO_2 coatings, such as magnetron sputtering [8], chemical vapor deposition [9, 10], hydrothermal growth [11–13], sol-gel method [14–16], and polymer-assisted deposition [17, 18]. Among these, the sol-gel method has been demonstrated as a facile wet-chemical technique to prepare thermochromic VO_2 coatings because of its low cost, scalable deposition, accommodation for various substrates, and convenience in terms of heteroatom doping and multilayer deposition [1, 4, 19]. In this chapter, a comprehensive review is presented on the recent progress in the sol-gel method to synthesize effective thermochromic VO_2 coatings. The fundamentals of the sol-gel method are first introduced to understand the chemistry involved in the synthesis of VO_2 coatings by the sol-gel method. Inorganic and organic sol-gel methods, as two commonly used methods for the synthesis of VO_2 coatings, are recommended, with emphasis on the preparation parameter–induced variations in phase structures, morphologies, and thermochromic properties. Moreover, two effective sol-gel strategies of doping and composite construction to optimize thermochromic properties of VO_2 coatings are elaborated.

10.2 Fundamentals of the Sol-Gel Method

The sol-gel method is a wet-chemical way to synthesize inorganic materials (e.g., metal oxides or hydroxides) through the hydrolysis and condensation of the precursor to form a sol and finally a network structure of gel [20–23]. A sol possesses the characteristics of a colloidal suspension, which is defined by the International Union of Pure and Applied Chemistry as the dispersion of one phase in another. In this definition, the molecules or polymolecular particles dispersed in a medium have at least in one direction a dimension roughly between 1 nm and 1 μm. Accordingly, not only the colloidal V_2O_5 suspension in water but also the covalent-type vanadium polymers in organic solvents in this paper can be classified as sols. During a sol-gel process, a gel can form in various ways. Small changes in conditions can lead to different gel structures even from the same precursor [24–26]. In general, a gel is defined as a nonfluid 3D network that extends through a fluid phase [23]. Four types of gels were classified by Flory in 1974: lamellar gels, covalent polymer networks, networks of physically aggregated polymers, and disordered particulate gels [27]. In 1996, considering the primary purpose of the sol-gel method to synthesize inorganic solids, Kakihana [28] presented a more useful classification with five different gel types, whose features are summarized in Table 10.1 [23]. These five key types of gels can provide a prospective guidance for the structural and morphological design of the final inorganic products.

10.3 Sol-Gel Process for VO_2 Coatings

In 1983, Greenberg [29] first used the sol-gel method to deposit VO_2 coatings and W-, Mo-, and Nb-doped VO_2 coatings on a glass substrate. The sol was obtained through the hydrolysis and condensation of vanadyl tri-isopropoxide, or $VO(OC_3H_7)_3$, in 2-propanol. After gel formation on a glass substrate, postannealing at 400°C was conducted in a reductive atmosphere to obtain the final VO_2 coatings. $WOCl_4$, $MoOCl_4$, and $Nb(C_2H_5O)_5$ were employed as the sources of W, Mo, and Nb dopants, respectively, to prepare the doped VO_2 coatings, whose temperature-induced metal–insulator transition (MIT) and IR switching properties were investigated in comparison to those of the undoped VO_2 coatings. Since then, VO_2 thermochromic coatings by the sol-gel technology have been frequently reported [30–35]. In general, the sol-gel method to prepare VO_2 coatings includes four fundamental steps [19]:

1. The formation of a stable and homogeneous sol through the hydrolysis and polycondensation of an active vanadium precursor in a liquid intermediate
2. The deposition of the sol on a substrate by dip coating or spin coating
3. Gel formation after a sufficient aging process
4. The formation of crystalline VO_2 coatings through an appropriate annealing process

Table 10.1 The features of the key five types of gels

Type of gel	Type of bonding	Composition	Schematic
Colloidal	Van der Waals or hydrogen bonding	Metal-oxide or hydroxide particles	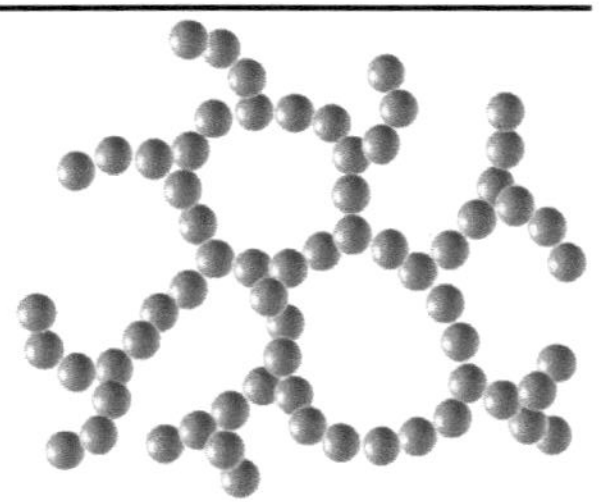
Metal-oxane polymers	Covalent bonding or intermolecular bonding	Inorganic polymers from the hydrolysis and condensation of molecular precursors	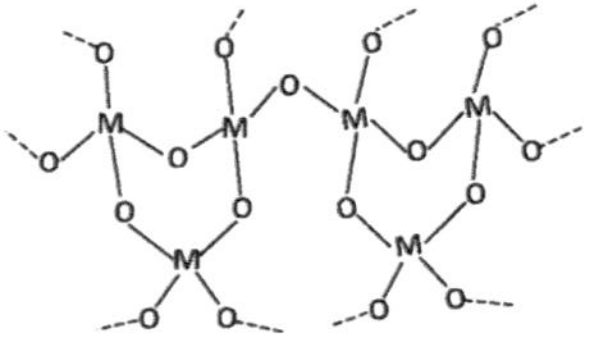
Metal complex	Weak intermolecular bonding	Concentrated metal complex solution	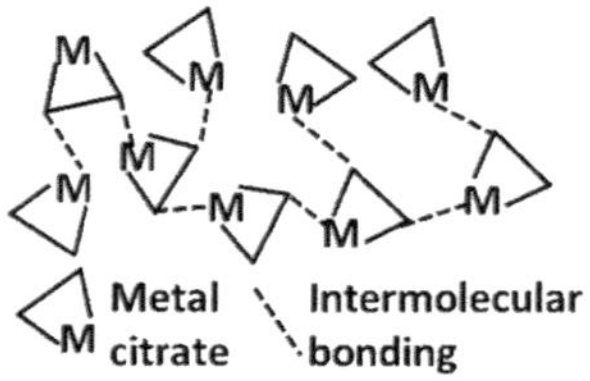

In situ polymerizable complex	Covalent and coordinate bonding	Polyesters of polyhydroxy alcohol and carboxylic acid with a metal complex	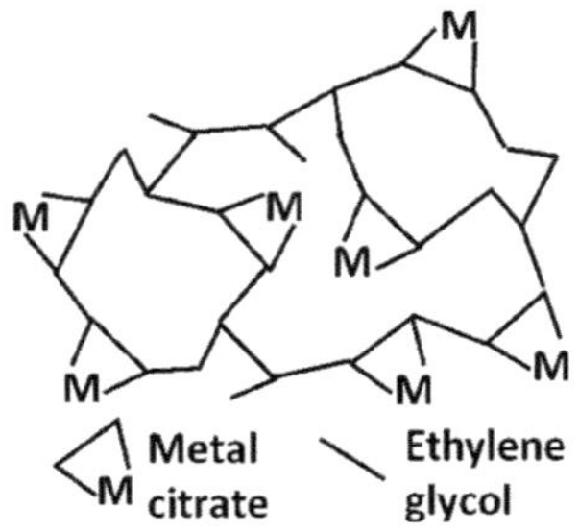
Coordinate and cross-linked polymer complex	Coordinate and intermolecular bonding	Coordinate polymers and metal salt solution	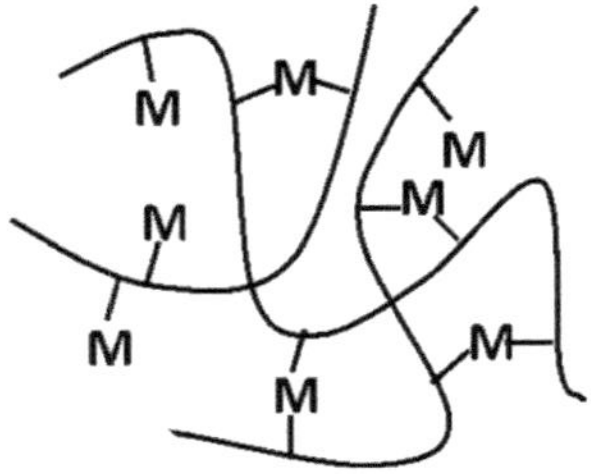

Vanadium alkoxides, vanadium oxides, vanadium salts, and molecular precursors, including V(acac)$_4$ and VO(acac)$_2$, are commonly used for the sol-gel synthesis of VO_2 thermochromic coatings [1, 10, 19]. According to the type of gel and the chemical reactions involved, the sol-gel method can be divided into inorganic [30] and organic [36]. For the organic sol-gel process, the sol forms through the hydrolysis and condensation of organic precursors, such as vanadium alkoxides, V(acac)$_4$, and VO(acac)$_2$. As a result, a vanadium-oxane polymer gel is obtained [26, 36, 37]. On the other hand, an inorganic process occurs on using inorganic V_2O_5 as the precursor, where a vanadium oxide colloidal gel forms in an aqueous medium [35, 38, 39]. It is worth mentioning that an organic process can be realized even with the V_2O_5 as the initial precursor, because of a transformation reaction from V_2O_5 to an organic vanadium alkoxide and the final formation of a vanadium-oxane polymer gel [24].

10.3.1 Inorganic Sol-Gel Method

V_2O_5 is the most commonly used inorganic precursor to prepare sol-gel VO_2 coatings because of its low cost and easy availability [19, 33]. In fact, the V_2O_5 sol is very stable and easy to form, simply by quenching the melted V_2O_5 in water. In the last decades, studies have focused on the effects of the procedure parameters on the quality and switching properties of the final VO_2 coatings, especially the postannealing treatment to reduce V_2O_5 into VO_2 [24, 35, 38].

In 1996, Yin et al. [30, 40] first reported the synthesis of VO_2 coatings by an inorganic sol-gel process by quenching melted V_2O_5 into water to form a V_2O_5 colloidal gel. The process included four steps: quenching, coating, drying, and vacuum annealing treatment. The effect of the drying process on the coating blistering was investigated. According to the results, the heating rate was the most important factor affecting the surface blistering. When the heating rate was less than 0.5°C min^{-1}, no blistering was observed on the coating surface. Moreover, the effects of film thickness, quenching temperature, and annealing time on the resistivity

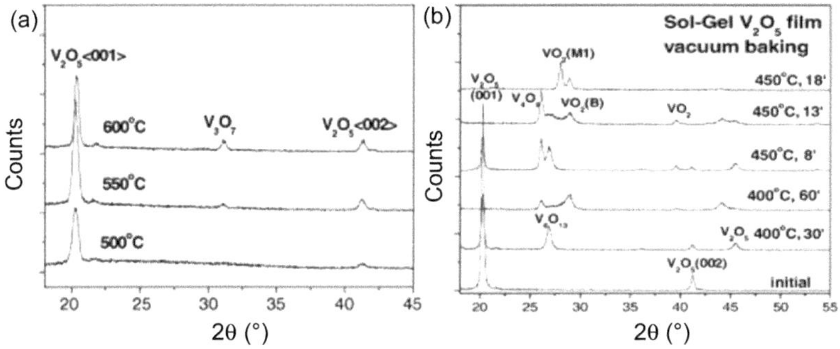

Figure 10.1 XRD patterns of sol-gel V_2O_5 films annealed in a temperature range of (a) 500°C–600°C for 20 min and (b) 400°C–450°C for a different amount of time by rapid thermal annealing (RTA). Cited from Ref. [38].

switching property were studied as well. It was found that lower quenching temperatures were more favorable for better resistivity and switching properties and the magnitude of the resistivity change increased with increasing film thickness and annealing time as well.

In 2002, Yuan et al. [38] prepared orientated V_2O_5 thin films on a SiO_2/Si substrate using a V_2O_5 sol to study the valence reduction process from V_2O_5 to VO_2 thin films. The initial V_2O_5 films were obtained by spin coating. Then, the films were treated by various rapid thermal annealing (RTA) conditions in the temperature range of 400°C–600°C and their X-ray diffraction (XRD) patterns were collected to study the valence reduction process. From the XRD results (Figs. 10.1a and 10.1b), it was demonstrated that the reduction process of V_2O_5 to VO_2 followed the sequence $V_2O_5 \rightarrow V_3O_7 \rightarrow V_4O_9 \rightarrow V_6O_{13} \rightarrow VO_2$, that is, from the phase V_nO_{2n+1} to VO_2 [38].

Liu et al. [32, 39] prepared thermochromic VO_2 coatings by postannealing treatment of V_2O_5 coatings on a polished fused quartz substrate. A thermodynamic calculation was conducted to direct experiments. The calculation results showed that V_2O_5 decomposed into VO_2 at 550°C under an atmospheric pressure of less than 0.06 Pa. Therefore, they annealed the initial V_2O_5 coatings at a pressure of 0.05 Pa at 450°C, 500°C, 550°C, and 600°C for 6 h

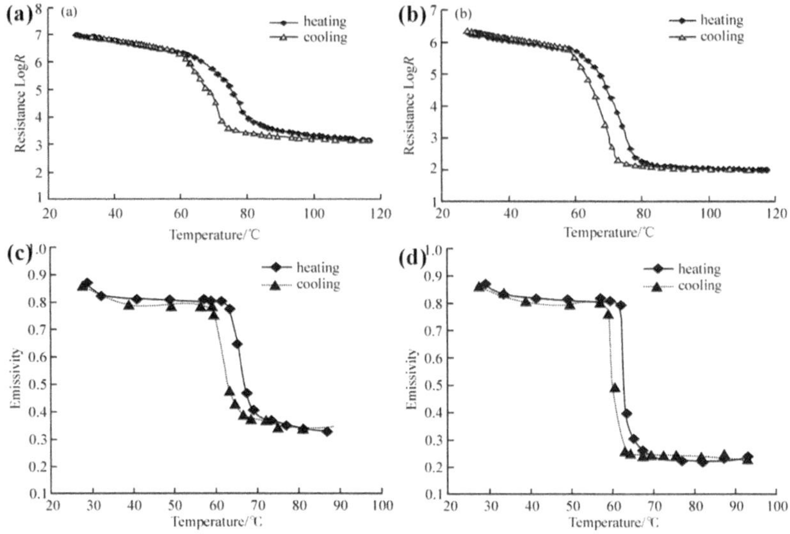

Figure 10.2 Hysteresis curves of the resistance and IR emissivity of the VO_2 coatings on quartz with thickness of (a), (c) 400 nm and (b), (d) 900 nm annealing at 550°C for 10 h under air pressure of 0.05 Pa. Cited from Ref. [39] and reproduced with permission from Springer

and for annealing times ranging from 2 h to 14 h at 550°C. V_2O_5 and V_3O_7 were still observed for the samples annealed at 450°C and 500°C for 6 h. The VO_2 phase appeared at 550°C, with the absence of V_2O_5. Pure polycrystalline VO_2 coatings formed at 600°C for 6 h. Considering the high-temperature-induced coarse surface at 600°C, the VO_2 coatings were synthesized at 550°C for various time. Pure VO_2 coatings were obtained by annealing the sample at 550°C for 10 h. Furthermore, the effect of film thickness on resistance switching and IR chromic properties was investigated. VO_2 coatings with thicknesses of 400 nm and 900 nm on fused quartz substrates were prepared by annealing at 550°C for 10 h under an air pressure of 0.05 Pa. As shown in Figs. 10.2c and 10.2d, the resistance of the 400 nm coatings dropped by ~3 orders of magnitude at 68°C and the half-maximum width of the hysteresis cycle was about 7°C, while the VO_2 coatings of 900 nm exhibited a resistance change of over 4 orders of magnitude at 68°C and a half-maximum width of only

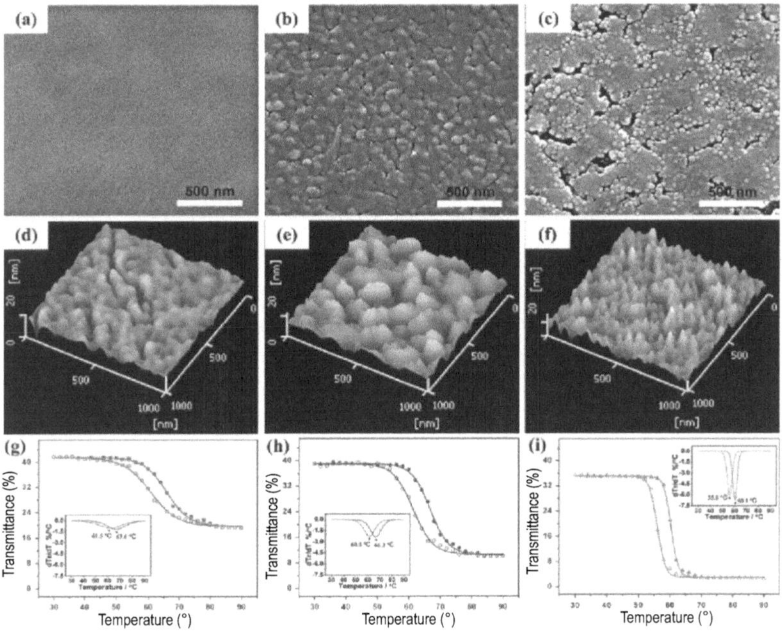

Figure 10.3 SEM images of annealed V_2O_5 coatings (~150 nm in thickness) at (a) 300°C, (b) 500°C, and (c) 700°C. AFM images of VO_2 films annealed at 500°C with thicknesses of (d) 30 nm, (e) 150 nm, and (f) 320 nm. Hysteresis loops of the IR ($\lambda = 25$ μm) transmittance against temperature for VO_2 films with thicknesses of (g) 30 nm, (h) 150 nm, and (i) 320 nm. The insets show the derivative of the temperature dependence of transmittance (dTr/dT). Cited from Ref. [34].

5°C. Accordingly, better IR chromic property (Figs. 10.2e and 10.2f) was observed on the thicker VO_2 coatings with a higher IR emittance variability of 0.6 and a narrower hysteresis half-maximum width of 3°C than on 400 nm VO_2 coatings (0.5 in IR emittance variability and a hysteresis half-maximum width of 4°C) [39].

Shi et al. [34] designed a series of experiments to investigate the effect of annealing temperature, annealing time, and film thickness on the grain size and morphology of the final VO_2 coatings. Dip coating was used to prepare the initial V_2O_5 coatings on a Si substrate, and postannealing treatment was conducted in a N_2 atmosphere. Figure 10.3a–c shows the SEM images of the coatings annealed at 300°C, 500°C, and 700°C, respectively. Extremely small grains were observed on the coatings annealed at 300°C, where only slight crystallization occurred. As the annealing temperature was increased to 500°C, a homogeneously distributed coating with the grain size of 120 nm formed. With a further increase in the annealing temperature, to 700°C, a film assembled by agglomerated nanoparticles with the grain size of 30 nm was prepared, which was explained by the recrystallization of the films at high temperatures. Subsequently, VO_2 films were prepared at 500°C for different annealing times (20 min, 40 min, 60 min, and 90 min). No obvious effect of the annealing time was observed on the grain size, morphology, and phase structure of the films, but annealing for a longer length of time proved to be favorable for the formation of a more stoichiometric VO_2 phase. The effect of thickness on surface morphology was observed. AFM images of 30 nm, 150 nm, and 320 nm thick films in Fig. 10.3d–f indicate that the grain size becomes smaller as the film thickness increases, along with the evolution of a compact and homogenous film. Moreover, the IR thermochromic properties of the VO_2 films of different thicknesses were compared, as shown in Fig. 10.3g–i. It was found that as the film thickness increased, the ability of the films to modulate the IR transmittance gradually improved and the transition temperature dropped.

Li et al. [35] further investigated the stability of VO_2 thermochromic coatings prepared by the inorganic sol-gel method on exposure to air. The as-prepared VO_2 thermochromic coatings were treated by annealing in air at different temperatures and for different lengths of time, and the morphologies and IR switching properties of the resulting coatings were compared. Figure 10.4a–g shows the SEM images of as-prepared VO_2 coatings and coatings annealed at 200°C, 250°C, 300°C, 350°C, 370°C, and 400°C in air for 20 min, respectively. The grain growth was observed for the VO_2 coatings obtained at 300°C. And the grainy structure became more obvious, with a reduction in the mesoporous structure. When the annealing temperature was over 300°C, the grain size continued to increase and the crystals varied from grainy to a bar-like structure,

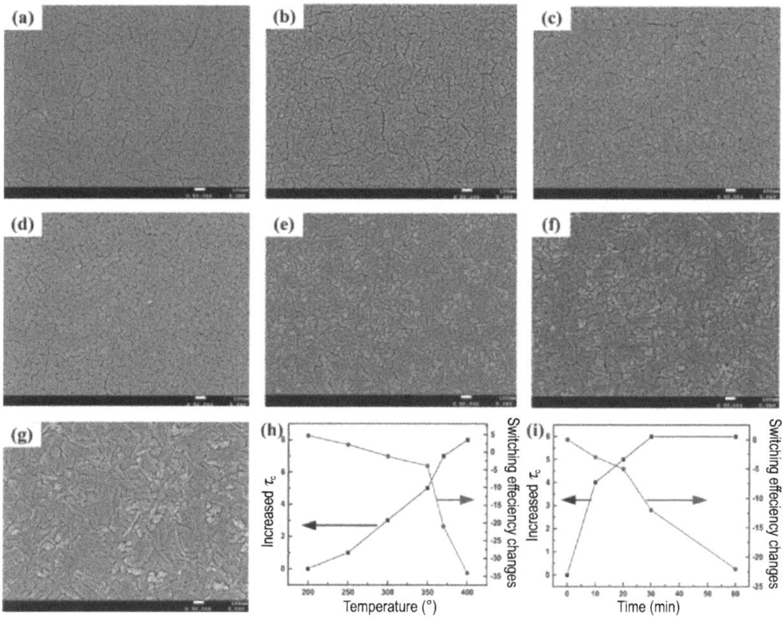

Figure 10.4 SEM images of (a) as-prepared VO_2 coatings and coatings after annealing at (b) 200°C, (c) 250°C, (d) 300°C, (e) 350°C, (f) 370°C, and (g) 400 °C in air for 20 min. (h) The air annealing temperature and (i) annealing time (350°C) evolution of the increased τ_c and the variation in the IR switching efficiency. Cited from Ref. [35].

resulting in more compact coatings. Notably, a new phase appeared when the annealing temperature was above 300°C, indicating the oxidation of VO_2 into V_2O_5 in air. The response to these changes in morphology and phase structure, the transition temperature (τ_c), and the IR switching efficiency varied. It was exhibited (Fig. 10.4h,i) that T_t increased and the IR switching efficiency decreased as the annealing temperature rose. Moreover, similar results were obtained with increasing annealing time at 350°C. These results indicate that VO_2 thermochromic coatings are quite stable in air below 200°C and further increase in temperature induces the oxidation of VO_2, along with a deterioration in the IR thermochromic properties.

10.3.2 Organic Sol-Gel Method

Generally, the chemistry of an organic sol-gel method involves the hydrolysis and condensation of an organic precursor to form a sol, the aging process for the formation a vanadium polymer gel, and the postannealing process to obtain crystalline VO_2 coatings [26, 36, 37]. Various vanadium precursors have been used in the organic sol-gel process to synthesize VO_2 coatings, such as vanadium alkoxides [29], VO(acac) [26], and inorganic V_2O_5, etc. [24].

A series of novel vanadium dioxide nanostructures, including metastable monoclinic VO_2(B) nanoneedles and nanocorals, orthorhombic VO_2(O) nanoparticles, and monoclinic VO_2(M) nanofacets, were successfully prepared by Wang and Chen [41] by controlling the aging time and the postannealing conditions in the sol-gel process with $VO(acac)_2$ as the precursor. Table 10.2 [41] exhibit the aging and post–heat treatment conditions to obtain the four VO_2 nanostructures.

Wu et al. [24] presented an organic sol-gel method to fabricate thermochromic VO_2 coatings on mica substrates by using low-cost V_2O_5 powders, isobutyl alcohol, and benzyl alcohol as the raw materials. This process was classified as an organic sol-gel process because of the generation of an organic vanadium precursor and

Table 10.2 The aging and post–heat treatment conditions to obtain various VO_2 nanostructures, including VO_2(B) nanocorals, nanoneedles, VO_2(O) nanoparticles, and VO_2(M) nanofacets

Aging treatment	Heat treatment	Products
#1 Aging for 24 h	550°C for 30 min + 600°C for 5 min in H_2/N_2	VO_2(B) nanocorals
#2 No aging	550°C for 30 min + 600°C for 5 min in H_2/N_2	VO_2(B) nanoneedles
#3 No aging	550°C for 30 min + 600°C for 5 min in vacuum	VO_2(O) nanoparticles
#4 No aging	#2 + 600°C for 4 h + air annealing thermal treatment in ambient atmosphere	VO_2(M) nanofacets

subsequent hydrolysis and condensation of the organic precursor to form an organic sol. The chemical reactions involved are listed as follows [24]:

1. Generation of an organic precursor

$$4C_4H_{10}O + V_2O_5 \rightarrow 2VO(i - OC_4H_9)_2 + 2H_2O$$

$$\text{Or} \quad 6C_4H_{10}O + V_2O_5 \rightarrow 2VO(i - OC_4H_9)_3 + 3H_2O$$

2. Hydrolysis reaction of the organic precursor solution

$$2VO(i - OC_4H_9)_3 + 2H_2O \rightarrow 2V(OH)O - (C_4H_9)_2 + 2C_4H_{10}O$$

3. Condensation of the organic hydrolysis product

$$2V(OH)O - (C_4H_9)_2 \rightarrow V_2O_3(OC_4H_9)_2 + H_2O$$

The effect of annealing temperature on the thermochromic properties of the as-deposited VO_2 coatings was investigated in the annealing temperature range of 440°C–540°C with an interval of 20°C. The thermo-optical hysteresis curves with inset $\mathrm{dTr}/\mathrm{d}T$-T plots for the VO_2 films annealed between 440°C and 540°C are exhibited in Fig. 10.5a–f, respectively. It was found that the coatings annealed at 500°C exhibited not only the narrowest hysteresis width ($\Delta\tau_c$), of 8°C, compared to those of the other coatings (~18°C) but also the highest IR modulation ΔT_{IR}, of 73%, and transmittance switching from 76% at 40°C to 3% at 90°C.

Yu et al. [26] studied the optical switching property of a VO_2 thin film fabricated using a $VO(acac)_2$ precursor via a sol-gel method by varying the thermal aging time of the precursor. They found that the aging process significantly improved the coating ability with enhanced adhesivity. Moreover, the morphology and optical switching property of the deposited films were affected. Figure 10.6a–d shows the SEM images of the VO_2 coatings evolved under different aging conditions. It was observed that the aging process significantly affected the morphology of the coatings. Nonaged coatings were composed of sparse particles, with a low density. In contrast, as the aging time increased, coral-like grains appeared and compact coatings were formed. Furthermore, the effects of

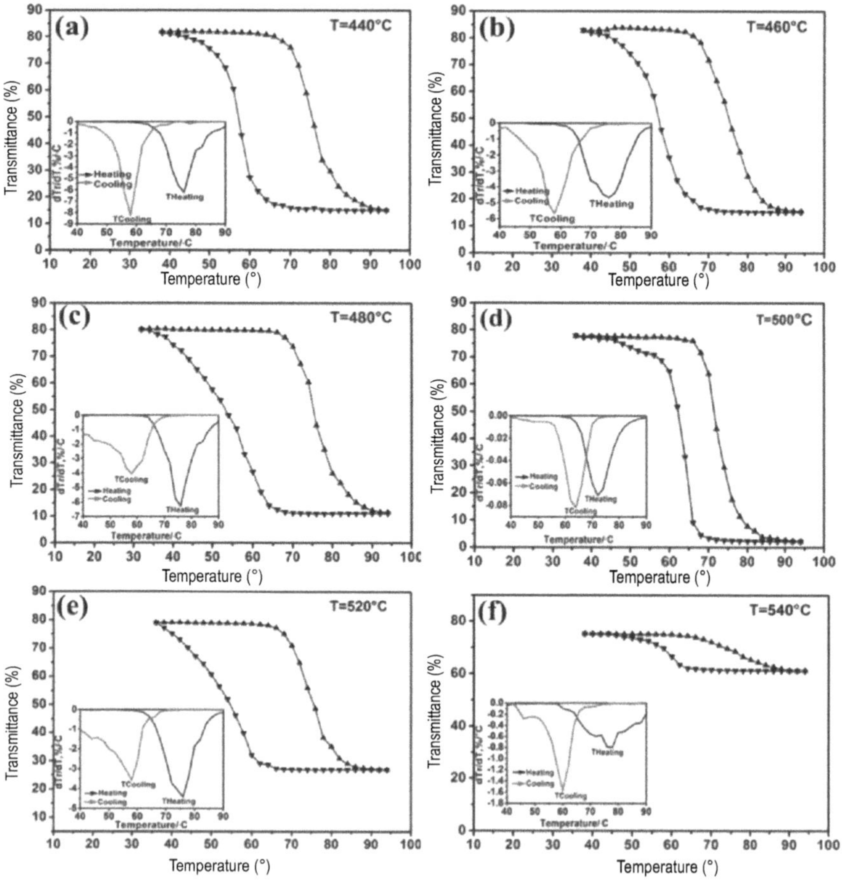

Figure 10.5 Hysteresis loop of transmittance at 4.3 μm versus temperature with inset dTr/dT-T plots for the VO_2 films annealed between 440°C and 540°C: (a) 440°C, (b) 460°C, (c) 480°C, (d) 500°C, (e) 520°C, and (f) 540°C. Cited from Ref. [24].

aging time on the optical switching properties of VO_2 coatings were investigated. As shown in Figs. 10.6e and 10.6f, the visible transmittance (T_{lum}) at 25°C decreased from ~90% to ~45% with the aging time increasing from 0 day to 7 days, while the IR switching efficiency (ΔT_{2000nm}) of the coatings aged for 7 days reached a maximum value of 51%.

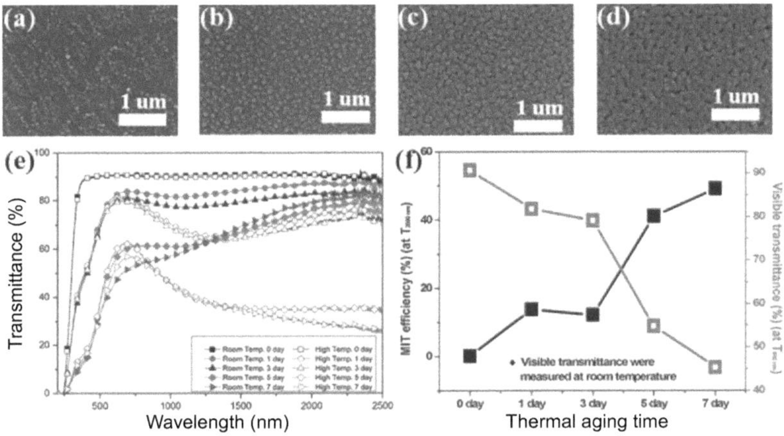

Figure 10.6 SEM images of postannealed VO_2 coatings fabricated with aging time variations: (a) Nonaged, (b) 3 days, (c) 5 days, and (d) 7 days. (e) The optical transmittance spectra and (f) the MIT efficiency and visible transmittance of VO_2 coatings evolved with aging time variations. The MIT efficiency was calculated by the difference at high temperature and room temperature at 2000 nm. Visible transmittances were measured by room temperature value at 600 nm. Cited from Ref. [26].

10.4 Sol-Gel Strategies for Improved Thermochromic Properties

Considering the primary aim of building fenestration, a smart thermochromic window should meet three key criteria: (i) high transparency in the visible spectrum for indoor natural light conservation, (ii) stable and optimized ability to modulate the incident IR light, and (iii) appropriate transition temperature close to room temperature. Great efforts have been made to optimize the thermochromic properties of VO_2 coatings, including morphology [25, 42] and structure designing [11–13], doping [6, 43–45], composites [16, 46, 47], and surface engineering [48]. Recent studies have demonstrated heteroatom doping to be an effective way to lower the phase transition temperature to room temperature [15, 44]. Moreover, the construction of a composite coating with a visible light antireflection effect has proved to boost the visible light transmittance without weakening the thermochromic properties [2–4].

10.4.1 Doping

The MIT phase transition temperature is closely related with the dimerization of the V–V chain VO_2 [49]. The element dopants can introduce electron or hole doping and internal strain/stress, which all affect the electronic structure and transition temperature of VO_2 [2, 19]. In the frame of classical thermodynamics, the dependence of transition temperature τ_c on the uniaxial stress σ can be understood following the Clausius–Clapeyron equation [6]:

$$d\tau_c/d\sigma = (\varepsilon_0 \tau_c^0)/\Delta H, \tag{10.1}$$

where ε_0 is the initial strain, τ_c^0 is the original phase transition temperature, and ΔH is the latent heat of the phase transition. Introduction of higher-valence-state cations (compared to V^{4+}) into the crystal lattice of VO_2 leads to extra electrons, which can induce a decrease in ΔH due to the W-induced dimerization of the V–V chain in the Rutile phase [5, 6], thereby reducing τ_c according to the relation $\tau_c = \Delta H/\Delta S$. Besides, substituting V^{4+} by cations of larger ionic radii or inserting cations at interstitial sites of the VO_2 lattice can introduce compressive strain and induce structural deformation, which reduces τ_c as well [5, 6, 50]. The sol-gel method has been demonstrated as a low-cost and feasible way to synthesize doped VO_2 coatings [1, 19]. VO_2 coatings doped by W^{6+} [44, 45], Mo^{6+} [15, 31, 51], Ti^{4+} [52], and Zr^{4+} [53] and codoped by W^{6+}/Mo^{6+} [50] and Mg^{2+}/W^{6+} [6] have been successfully synthesized by the sol-gel method.

W^{6+} is known to reduce the thermochromic switching temperature of VO_2 by the greatest extent per atomic percentage (20°C–26°C per at.%). Liang et al. [45] reported the synthesis of self-cleaning W-doped thermochromic VO_2 coatings on a fused quartz substrate via an organic sol-gel process with the addition of tungsten chloride as the tungsten source. The effects of doping with different levels (0–1.5 at.%) of tungsten were investigated on the thermochromic properties of VO_2 coatings. Figure 10.7 shows the optical transmittance spectra of VO_2 coatings doped with various

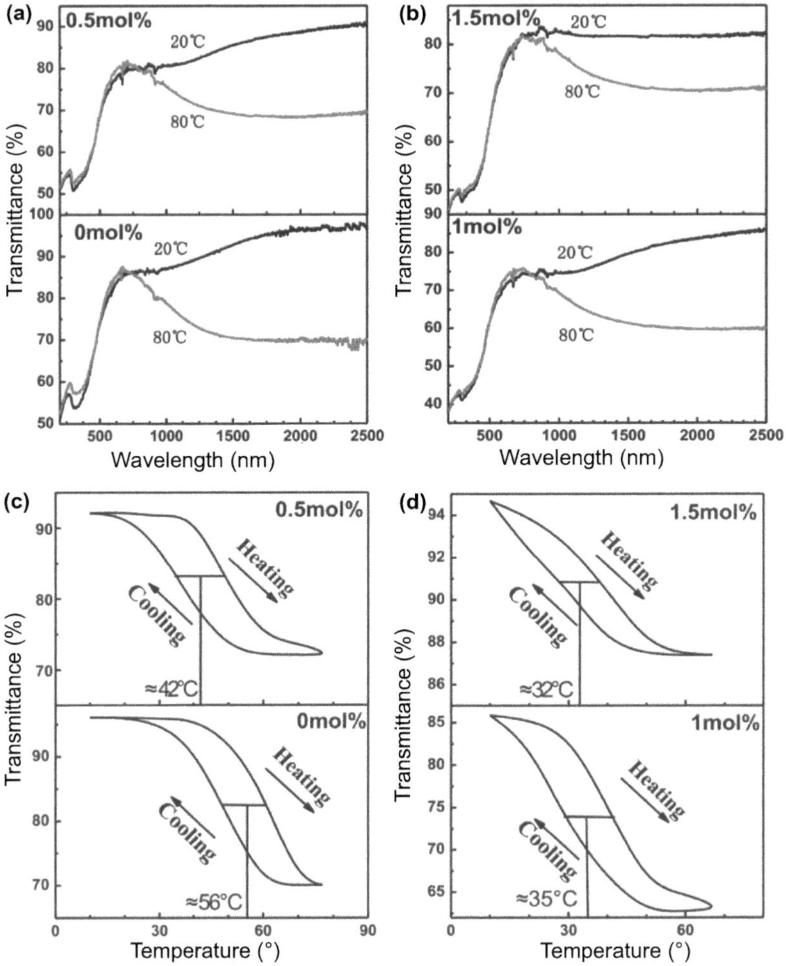

Figure 10.7 Thermochromic optical properties of W-doped VO_2 coatings with doping levels of 0–1.5 at.%. (a), (b) the optical transmittance spectra in the wavelength range of 200–2500 nm, and (c, d) the corresponding thermal hysteresis loops at the wavelength of 2000 nm. Cited from Ref. [45].

levels of W in the wavelength range of 250–2500 nm and the corresponding thermal hysteresis loops at the wavelength of 2000 nm. It was observed that W doping exhibited a significant effect on the thermochromic properties of VO_2 coatings, especially the transition temperature as the doping level increased.

Table 10.3 [45] summarizes the thermochromic optical properties of W-doped VO_2 coatings. The table shows that high values of visible light transmittance T_{lum} of 80.6% in the semiconductor state and 79.2% in the metallic state were realized for undoped VO_2 coatings. T_{lum} decreased as the doping level increased, and reached a minimum value of 71.6%/70.1% at W doping of 1 at.%. An increase in T_{lum} for VO_2 coating with 1.5 at.% W doping was caused by a reduction in the thickness from 392 nm to 259 nm. At a wavelength of 2000 nm, both solar light modulation ΔT_{sol} and IR light modulation ΔT_{2000nm} were reduced as the doping level increased. As the doping level reached 1.5 at.%, the ΔT_{sol} became 6.1% and the ΔT_{2000nm} became 10%, compared with those of undoped VO_2 (9.1% and 26%, respectively). It is worth noting that the phase transition temperature τ_c significantly dropped as the doping level increased, reducing almost 20°C per W at.% from the initial 56°C of the undoped VO_2. Phase transition temperatures of 42°C, 35°C, and 32°C were realized with W doping levels of 0.5 at.%, 1 at.%, and 1.5 at.%, respectively.

Hu et al. [44] investigated the thermochromic properties of porous W-doped VO_2 coatings prepared from an aqueous vanadium solution/polyvinylpyrrolidone (PVP) sol. Ammonium tungstate hydrate ($H_{40}N_{10}O_{41}W_{12} \cdot xH_2O$) was selected as the W doping source. The transmittance spectra at 20°C (solid lines) and 90°C (dashed lines) in the wavelength of 250–2500 nm and the corresponding inset SEM images and d(Tr)/d(T)-T plots at the wavelength of 2000 nm for porous W-doped VO_2 coatings on silica substrates with different doping levels are exhibited in Fig. 10.8. All the W-doped VO_2 coatings showed similar porous structures with favorable holes for visible light transmittance. It was found that all films exhibited significant optical modulation in the IR region. The IR transmittance change decreased with an increase in the W doping level. However, the visible light transmittance was hardly affected by the increased W doping level, remaining at ~60%. According to the

Table 10.3 Thermochromic properties of VO_2 coatings with different levels of W doping

W (at.%)	Thickness (nm)	T_{lum} (%)	ΔT_{lum} (%)	T_{sol} (%)	ΔT_{sol} (%)	ΔT_{2000nm} (%)	$\Delta \tau_c$ (°C)	τ_c (°C)
0	226	80.6/79.2	−1.5	81.4/72.3	9.1	26	22	56
0.5	234	78.9/77.8	−1.0	79.5/71.2	8.3	20	15	42
1.0	392	71.6/70.1	−1.5	71.7/63.2	8.6	22.7	16	35
1.5	259	74.5/74.3	-0.2	76.7/70.6	6.1	10	12	32

Table 10.4 Thermochromic optical parameters of W-doped VO_2 coatings with increased doping levels from 0 at.% to 2 at.%

W (at. %)	$T_{vis-max}$ (%)	T_{lum} (%)	ΔT_{lum} (%)	T_{sol} (%)	ΔT_{sol} (%)	ΔT_{2000nm} (%)	$\Delta \tau_c$ (°C)	τ_c (°C)
0	57.3/58.8	44.4/45	-0.6	50.9/41.5	9.4	52.9	20	55
0.5	63.1/64.2	49.3/51.6	-2.3	55.7/48.5	7.2	45.9	15	45
1	57.5/58.9	44.8/47.4	-2.6	50.2/44.2	6	38.5	12	35
2	58.4/59.2	44.6/45.7	-1.1	49.4/42.5	6.9	37	15	28

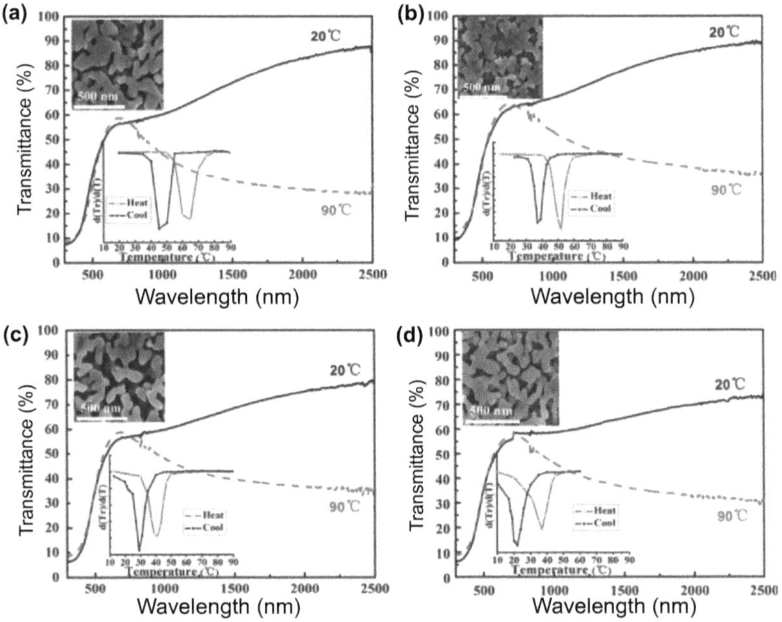

Figure 10.8 Transmittance spectra at 20°C (solid lines) and 90°C (dashed lines) in the wavelength of 250–2500 nm and the corresponding inset SEM images and d(Tr)/d(T)-T plots at the wavelength of 2000 nm for porous W-doped VO_2 coatings on silica substrates with different doping levels: (a) 0 at.%, (b) 0.5 at.%, (c) 1 at.%, and (d) 2 at.%, respectively. Cited from Ref. [44].

transmission spectra, the IR modulation ΔT_{2000nm} at the wavelength of 2000 nm and solar spectrum modulation ΔT_{sol} are given in Table 10.4 [44]. Both decreased with increasing W doping levels. A significant effect of W doping was observed on the phase transition temperature. Compared to the undoped porous VO_2 coatings with a transition temperature of 55°C, the doped ones, with W doping levels of 0.5 at.%, 1 at.%, and 2 at.%, exhibited respective transition temperatures of 45°C, 35°C, and 28°C, which was close to room temperature (20°C–30°C).

Although a significant decrease in the critical transition temperature has been achieved even close to room temperature by doping with high-valence cations with large ionic radii, such as W^{6+} and Mo^{6+}, a deterioration in visible transmittance and solar modulation is observed [44, 45]. Therefore, a dopant that can reduce the transition temperature without sacrificing the luminous transmittance is eagerly awaited. According to the Clausius–Clapeyron equation (Eq. 10.1), Zr^{4+}, with a larger ion radius (r = 0.71 Å and coordination number [CN] = 6), rather than V^{4+} (r = 0.58 Å and CN = 6) [54], has the probability to lower the critical transition temperature by causing lattice distortion in VO_2. In addition, Zr dopants could widen the optical gaps of VO_2, resulting in enhanced visible transmittance [5]. The positive effect of Zr^{4+} doping has been demonstrated on the thermochromic optimization of VO_2 coatings. Lu et al. [53] prepared Zr-doped VO_2 coatings on a glass substrate through the sol-gel method by using $VO(acac)_2$ as the vanadium precursor and $Zr(NO_3)_4$ as the Zr^{4+} source. The effect of Zr doping on the metal–semiconductor transition and thermochromic transmittance switching was investigated. It was found that as the doping level of Zr^{4+} increased up to 2 wt.%, the critical temperature of phase transition decreased to 50°C, with little variation in the visible light transmittance [53].

Ti^{4+} is another kind of dopant for thermochromic VO_2 coatings, which can increase the critical phase transition temperature but does not change the transmission/absorption characteristics of the VO_2 coatings [52, 55]. Hu et al. [52] used the hybrid of an inorganic V_2O_5 sol and an organic titanium sol to synthesize Ti-doped VO_2 coatings with different Ti concentrations. Figure 10.9a shows the IR transmittance spectra of an undoped VO_2 film and a VO_2 film doped with 2.8 at.% Ti at 40°C and 90°C. Little difference was found in the transmittance spectra of the doped and undoped VO_2 coatings, which indicates the zero effect of Ti doping on the IR transmittance switching of VO_2 coatings. Besides, as shown in Fig. 10.9b, the critical transition temperature increases and the hysteresis width decreases as the Ti doping concentration increases. The special functions of Ti doping provide a way to adjust the critical phase transition temperature of VO_2 in the range above the original 68°C without influencing the IR transmittance switching properties.

Considering the different effects of dopants, codoping may provide a way to satisfy both decreasing the τ_c and maintaining an acceptable T_{lum}, favoring access to VO_2 coatings with sufficient

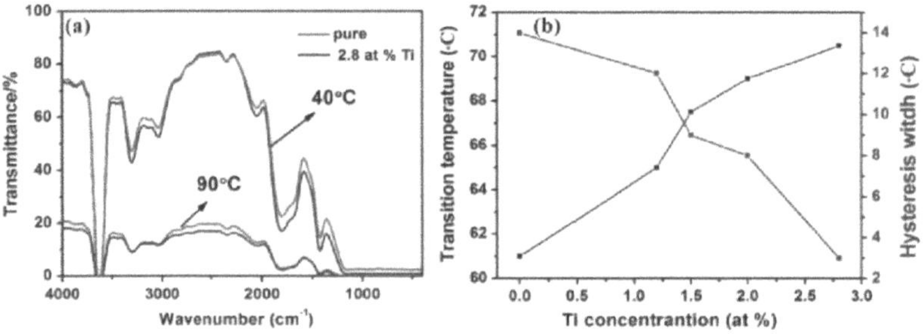

Figure 10.9 (a) IR transmittance spectra of undoped VO_2 coating and VO_2 coating doped with 2.8 at.% Ti at 40°C and 90°C. (b) Variation in the transition temperature and hysteresis width of VO_2 films as a function of Ti doping concentration. Cited from Ref. [52].

luminous transmittance and solar spectrum modulation at rather low transition temperatures.

Wang et al. [6] prepared Mg/W-codoped VO_2 coatings on a silica substrate through a sol-gel method with PVP as a cross-linking reagent. In this study, the large τ_c-reducing rate of W and the superior T_{lum}-enhancing ability of Mg were simultaneously demonstrated on the Mg/W-codoped VO_2 coatings, resulting in a good combination of low τ_c of ~35°C and a high T_{lum} of 81.3%. The UV-vis-NIR transmittance spectra of Mg/W-codoped VO_2 coatings were recorded at 15°C and 90°C (Fig. 10.10a). A decrease in the transmission was observed for all of the samples above the wavelength of 1000 nm upon heating from 15°C to 90°C, in accordance with the intrinsic MIT characteristics of VO_2. Among all the samples, the single Mg-doped $V_{0.95}Mg_{0.05}O_2$ exhibited the highest T_{sol} in the entire spectrum (250–2500 nm) at both temperatures, especially in the visible range, indicating the ability of Mg to enhance T_{lum}. On the other hand, the single W-doped $V_{0.98}W_{0.02}O_2$ showed the lowest phase transition temperature (27.05°C) and the lowest T_{sol} (250–2500 nm) at the same time, revealing the weakness of W as a dopant. On doping with 2 at.% W, the T_{lum} increased with an increase the Mg doping concentration, and the optimal T_{lum} was realized on $V_{0.94}W_{0.02}Mg_{0.04}O_2$, with the lowest visible-light consumption by reflection (%R) and absorption (%A) (Fig. 10.10b). Figure 10.10c show hysteresis loops for the temperature-dependent transmittance

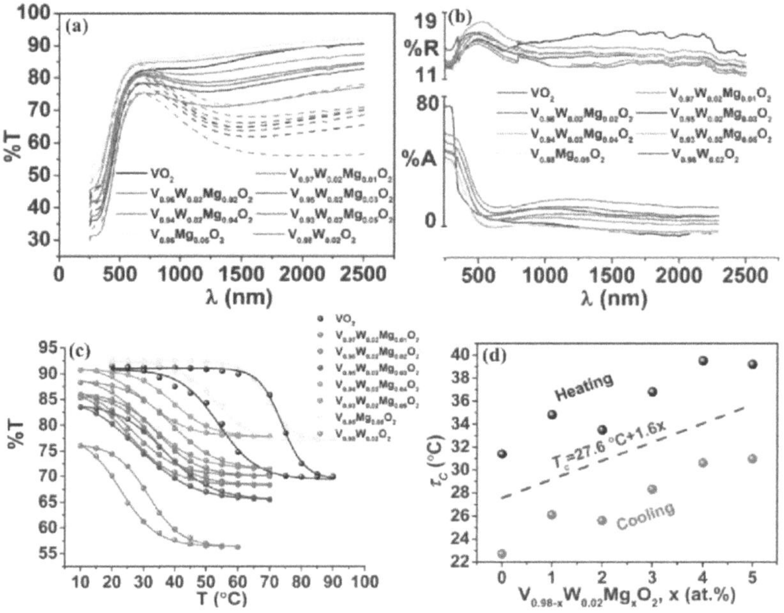

Figure 10.10 (a) Transmittance spectra and (b) recorded reflection and absorption spectra of pristine and Mg/W-codoped VO_2 coatings in the UV-vis-NIR range at 15°C (solid lines) and 90 °C (dashed lines). (c) Hysteresis loop for the temperature-dependent transmittance of the pristine and Mg/W-codoped VO_2 coatings at a wavelength of 2500 nm. (d) Dependence of τ_c on Mg doping concentration on the basis of doping with 2 at.% W with a fitted linear relationship. Cited from Ref. [6].

of pristine and Mg/W-codoped VO_2 coatings at a wavelength of 2500 nm, and the corresponding dependence of τ_c on Mg doping concentration was investigated. As shown in Fig. 10.10d, the τ_c exhibited a tendency to increase, where continuous Mg doping in the presence of a 2 at.% W dopant resulted in a slight increase in τ_c from 27.05°C ($V_{0.98}W_{0.02}O_2$) to 35.08 °C ($V_{0.93}W_{0.02}Mg_{0.05}O_2$) at a rate of 1.6°C/at.%.

In addition, Mo/W [50, 56] and Ti/W [57] codoping has been carried out to tune the τ_c of VO_2. Mo/W-codoped VO_2 thin films exhibited an interesting nonlinear effect between co- and single-doped materials [56]. The co-doped VO_2 films showed higher τ_c-lowering efficiency than those single-element-doped films [56].

Ti/W codoping induced a higher τ_c in the material than single W doping because of the decrease in the carrier density arising from the electron transfer from the donor W to the acceptor Ti, while the transmission was largely compromised compared with that of pure VO_2 [57]. Therefore, codoping with proper metal ions can simultaneously realize low τ_c and high luminous transmittance through a synergetic effect.

10.4.2 VO_2-Based Composite Coatings

Positive ways to enhance the visible light transmittance of VO_2 thermochromic coatings are the need of the hour, though improvements can be made by codoping. The introduction of antireflection coatings (ARCs) on the surface of VO_2 coatings by using SiO_2, Al_2O_3, TiO_2, etc., has attracted a lot of attention because of the impressive improvement they confer in terms of the visible light transmittance of VO_2 coatings [1, 2, 4]. Zhang et al. [16] presented a way to prepare a mesoporous SiO_2/VO_2 composite coating on a quartz substrate through a low-cost sol-gel method. The preparation procedure is exhibited in Scheme 10.1, where the VO_2 film was first prepared by an organic sol-gel method by using ammonium

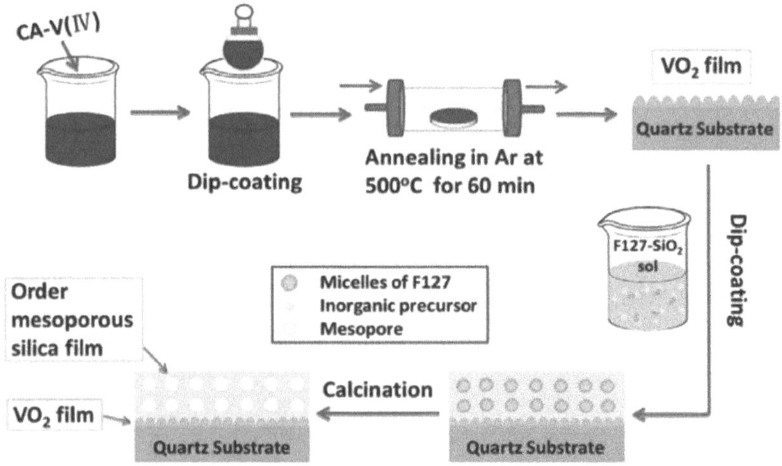

Scheme 10.1 The preparation procedure of mesoporous SiO_2/VO_2 coatings on quartz substrates. Cited from Ref. [16].

Table 10.5 The thermochromic properties of VO_2/Si-Al gel nanocomposite coatings with different solid content ratios of VO_2

	T_{lum} (%)		T_{sol} (%)		
Content of VO_2 (wt.%)	**20°C**	**90°C**	**20°C**	**90°C**	**ΔT_{sol} (%)**
2.5	80.3	81.9	76.7	76.7	0
5	69.8	69.2	67.1	61.6	5.5
10	59.5	57.1	59.2	51.3	7.9
12.5	55.6	53.4	56.9	50.1	6.8
15	56	56.3	57.1	55.5	1.6

citrato-oxovanadate (CA-V(IV)) as a precursor. After the formation of the pristine VO_2 coating, a block copolymer Pluronic F127-SiO_2 sol was dip-coated on it. Mesoporous SiO_2/VO_2 coatings were obtained after the removal of F127 through sufficient calcination. At refining conditions with the proper refractive index (1.299) and porosity of SiO_2, the integral visible transmittance at 25°C ($T_{vis,L}$) and 90°C ($T_{vis,H}$) increased to 80% and 78.9%, compared with 69.8% and 67.6%, respectively, of pure VO_2 coating. Simultaneously, the high IR switching ability ΔT_{2000nm} of 29% and solar spectrum modulation ΔT_{sol} of 10.2% were effectively maintained.

Liu et al. [47] prepared VO_2/Si-Al gel nanocomposite coatings by a sol-gel method and investigated their thermochromic properties. Table 10.5 lists the thermochromic properties of VO_2/Si-Al gel nanocomposite coatings with different solid content ratios of VO_2. The table indicates that both luminous transmittance T_{lum} and integral solar light transmittance T_{sol} decreased with an increasing VO_2 content ratio because of the poor transmittance of VO_2 in the visible light range. The solar spectrum modulation ΔT_{sol} was enhanced as the content of VO_2 increased, and reached a maximum value of 7.9% at the VO_2 content of 10 wt.%, after which the ΔT_{sol} dropped against the increase of VO_2 content, caused by the unfavorable optical transparency and large particle agglomeration. The thermochromic properties of VO_2/Si-Al nanocomposite coatings of different thicknesses are summarized in Table 10.6. It is seen that both T_{lum} and T_{sol} decreased as the VO_2 coating thickness increased from 1 μm to 3 μm, with the ΔT_{sol} continuously improving. A ΔT_{sol} as high as 12% was realized for the coating with a thickness of 3 μm,

Table 10.6 The thermochromic properties of VO_2/Si-Al nanocomposite coatings of different thicknesses

	T_{lum} (%)		T_{sol} (%)		
Thickness (μm)	20°C	90°C	20°C	90°C	ΔT_{sol} (%)
1	78.7	79.4	74.9	74.4	0.5
1.4	71.1	71.2	68.1	65.2	2.9
2	72	69.4	68.8	63.4	5.4
3	63.7	54.4	61.9	49.9	12

and the corresponding T_{lum} of 63.7% at 20°C and 54.4% at 90°C remained.

Suzuki et al. [58] produced a composite VO_2-based film by mixing polylactic acid with VO_2-coated SiO_2 monodispersed particles via a sol-gel process. VO_2/mica pigments were fabricated by He and coworkers [59] through an aqueous sol-gel process. The composite coatings showed good thermochromic properties. Efficient use of vanadium raw materials and good modulation of their optical performance were demonstrated in VO_2-coated matrix particles [59]. Other VO_2-SiO_2 composite coatings via the sol-gel method were reported by Chen and coworkers [60]. Precursors of $VO(OR)_3$ and $Si(OR)_4$ in Si/V volumetric mixing ratios ranging from 0 to 1 were used, followed by spin coating and heat treatment at 600°C for 15 min in reducing atmospheres. The as-obtained VO_2-SiO_2 coatings showed a higher luminous transmittance than that of the pristine VO_2 film (33%). The highest T_{lum}, of 75%, was realized for the VO_2-SiO_2 coating at the Si/V ratio of 0.05. The presence of SiO_2 in the VO_2 coatings not only enhanced the luminous transmittance but also maintained a high IR modulation, ΔT_{3200nm}, of 50%. Therefore, high optical modulation and sufficient luminous transmittance can be simultaneous achieved by constructing VO_2-based composite coatings via the sol-gel method.

Besides ARC composites, new composites have been constructed by uniformly dispersing highly crystalline VO_2 nanoparticles in a flexible and highly transparent polymer gel matrix [61]. Scheme 10.2 shows the schematic synthesis procedure of flexible VO_2/polymer composites and their optical switching mechanism, in which VO_2 nanoparticles in a polymer gel matrix perform thermal optical

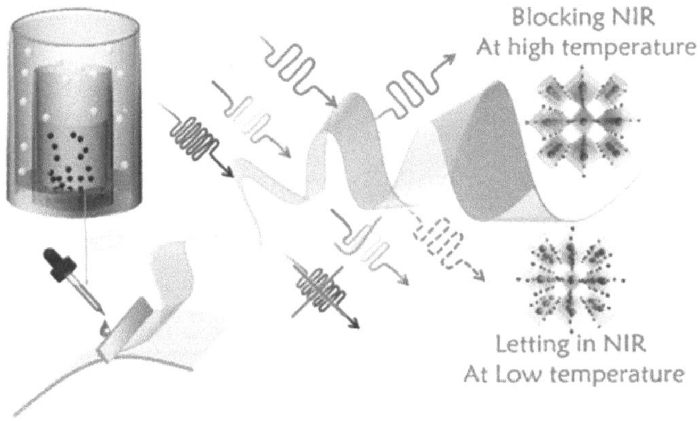

Scheme 10.2 The schematic synthesis procedure of flexible VO_2/polymer composites and their optical switching mechanism. Cited from Ref. [61].

switching of the solar IR heat, while the transparent polymer still maintains high visible light transmittance [61]. In this way, VO_2 nanoparticles with a high optical switching performance can be rationally synthesized by simply tuning the weight ratio of polymer/VO_2 and thickness of the flexible films. Guo et al. [61] first synthesized highly crystalline ultrasmall (~30 nm) VO_2(M) nanoparticles by a facile hydrothermal method. The as-obtained VO_2(M) nanoparticles were dispersed in a polyacrylonitrile solution to reach a uniform sol. By dip coating, a flexible VO_2/PAN composite film was finally peeled from a polyethylene terephthalate substrate (Fig. 10.11e). Figure 10.11a–d shows scanning electron microscope and transmission electron microscope images of the VO_2(M) nanoparticles, demonstrating the small size and fine crystalline M phase. Figure 10.11f exhibits the transmittance spectra of flexible VO_2/PAN composite films of different thicknesses and different VO_2 particle sizes. The integrated luminous transmittance (T_{lum}, 380–780 nm) of the flexible VO_2/PAN composite film with the thickness of ~1.1 μm (1#) was 54.26%, and the solar modulating ability (ΔT_{sol}, 300–2500 nm) was 12.34% (T_{sol} ~63.42% at 30°C and ~51.08% at 90°C, respectively), indicating a desirable thermochromic performance. The thicker films displayed a better solar modulating ability, with a ΔT_{sol} of ~14.34% for 2# (~2.3 μm thick) and of ~13.25% for 3# (~4 μm thick), but the degradation

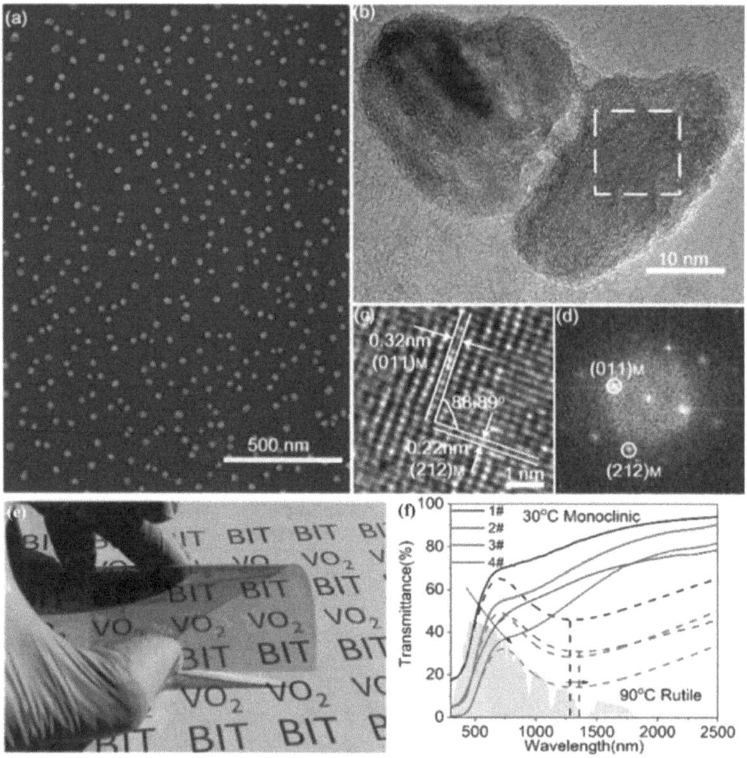

Figure 10.11 (a) SEM, (b) TEM, (c) HRTEM images, and (d) SAED patterns of the VO_2(M) nanoparticles. (e) Picture of the flexible VO_2/PAN composite film and (f) transmittance spectra of flexible VO_2/PAN composite films (1#, thickness ~1.1 μm; 2#, thickness ~2.3 μm; 3#, thickness ~4 μm with a VO_2 nanoparticle size of ~30 nm; and 4#, thickness ~2.7 μm with a VO_2 nanoparticle size of ~79 nm). The real lines denote the transmittance spectra measured at 30°C, and the dashed lines show the transmittance spectra measured at 90°C. Cited from Ref. [61].

of integrated luminous transmittance was observed with a T_{lum} of ~37.19% and ~20.96%, respectively. For 4# (~2.7 μm thick) with a VO_2 nanoparticle size of ~79 nm, after annealing treatment, a weakened thermochromic performance (T_{lum}, ~31.49%; ΔT_{sol}, ~9.38%) was observed because of its large-sized particle scattering solar light.

10.5 Conclusion

Literature demonstrates that the sol-gel method is a facile wet-chemical way to prepare thermochromic VO_2 coatings, which is applicable to various substrates and convenient for heteroatom doping and composite coating construction. Three typical types of vanadium gels were discussed: a colloidal gel from melting-quenching of V_2O_5 in water, a vanadium-oxane polymer complex formed by the hydrolysis and condensation of the organic precursors such as VO(acac) and vanadium alkoxides, and a coordinated and cross-linked polymer complex obtained by a vanadium solution with the addition of a polymer as the cross-link reagent. VO_2 coatings with different morphologies and thermochromic properties can be synthesized by controlling the preparation conditions, including types of gels, aging time, and postannealing parameters (temperature, time, atmosphere, gas flow rate, etc.). The key to improving the thermochromic properties of VO_2 coatings lies in the reduction of the critical phase transition temperature to room temperature, a high IR transmittance switching ability, and high solar energy modulation, while sufficient visible light transmittance is maintained for daily life. Ionic doping with W^{6+}, Mo^{6+}, Zr^{4+}, etc., and building up the composite construction with ARCs, including SiO_2, TiO_2, and Al_2O_3, have been the main strategies to overcome the shortcomings of VO_2 coatings. W^{6+} doping has exhibited the most positive effect on lowering τ_c, at the rate of over 20°C/at.%, but induces serious deterioration of visible light transmittance and solar modulation ability. Mg^{2+}/W^{6+}-codoped VO_2 coatings by a sol-gel approach achieved the excellent combination of the large τ_c-reducing rate of W^{6+} and the superior T_{lum}-enhancing ability of Mg^{2+}, resulting in a low τ_c of ~35°C and a high T_{lum} of 81.3% at the same time. VO_2-based composite coatings with ARCs retained the best possible optical modulation ability while enhancing transmittance in the visible light range. Mesoporous SiO_2/VO_2 coatings realized an enhanced T_{lum} of 80% (20°C) and 78.9% (90°C) from 69.8% (20°C) and 67.6% (90°C) of the pure VO_2 coating, respectively, with a still high ΔT_{2000nm} of 29% and ΔT_{sol} of 10.2%. Besides, flexible composite films constructed by dispersing

VO_2 nanoparticles in a highly transparent polymer gel matrix have proved to be another effective way toward thermochromic coatings with a fine-tuned solar light modulating ability and considerable visible light transmittance. The sol-gel method has proved to be an effective and low-cost way to develop thermochromic VO_2 coatings with rather satisfactory optical switching properties for practical application in smart thermochromic windows.

References

1. Gao, Y., Luo, H., Zhang, Z., et al. (2012). Nanoceramic VO_2 thermochromic smart glass: a review on progress in solution processing, *Nano Energy*, **1**, pp. 221–246.
2. Wang, S., Liu, M., Kong, L., et al. (2016). Recent progress in VO_2 smart coatings: strategies to improve the thermochromic properties, *Prog. Mater. Sci.*, **81**, pp. 1–54.
3. Cui, Y., Ke, Y., Liu, C., et al. (2018). Thermochromic VO_2 for energy-efficient smart windows, *Joule*, **2**, pp. 1–40.
4. Xu, F., Cao, X., Luo, H., et al. (2018). Recent advances in VO_2-based thermochromic composites for smart windows, *J. Mater. Chem. C*, **6**, pp. 1903–1919.
5. Shen, N., Chen, S., Chen, Z., et al. (2014). The synthesis and performance of Zr-doped and W–Zr-codoped VO_2 nanoparticles and derived flexible foils, *J. Mater. Chem. A*, **2**, pp. 15087–15093.
6. Wang, N., Liu, S., Zeng, X. T., et al. (2015). Mg/W-codoped vanadium dioxide thin films with enhanced visible transmittance and low phase transition temperature, *J. Mater. Chem. C*, **3**, pp. 6771–6777.
7. Chang, T.-C., Cao, X., Bao, S.-H., et al. (2018). Review on thermochromic vanadium dioxide based smart coatings: from lab to commercial application, *Adv. Manuf.*, **6**, pp. 1–19.
8. Chang, T., Cao, X., Li, N., et al. (2017). Facile and low-temperature fabrication of thermochromic Cr_2O_3/VO_2 smart coatings: enhanced solar modulation ability, high luminous transmittance and UV-shielding function, *ACS Appl. Mater. Interfaces*, **9**, pp. 26029–26037.
9. Top, I., Binions, R., Warwick, M. E. A., et al. (2018). VO_2/TiO_2 bilayer films for energy efficient windows with multifunctional properties, *J. Mater. Chem. C*, **6**, pp. 4485–4493.

10. Wang, S., Owusu, K. A., Mai, L., et al. (2018). Vanadium dioxide for energy conservation and energy storage applications: synthesis and performance improvement, *Appl. Energy*, **211**, pp. 200–217.
11. Zhang, J. S., Zhao, Z. J., Li, J. B., et al. (2017). Evolution of structural and electrical properties of oxygen-deficient VO_2 under low temperature heating process, *ACS Appl. Mater. Interfaces*, **9**, pp. 27135–27141.
12. Zhang, J. S., Jin, H. B., Chen, Z., et al. (2015). Self-assembling VO_2 nanonet with high switching performance at wafer-scale, *Chem. Mater.*, **27**, pp. 7419–7424.
13. Guo, D., Zhao, Z., Li, J., et al. (2017). Symmetric confined growth of superstructured vanadium dioxide nanonet with a regular geometrical pattern by a solution approach, *Cryst. Growth Des.*, **17**, pp. 5838–5844.
14. Zhang, C., Cao, W., Adedeji, A. V., et al. (2014). Preparation and properties of VO_2 thin films by a novel sol–gel process, *J. Sol-Gel Sci. Technol.*, **69**, pp. 320–324.
15. Khan, G. R., Asokan, K. and Ahmad, B. (2017). Room temperature tunability of Mo-doped VO_2 nanofilms across semiconductor to metal phase transition, *Thin Solid Films*, **625**, pp. 155–162.
16. Zhang, J., Wang, J., Yang, C., et al. (2017). Mesoporous SiO_2/VO_2 double-layer thermochromic coating with improved visible transmittance for smart window, *Sol. Energy Mater. Sol. Cells*, **162**, pp. 134–141.
17. Singh, A. K., Kiruthika, S., Mondal, I., et al. (2017). Fabrication of solar and electrically adjustable large area smart windows for indoor light and heat modulation, *J. Mater. Chem. C*, **5**, pp. 5917–5922.
18. Kim, Y., Yu, S., Park, J., et al. (2018). High-throughput roll-to-roll fabrication of flexible thermochromic coatings for smart windows with VO_2 nanoparticles, *J. Mater. Chem. C*, **6**, pp. 3451–3458.
19. Seyfouri, M. M. and Binions, R. (2017). Sol-gel approaches to thermochromic vanadium dioxide coating for smart glazing application, *Sol. Energy Mater. Sol. Cells*, **159**, pp. 52–65.
20. Zheludkevich, M. L., Salvado, I. M. and Ferreira, M. G. S. (2005). Sol–gel coatings for corrosion protection of metals, *J. Mater. Chem.*, **15**, pp. 5099–5111.
21. Avnir, D., Coradin, T., Lev, O., et al. (2006). Recent bio-applications of sol–gel materials, *J. Mater. Chem.*, **16**, pp. 1013–1030.
22. Chen, H., Zhang, G., Wei, Z., et al. (2010). Layer-by-layer assembly of sol–gel oxide "glued" montmorillonite-zirconia multilayers, *J. Mater. Chem.*, **20**, pp. 4925–4936.

23. Danks, A. E., Hall, S. R. and Schnepp, Z. (2016). The evolution of 'sol–gel' chemistry as a technique for materials synthesis, *Mater. Horiz.*, **3**, pp. 91–112.
24. Wu, J., Huang, W., Shi, Q., et al. (2013). Effect of annealing temperature on thermochromic properties of vanadium dioxide thin films deposited by organic sol–gel method, *Appl. Surf. Sci.*, **268**, pp. 556–560.
25. Xu, Y., Huang, W., Shi, Q., et al. (2013). Shape-dependent thermochromic phenomenon in porous nano-structured VO_2 films, *Mater. Res. Bull.*, **48**, pp. 4146–4149.
26. Yu, J. H., Nam, S. H., Kim, D., et al. (2016). Thermal aging effect of vanadyl acetylacetonate precursor for deposition of VO_2 thin films with thermochromic properties, *Mater. Res. Bull.*, **82**, pp. 11–15.
27. Flory, P. J. (1974). Introductory lecture: gels and gelling processes, *Faraday Discuss. Chem. Soc.*, **57**, pp. 7–18.
28. Kakihana, M. (1996). "Sol-gel" preparation of high temperature superconducting oxides, *J. Sol-Gel Sci. Technol.*, **6**, pp. 7–55.
29. Greenberg, C. B. (1983). Undoped and doped VO_2 films grown from $VO(OC_3H_7)_3$, *Thin Solid Films*, **110**, pp. 73–82.
30. Yin, D., Xu, N., Zhang, J., et al. (1996). High quality vanadium dioxide films prepared by an inorganic sol-gel method, *Mater. Res. Bull.*, **31**, pp. 335–340.
31. Hanlon, T. J., Coath, J. A. and Richardson, M. A. (2003). Molybdenum-doped vanadium dioxide coatings on glass produced by the aqueous sol-gel method, *Thin Solid Films*, **436**, pp. 269–272.
32. Liu, D. Q., Zheng, W. W., Cheng, H. F., et al. (2009). Thermochromic VO_2 thin film prepared by post annealing treatment of V_2O_5 thin film, *Adv. Mater. Res.*, **79–82**, pp. 747–750.
33. Wang, N., Magdassi, S., Mandler, D., et al. (2013). Simple sol–gel process and one-step annealing of vanadium dioxide thin films: synthesis and thermochromic properties, *Thin Solid Films*, **534**, pp. 594–598.
34. Shi, Q., Huang, W., Lu, T., et al. (2014). In situ growth of sol-gel-derived nano-VO_2 film and its phase transition characteristics, *J. Nanopart. Res.*, **16**, pp. 1–9.
35. Li, D., Huang, W., Song, L., et al. (2015). Thermal stability of VO_2 thin films deposited by sol–gel method, *J. Sol-Gel Sci. Technol.*, **75**, pp. 189–197.
36. Berezina, O., Kirienko, D., Pergament, A., et al. (2015). Vanadium oxide thin films and fibers obtained by acetylacetonate sol–gel method, *Thin Solid Films*, **574**, pp. 15–19.

37. Pan, M., Zhong, H., Wang, S., et al. (2004). Properties of VO_2 thin film prepared with precursor VO(acac)$_2$, *J. Cryst. Growth*, **265**, pp. 121–126.

38. Yuan, N., Li, J., Lin, C. (2002). Valence reduction process from sol-gel V_2O_5 to VO_2 thin films, *Appl. Surf. Sci.*, **191**, pp. 176–180.

39. Liu, D., Cheng, H., Zheng, W., et al. (2012). Infrared thermochromic properties of VO_2 thin films prepared through aqueous sol-gel process, *J. Wuhan Univ. Technol.-Mater. Sci. Ed.*, **27**, pp. 861–865.

40. Dachuan, Y., Niankan, X., Jingyu, Z., et al. (1996). Vanadium dioxide films with good electrical switching property, *J. Phys. D: Appl. Phys.*, **29**, p. 1051.

41. Wang, Y. T. and Chen, C. H. (2013). Facile growth of thermochromic VO_2 nanostructures with greatly varied phases and morphologies, *Inorg. Chem.*, **52**, pp. 2550–2555.

42. Xu, Y., Huang, W., Shi, Q., et al. (2013). Porous nano-structured VO_2 films with different surfactants: synthesis mechanisms, characterization, and applications, *J. Mater. Sci.: Mater. Electron.*, **24**, pp. 3823–3829.

43. Zhang, R., Jin, H.-B., Guo, D., et al. (2016). The role of Fe dopants in phase stability and electric switching properties of Fe-doped VO_2, *Ceram. Int.*, **42**, pp. 18764–18770.

44. Hu, L., Tao, H., Chen, G., et al. (2016). Porous W-doped VO_2 films with simultaneously enhanced visible transparency and thermochromic properties, *J. Sol-Gel Sci. Technol.*, **77**, pp. 85–93.

45. Liang, Z., Zhao, L., Meng, W., et al. (2017). Tungsten-doped vanadium dioxide thin films as smart windows with self-cleaning and energy-saving functions, *J. Alloys Compd.*, **694**, pp. 124–131.

46. Liu, C., Wang, N. and Long, Y. (2013). Multifunctional overcoats on vanadium dioxide thermochromic thin films with enhanced luminous transmission and solar modulation, hydrophobicity and anti-oxidation, *Appl. Surf. Sci.*, **283**, pp. 222–226.

47. Liu, C., Cao, X., Kamyshny, A., et al. (2014). VO_2/Si-Al gel nanocomposite thermochromic smart foils: largely enhanced luminous transmittance and solar modulation, *J. Colloid Interface Sci.*, **427**, pp. 49–53.

48. Wang, N., Peh, Y. K., Magdassi, S., et al. (2018). Surface engineering on continuous VO_2 thin films to improve thermochromic properties: top-down acid etching and bottom-up self-patterning, *J. Colloid Interface Sci.*, **512**, pp. 529–535.

49. Goodenough, J. B. (1971). The two components of the crystallographic transition in VO_2, *J. Solid State Chem.*, **3**, pp. 490–500.

50. Xu, Y., Huang, W., Zhang, Y., et al. (2012). Synthesis and properties of Mo and W ions Co-doped porous nano-structured VO_2 films by sol–gel process, *J. Sol-Gel Sci. Technol.*, **64**, pp. 493–499.
51. Mai, L. Q., Hu, B., Hu, T., et al. (2006). Electrical property of Mo-doped VO_2 nanowire array film by melting-quenching sol-gel method, *J. Phys. Chem. B*, **110**, pp. 19083–19086.
52. Hu, Y., Shi, Q., Huang, W., et al. (2015). Preparation and phase transition properties of Ti-doped VO_2 films by sol–gel process, *J. Sol-Gel Sci. Technol.*, **78**, pp. 19–25.
53. Lu, W., Zhao, G., Song, B., et al. (2017). Preparation and thermochromic properties of sol-gel-derived Zr-doped VO_2 films, *Surf. Coat. Technol.*, **320**, pp. 311–314.
54. Shannon, R. D. and Prewitt, C. T. (1969). Effective ionic radii in oxides and fluorides, *Acta Cryst.*, **B25**, pp. 925–946.
55. Du, J., Gao, Y., Luo, H., et al. (2011). Significant changes in phase-transition hysteresis for Ti-doped VO_2 films prepared by polymer-assisted deposition, *Sol. Energy Mater. Sol. Cells*, **95**, pp. 469–475.
56. Jiazhen, Y., Yue, Z., Wanxia, H., et al. (2008). Effect of Mo-W Co-doping on semiconductor-metal phase transition temperature of vanadium dioxide film, *Thin Solid Films*, **516**, pp. 8554–8558.
57. Soltani, M., Chaker, M., Haddad, E., et al. (2004). Effects of Ti–W codoping on the optical and electrical switching of vanadium dioxide thin films grown by a reactive pulsed laser deposition, *Appl. Phys. Lett.*, **85**, pp. 1958–1960.
58. Suzuki, H., Yamaguchi, K. and Miyazaki, H. (2007). Fabrication of thermochromic composite using monodispersed VO_2 coated SiO_2 nanoparticles prepared by modified chemical solution deposition, *Compos. Sci. Technol.*, **67**, pp. 3487–3490.
59. He, P., Huang, W.-X., Yan, J.-Z., et al. (2011). Preparation and thermochromic property of VO_2/mica pigments, *Mater. Res. Bull.*, **46**, pp. 966–969.
60. Chen, H. K., Hung, H. C., Yang, C. K., et al. (2004). The preparation and characterization of transparent nano-sized thermochromic VO_2-SiO_2 films from the sol–gel process, *J. Non-Cryst. Solids*, **347**, pp. 138–143.
61. Guo, D., Ling, C., Wang, C., et al. (2018). Hydrothermal one-step synthesis of highly dispersed M-phase VO_2 nanocrystals and application to flexible thermochromic film, *ACS Appl. Mater. Interfaces*, **10**, pp. 28627–28634.

Chapter 11

VO_2-Based Smart Coatings with Long-Term Durability: Review and Perspective

Xun Cao and Tianci Chang
Shanghai Institute of Ceramics, Chinese Academy of Sciences, Heshuo Rode No. 588, Shanghai 201899, China
cxun@mail.sic.ac.cn

11.1 Introduction

Vanadium dioxide (VO_2) is an archetypal phase transition material and undergoes a reversible semiconducting–metallic transition at the critical temperature ($T_c \approx 68°C$) [1]. For temperatures below T_c, VO_2 is in a semiconducting state and shows a relatively high transmittance for solar radiation; when the temperature goes over T_c, VO_2 is in a metallic state with a low solar transmittance [2]. On the basis of this thermochromic feature, VO_2 has been widely investigated as "smart coatings" for building fenestration to realize solar control and the resultant energy-efficient function.

Three challenges limit the practical application of VO_2-based thermochromic smart coatings over relatively long periods [3–5]:

Vanadium Dioxide-Based Thermochromic Smart Windows
Edited by Yi Long and Yanfeng Gao

ISBN 978-981-4877-06-0 (Hardcover), 978-1-003-17731-9 (eBook)
www.jennystanford.com

(i) the solar modulation ability, denoted as $\Delta T_{sol} = T_{sol,lt} - T_{sol,ht}$, is usually less than 10% and not sufficient for an energy-efficient effect; (ii) the integrated luminous transmittance (T_{lum}) is less than 40% for coatings with a desirable ΔT_{sol}, which seems not acceptable for the viewing purpose of customers; (iii) the critical temperature T_c of phase transition is about 68°C, which is too high for such coatings to be used in most areas of the world. To solve these obstacles, many modification methods have been proposed and investigated by researchers, such as doping [6–9], designing multilayer structures [10–12], fabricating nanostructures [13–15], and preparing nanocomposites [16–18].

With the development of this field and the deeper insight into the VO_2 material, another important issue has caught the attention of researchers. Vanadium oxides are a big family and include various forms, such as VO, V_2O_3, V_4O_7, VO_2, V_6O_{13}, and V_2O_5. Among these oxides, VO_2 is not the most stable phase and probably transforms into other vanadium oxides, such as the thermodynamically stable phase V_2O_5, or their hydroxides, especially in a moist environment [4, 5, 19]. This can seriously degrade the thermochromic performance of VO_2-based smart coatings. Durability is a major challenge to the commercialization of VO_2-based smart coatings. Irrespective of whether excellent thermochromic performance have been achieved by the above methods, if the performance cannot be preserved for a long-term period, all these modification methods seem to be useless and meaningless.

In this chapter, we are going to review the strategy to improve the durability of VO_2-based smart coatings, which usually contain two forms, VO_2 nanoparticles and VO_2 films. Meanwhile, development trends and perspectives in this field will be discussed for accelerating the commercialization process of VO_2-based smart coatings.

11.2 Enhanced Durability of VO_2 Nanoparticles

VO_2 nanoparticles are an important form of VO_2-based smart coatings, which can be dispersed in polymer matrixes and used as flexible foils. Unfortunately, due to the relatively small size and large specific surface area, VO_2 nanoparticles are very likely to react with moisture and oxygen in the air and show fast deterioration. Therefore, there is a strong demand to improve the stability of VO_2 nanoparticles for their application as thermochromic smart coatings.

Preparing core-shell structures is a common way to improve the durability of VO_2 nanoparticles, where the VO_2 nanoparticle is the "core" and there is a barrier "shell," which is chemically stable and can protect the VO_2 nanoparticles from moisture and oxygen. Many shell materials have been investigated, such as SiO_2, ZnO, MgF_2, and AlO_x [20–26].

SiO_2 is the most studied material for shells for VO_2 nanoparticles [20–23]. SiO_2 can not only enhance the stability of VO_2 but also provides the possibility for additional surface modification and good incorporation into the polymer matrix [20]. A VO_2@SiO_2 core-shell structure synthesized by a solution method has been reported by Gao et al. [20] (Fig. 11.1a), and its transmission electron microscope (TEM) image is shown in Fig. 11.1b. The SiO_2 shell improves the antioxidation and antiacid properties of the VO_2 nanoparticles. Uncoated and SiO_2-coated VO_2 nanoparticles were annealed at 300°C in air for 2 h. Peaks belonging to V_2O_5 were detected in the X-ray diffraction (XRD) pattern of the uncoated sample but not of the VO_2@SiO_2 sample (Fig. 11.1c).

Due to a high stability, Al_2O_3 is another promising material for shells to protect materials from corrosion. Tong et al. [26] synthesized two kinds of core-shell structures: as prepared VO_2@Al-O and annealed VO_2@Al-O, and different durability tests have been carried out. Compared with bare VO_2 nanoparticles and the unannealed VO_2@Al-O core-shell structure, the annealed VO_2@Al-O core-shell structure shows the best stability. With protection from the Al-O shell, the VO_2 core remained stable at a high temperature (350°C in air) and in H_2O_2 solution. Especially, thermochromic films made of annealed VO_2@Al-O core-shell retained most of their thermochromic performance in a damp heating environment ($T = 60$°C, RH = 90%) for 20 days.

ZnO has also been investigated as a shell, and VO_2@ZnO core-shell structures have been prepared by Chen et al. [24] (Fig. 11.2a–c). The ZnO shell shows great influence on VO_2 nanoparticles. On one hand, due to the suitable refractive index of ZnO, the $T_{lum,lt}$

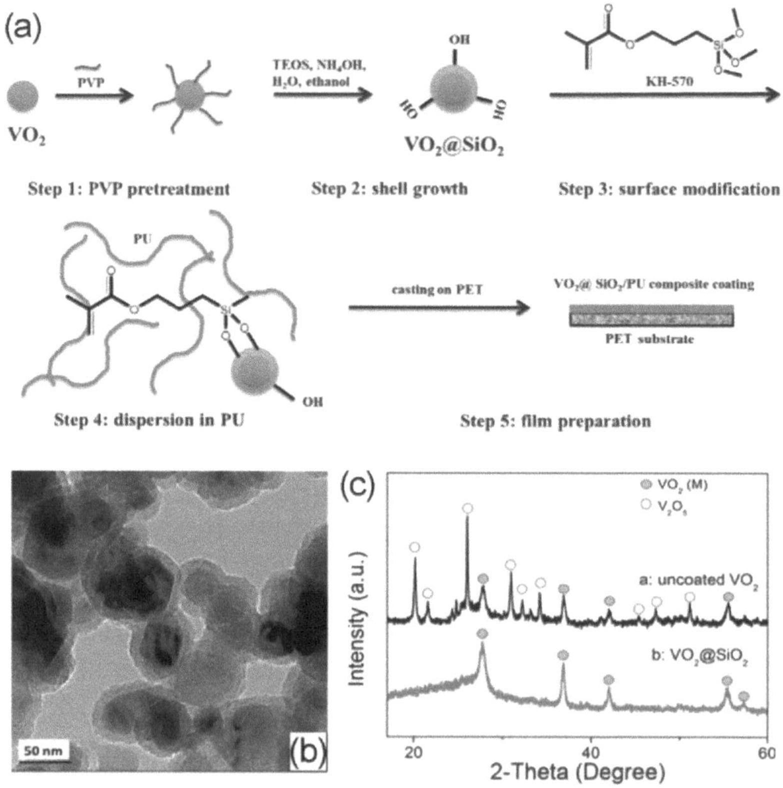

Figure 11.1 $VO_2@SiO_2$ core-shell structures for enhanced durability. (a) Preparation procedure for $VO_2@SiO_2$ nanoparticles and flexible composite films. (b) TEM images of $VO_2@SiO_2$ nanoparticles. (c) XRD patterns of the uncoated VO_2 and $VO_2@SiO_2$ nanoparticles after heat treatment at 300°C for 2 h in air.

and ΔT_{sol} of $VO_2@ZnO$ increased to 51% and 19.1%, while the $T_{lum,lt}$ and ΔT_{sol} of uncoated VO_2 is 38.9% and 17.2%, respectively. On the other hand, ZnO can largely improve the stability of VO_2 nanoparticles. An accelerating tests has been carried out in the environment with 60°C and 90% relative humidity to evaluate the durability of the samples. For an uncoated VO_2 film, the thermochromism vanishes after a 30 h treatment. Compared with the uncoated VO_2, $VO_2@ZnO$ shows striking durability (Figs. 11.2d and 11.2e).

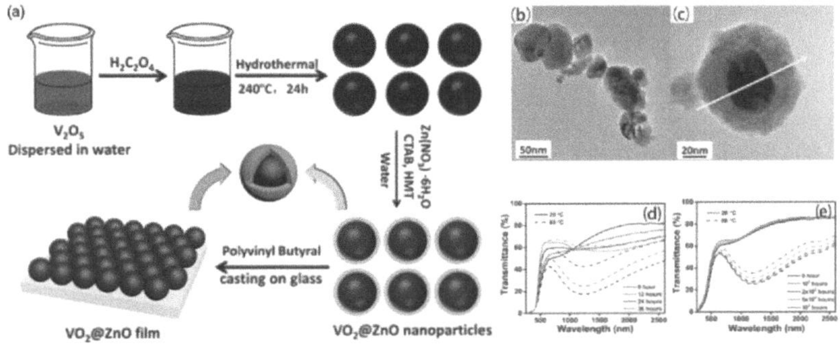

Figure 11.2 VO_2@ZnO core-shell structures for enhanced durability. (a) Experimental flow chart for the synthesis of VO_2@ZnO core-shell structure nanoparticles and a VO_2@ZnO film. (b, c) TEM image of uncoated VO_2 nanoparticles and a VO_2@ZnO core-shell structure nanoparticle. (d, e) Optical transmittance spectra of uncoated VO_2 and VO_2@ZnO at a constant temperature (60°C) and humidity (90%).

11.3 Protective Layers for VO_2 Thin Films

VO_2 thin films deposited directly on glass by deposition technologies such as magnetron sputtering (MS) and chemical vapor deposition are another form of VO_2-based smart coatings. For VO_2 thin films, preparing protective layers on the top surface is an effective way to improve their durability. Like the shell material for VO_2 nanoparticles, materials utilized as protective layers require high transparency and stability. Protective layers such as Al_2O_3, WO_3, SiO_2, and SiN_x have been investigated to mitigate the deterioration process of VO_2 films [11, 27–32].

Ji et al. deposited Al oxide top layers by MS to protect the underlying VO_2 films [27, 33]. Enhanced durability of the bilayer structure has been demonstrated by heating the sample in dry air at 300°C and in 95% relative humidity at 60°C. Wang et al. prepared Al_2O_3 protective layers by atomic layer deposition to protect VO_2 films [28]. They found that when the sample was exposed to damp heat, the protective effect offered by the Al_2O_3 layers decreased, maybe due to their corrosion by water.

Pan et al. [29] investigated the influence of combined protective layers on the deterioration process of VO_2 films, and the proposed $VO_2/TiO_2/ZnO$ protective coating is a promising material in terms of both durability and thermochromic performance. A multilayered $SiN_x/NiCrO_x/SiN_x/VO_x/SiN_x/NiCrO_x/SiN_x$ structure was reported by Zhan et al. [32] and its relative thermal stability investigated.

For application as a smart coating, besides showing robust endurance against environmental damage, a VO_2-based coating has to undergo a phase transition between semiconducting and metallic states. Therefore, stable fatigue performance is also an important property for thermochromic films. Long et al. [30] first proposed a long-time fatigue property of the $SiN_x/VO_2/SiN_x$ structure. The sample undergoes a phase transition in the heating-cooling process per 30 min to guarantee temperature stability and provide adequate time for detecting, and the detector records the transmittance of the beam at different states of the sample (Fig. 11.3a). The cycle occurs 48 times per 24 h, and the total number of cycles executed is over 35,000, indicating that our fatigue test lasted for more than 2 years. In this study, the contrast at 2500 nm is considered as the index to evaluate the durability of the sample, and the efficient fatigue life time is the point when the contrast is decreased by 10% of the pristine value. For the single-layer VO_2 film, the efficient fatigue life time is found to be 1.6×10^3 cycles (Fig. 11.3b). And the efficient fatigue life time is over 1×10^4 cycles for the $SiN_x/VO_2/SiN_x$ structure—when the retention percentage is 98% at least—6 times longer than that for the VO_2 single layer (Fig. 11.3c).

In a newly reported work by Chang et al., a naturally hydrophobic and stable HfO_2 layer was used to mitigate the deterioration process of VO_2 films [34]. In this work, on modifying the thickness of HfO_2, the VO_2/HfO_2 structure shows a largely improved thermochromic performance, with $T_{\text{lum,lt}} = 55.8\%$, $T_{\text{lum,ht}} = 43.6\%$, and $\Delta T_{\text{sol}} = 15.9\%$. Four types of samples have been prepared with different encapsulated surfaces and cross sections. After durability tests in boiling water and under high-temperature heating and an accelerating examination, thermochromic performance of VO_2 films were examined and the cross sections of VO_2 films have been demonstrated to play an important role in the deterioration process (Fig. 11.4a). By deliberating incorporating HfO_2 for an

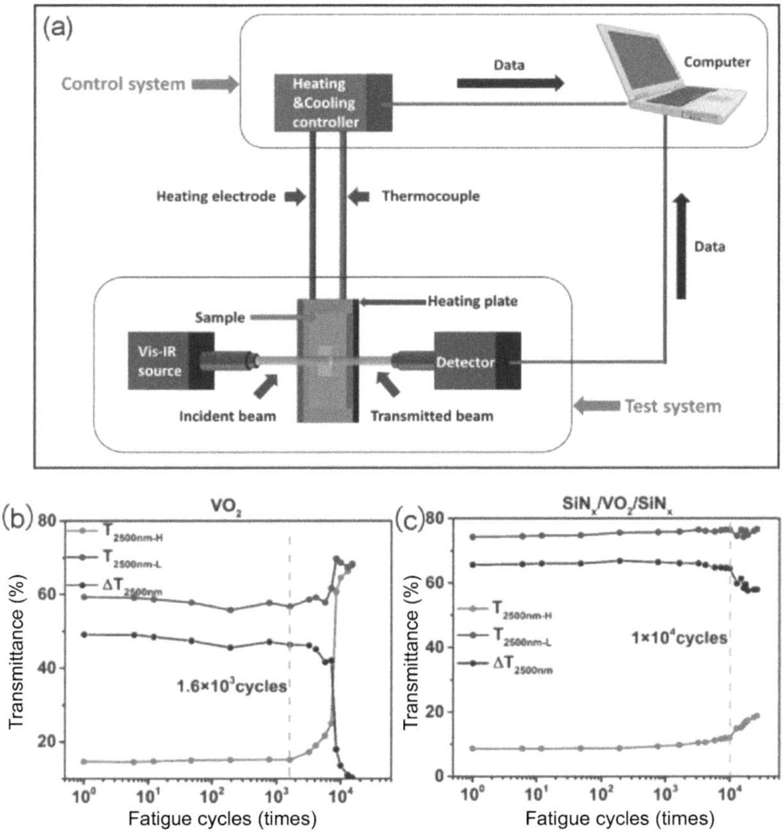

Figure 11.3 $SiN_x/VO_2/SiN_x$ multilayer structures for enhanced durability. (a) Schematic description of a self-made fatigue testing facility. (b, c) The variation in transmittance at a 2500 nm wavelength accompanied by an increment in the thermochromic fatigue cycles for (b) VO_2 and (c) $SiN_x/VO_2/SiN_x$.

encapsulation structure, the degradation process of VO_2 can be effectively suppressed. The proposed films can retain stable phase transition performances under high relative humidity (90%) and temperature (60°C) for over 100 days (Fig. 11.4b,c), which is equal to ~16 years in the real environment calculated by the Hellberg–Peck model.

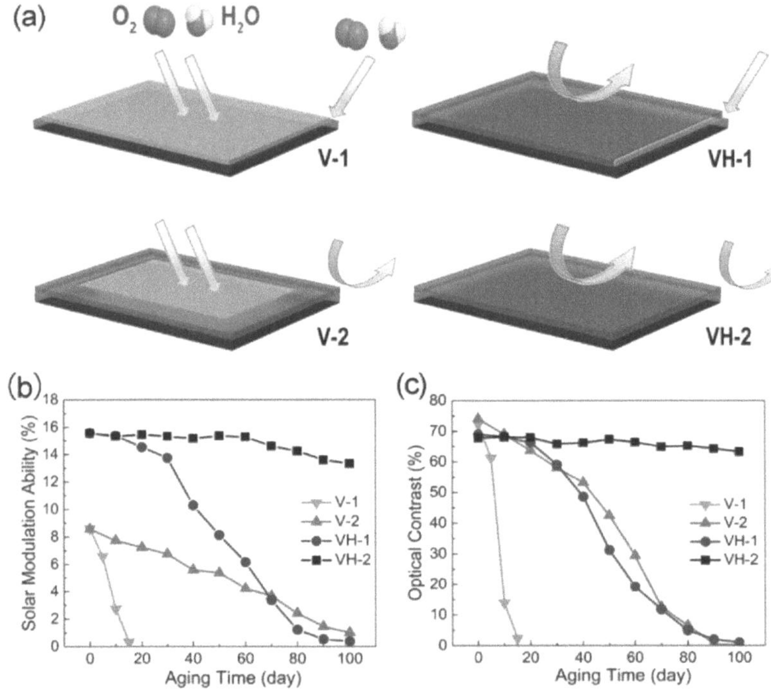

Figure 11.4 Environmental stability test of different samples of VO_2/HfO_2. (a) Schematic illustration of the possible effect of oxygen and moisture on different samples. (b) Solar modulation abilities versus increase in the aging time for different samples. (c) Optical contrast at 2500 nm in the infrared region versus increase in the aging time for different samples.

11.4 Summary and the Future

To improve the durability of VO_2-based smart coatings for long-term use, some research has been carried out. For VO_2 nanoparticles, it is usually to synthesize core-shell structures for enhanced durability. As for continuous VO_2 films, designing protective layers is an effective way for long-term stability. Although there are several works on the durability of VO_2-based smart coatings, more attention needs to be paid in future studies.

The following opinions should be taken into consideration in future work, in order to improve the stability of VO_2 for commercial applications:

- The deterioration process of VO_2 in different operating environments is not clear. Discovering the deterioration process and relative mechanism of VO_2 will help focus on enhanced durability.
- In the phase transition process, VO_2 shows different lattice parameters in the monoclinic and rutile phases. According to the study, there is a volume expansion of 0.3% in the cooling process of VO_2 from the rutile phase to the monoclinic phase. The volume change may affect current protective methods for VO_2, especially for the VO_2 core-shell structures, which require attention in future work.
- To test the durability of VO_2-based smart coatings, accelerating tests in extreme conditions were usually used in previous works. However, we cannot obtain the expected service life of VO_2 in the real environment on the basis of the data from the accelerating tests. Therefore, reliable relationships need to be built up between the accelerating tests and the real environments for an objective evaluation of the durability of VO_2-based smart coatings.

References

1. Morin, F. J. (1959). Oxides which show a metal-to-insulator transition at the neel temperature, *Phys. Rev. Lett.*, **3**, pp. 34–36.
2. Babulanam, S. M., Eriksson, T. S., Niklasson, G. A. and Granqvist, C. G. (1987). Thermochromic VO_2 films for energy-efficient windows, *Sol. Energy Mater.*, **16**, pp. 347–363.
3. Li, S. Y., Niklasson, G. A. and Granqvist, C. G. (2012). Thermochromic fenestration with VO_2-based materials: three challenges and how they can be met, *Thin Solid Films*, **520**, pp. 3823–3828.
4. Chang, T., Cao, X., Bao, S., Ji, S., Luo, H. and Jin, P. (2018). Review on thermochromic vanadium dioxide based smart coatings: from lab to commercial application, *Adv. Manuf.*, **6**, pp. 1–19.

5. Wang, S. F., Liu, M. S., Kong, L. B., Long, Y., Jiang, X. C. and Yu, A. B. (2016). Recent progress in VO_2 smart coatings: strategies to improve the thermochromic properties, *Prog. Mater. Sci.*, **81**, pp. 1–54.
6. Ji, C., Wu, Z., Wu, X., Wang, J., Gou, J., Huang, Z., Zhou, H., Yao, W. and Jiang, Y. (2018). Al-doped VO_2 films as smart window coatings: reduced phase transition temperature and improved thermochromic performance, *Sol. Energy Mater. Sol. Cells*, **176**, pp. 174–180.
7. Li, S.-Y., Mlyuka, N. R., Primetzhofer, D., Hallén, A., Possnert, G., Niklasson, G. A. and Granqvist, C. G. (2013). Bandgap widening in thermochromic Mg-doped VO_2 thin films: quantitative data based on optical absorption, *Appl. Phys. Lett.*, **103**, p. 161907.
8. Dai, L., Chen, S., Liu, J., Gao, Y., Zhou, J., Chen, Z., Cao, C., Luo, H. and Kanehira, M. (2013). F-doped VO_2 nanoparticles for thermochromic energy-saving foils with modified color and enhanced solar-heat shielding ability, *Phys. Chem. Chem. Phys.*, **15**, pp. 11723–11729.
9. Chen, R., Miao, L., Cheng, H. L., Nishibori, E., Liu, C. Y., Asaka, T., Iwamoto, Y., Takata, M. and Tanemura, S. (2015). One-step hydrothermal synthesis of $V_{1-x}W_xO_2$ (M/R) nanorods with superior doping efficiency and thermochromic properties, *J. Mater. Chem. A*, **3**, pp. 3726–3738.
10. Mlyuka, N. R., Niklasson, G. A. and Granqvist, C. G. (2009). Thermochromic multilayer films of VO_2 and TiO_2 with enhanced transmittance, *Sol. Energy Mater. Sol. Cells*, **93**, pp. 1685–1687.
11. Chang, T. C., Cao, X., Dedon, L. R., Long, S. W., Huang, A. B., Shao, Z. W., Li, N., Luo, H. J. and Jin, P. (2018). Optical design and stability study for ultrahigh-performance and long-lived vanadium dioxide-based thermochromic coatings, *Nano Energy*, **44**, pp. 256–264.
12. Powell, M. J., Quesada-Cabrera, R., Taylor, A., Teixeira, D., Papakonstantinou, I., Palgrave, R. G., Sankar, G. and Parkin, I. P. (2016). Intelligent multifunctional $VO_2/SiO_2/TiO_2$ coatings for self-cleaning, energy-saving window panels, *Chem. Mater.*, **28**, pp. 1369–1376.
13. Ke, Y. J., Wen, X. L., Zhao, D. Y., Che, R. C., Xiong, Q. H. and Long, Y. (2017). Controllable fabrication of two-dimensional patterned VO_2 nanoparticle, nanodome, and nanonet arrays with tunable temperature-dependent localized surface plasmon resonance, *ACS Nano*, **11**, pp. 7542–7551.
14. Lu, Q., Liu, C., Wang, N., Magdassi, S., Mandler, D. and Long, Y. (2016). Periodic micro-patterned VO_2 thermochromic films by mesh printing, *J. Mater. Chem. C*, **4**, pp. 8385–8391.

15. Zhou, M., Bao, J., Tao, M., Zhu, R., Lin, Y., Zhang, X. and Xie, Y. (2013). Periodic porous thermochromic VO_2 (M) films with enhanced visible transmittance, *Chem. Commun. (Camb)*, **49**, pp. 6021–6023.

16. Li, W., Ji, S., Qian, K. and Jin, P. (2015). Preparation and characterization of VO_2–$BaSO_4$ composite films with enhanced optical properties in thermochromic field, *Ceram. Int.*, **41**, pp. 5049–5056.

17. Luo, Y., Li, M., Li, K., Xu, C., Xu, S. and Li, G. (2019). Optimized thermochromic properties of $VO_2/CoFe_2O_4$ nanocomposite films via band gap engineering, *Sol. Energy Mater. Sol. Cells*, **191**, pp. 258–265.

18. Li, W., Ji, S., Qian, K. and Jin, P. (2015). Preparation and characterization of VO_2 (M)-SnO_2 thermochromic films for application as energy-saving smart coatings, *J. Colloid Interface Sci.*, **456**, pp. 166–173.

19. Cui, Y., Ke, Y., Liu, C., Chen, Z., Wang, N., Zhang, L., Zhou, Y., Wang, S., Gao, Y. and Long, Y. (2018). Thermochromic VO_2 for energy-efficient smart windows, *Joule*, **2**, pp. 1707–1746.

20. Gao, Y., Wang, S., Luo, H., Dai, L., Cao, C., Liu, Y., Chen, Z. and Kanehira, M. (2012). Enhanced chemical stability of VO_2 nanoparticles by the formation of SiO_2/VO_2 core/shell structures and the application to transparent and flexible VO_2-based composite foils with excellent thermochromic properties for solar heat control, *Energy Environ. Sci.*, **5**, p. 6104.

21. Zhou, Y., Huang, A., Li, Y., Ji, S., Gao, Y. and Jin, P. (2013). Surface plasmon resonance induced excellent solar control for VO_2@SiO_2 nanorods-based thermochromic foils, *Nanoscale*, **5**, pp. 9208–9213.

22. Huang, A., Zhou, Y., Li, Y., Ji, S., Luo, H. and Jin, P. (2013). Preparation of $V_xW_{1-x}O_2$(M)@SiO_2 ultrathin nanostructures with high optical performance and optimization for smart windows by etching, *J. Mater. Chem. A*, **1**, p. 12545.

23. Zhu, J., Zhou, Y., Wang, B., Zheng, J., Ji, S., Yao, H., Luo, H. and Jin, P. (2015). Vanadium dioxide nanoparticle-based thermochromic smart coating: high luminous transmittance, excellent solar regulation efficiency, and near room temperature phase transition, *ACS Appl. Mater. Interfaces*, **7**, pp. 27796–27803.

24. Chen, Y., Zeng, X., Zhu, J., Li, R., Yao, H., Cao, X., Ji, S. and Jin, P. (2017). High performance and enhanced durability of thermochromic films using VO_2@ZnO core–shell nanoparticles, *ACS Appl. Mater. Interfaces*, **9**, pp. 27784–27791.

25. Zhao, S., Tao, Y., Chen, Y., Zhou, Y., Li, R., Xie, L., Huang, A., Jin, P. and Ji, S. (2019). Room-temperature synthesis of inorganic-organic hybrid coated VO_2 nanoparticles for enhanced durability and flexible temperature-responsive near-infrared modulator application, *ACS Appl. Mater. Interfaces*, **11**, pp. 10254–10261.
26. Tong, K., Li, R., Zhu, J., Yao, H., Zhou, H., Zeng, X., Ji, S. and Jin, P. (2017). Preparation of VO_2/Al-O core-shell structure with enhanced weathering resistance for smart window, *Ceram. Int.*, **43**, pp. 4055–4061.
27. Ji, Y., Li, S., Niklasson, G. A. and Granqvist, C. G. (2014). Durability of thermochromic VO_2 thin films under heating and humidity: effect of Al oxide top coatings, *Thin Solid Films*, **562**, pp. 568—573.
28. Wang, X., Cao, Y., Yang, C., Yan, L. and Li, Y. (2016). Vanadium dioxide film protected with an atomic-layer-deposited Al_2O_3 thin film, *J. Vac. Sci. Technol., A*, **34**, p. 01A106.
29. Pan, G. T., Yang, Y. L., Chong, S. H., Arjun, N., Yang, T. C. K. and Lai, Y. C. (2017). The durability study of thermochromic vanadium dioxide films with the addition of barrier coatings, *Vacuum*, **145**, pp. 158–168.
30. Long, S., Cao, X., Li, N., Xin, Y., Sun, G., Chang, T., Bao, S. and Jin, P. (2019). Application-oriented VO_2 thermochromic coatings with composite structures: optimized optical performance and robust fatigue properties, *Sol. Energy Mater. Sol. Cells*, **189**, pp. 138–148.
31. Long, S. W., Zhou, H. J., Bao, S. H., Xin, Y. C., Cao, X. and Jin, P. (2016). Thermochromic multilayer films of $WO_3/VO_2/WO_3$ sandwich structure with enhanced luminous transmittance and durability, *RSC Adv.*, **6**, pp. 106435–106442.
32. Zhan, Y., Xiao, X., Lu, Y., Cao, Z., Qi, S., Huan, C., Cheng, H., Shi, J. and Xu, G. (2018). Enhanced thermal stability and thermochromic properties of VO_x-based thin films by room-temperature magnetron sputtering, *Sol. Energy Mater. Sol. Cells*, **174**, pp. 102–111.
33. Ji, Y.-X., Niklasson, G. A. and Granqvist, C. G. (2014). Durability of VO_2-based thin films at elevated temperature: towards thermochromic fenestration, *J. Phys. Conf. Ser.*, **559**, p. 012005.
34. Chang, T., Cao, X., Li, N., Long, S., Zhu, Y., Huang, J., Luo, H. and Jin, P. (2019). Mitigating deterioration of vanadium dioxide thermochromic films by interfacial encapsulation, *Matter*, **1**, pp. 734–744.

Chapter 12

Conclusions and Perspectives

Yi Long,[a] Yanfeng Gao,[b] and Xuchuan Jiang[c]
[a]*School of Materials Science and Engineering, Nanyang Technological University, 50 Nanyang Avenue, Singapore 639798, Singapore*
[b]*State Key Laboratory of High Performance Ceramics and Superfine Microstructure, Shanghai Institute of Ceramics, Chinese Academy of Sciences, Shanghai 200050, China*
[c]*Department of Chemical Engineering, Monash University, Wellington Road, Clayton 3800, Australia*
longyi@ntu.edu.sg, yfgao@mail.sic.ac.cn, xuchuan.jiang@monash.edu

A smart window is a promising technology for reducing the energy consumption of a building and enhancing the comfort level of a house. Thermochromic windows are gaining attention due to their rational stimulus, constant luminous transmission in different seasons, and automatic regulation of indoor solar transmission (particularly the infrared [IR] range), together with the relatively simple structure, facile fabrication, and low cost. In this book, the progress in vanadium dioxide (VO_2)-based smart windows is overviewed, from the band structure designing at the electronic and atomic scales to morphology engineering from nano- to micro-scales (Fig. 12.1). The approaches to improve the thermochromic performance of VO_2-based smart windows are discussed, such as creating nanocomposites, applying multifunctional antireflection coatings, controlling porosity, biomimetic patterning, and creating grids (Fig. 12.1c). The approaches for fabricating VO_2 are elaborated,

Vanadium Dioxide-Based Thermochromic Smart Windows
Edited by Yi Long and Yanfeng Gao

ISBN 978-981-4877-06-0 (Hardcover), 978-1-003-17731-9 (eBook)
www.jennystanford.com

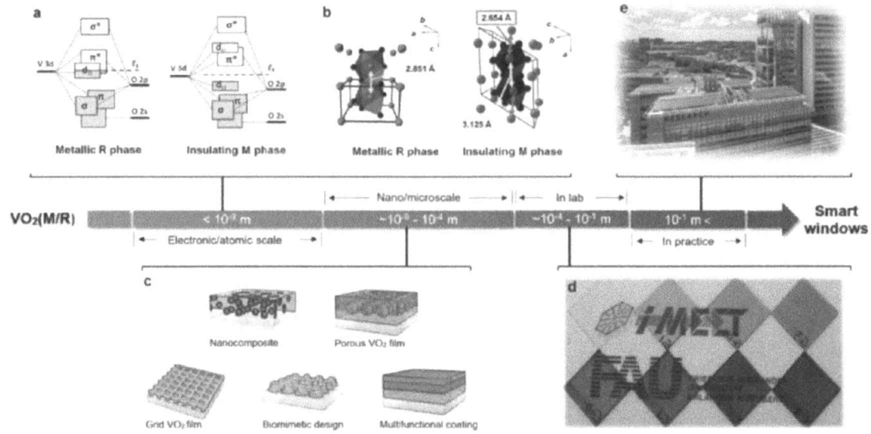

Figure 12.1 Overview of the development of VO_2 smart windows from the atomic scale to the architecture scale. (a) Molecular orbital diagrams showing the band structures of the metallic R and insulating M phases of VO_2. (b) Schematic of the atomic structures of the metallic tetragonal phase R and the insulating monoclinic phase M, with the V–V distances in each crystal structure highlighted. (c) Nano-/microengineering toward performance enhancement. (d) Photographs of VO_2-based thermochromic samples in the lab. (e) Photographs of architectural windows in practice. Reproduced from [1], Copyright © 2018 Elsevier.

including physical/chemical vapor deposition and hydrothermal and sol gel processes. In this chapter, challenges and perspectives are discussed. We hope that this book inspires more innovative researchers and accelerates the technical transfer from the lab to the industry (Fig. 12.1d,e) [1].

We have summarized some of the best-reported results in VO_2-based thermochromic smart windows in Fig. 12.2 by classifying different categories, with the pristine single layer as a benchmark [1]. Both simulation and experiment results in terms of τ_c, ΔT_{sol}, and T_{lum} are included as these three parameters are of prime importance in the applications. The constant goal in thermochromic smart windows is to reduce τ_c and enhance both ΔT_{sol} and T_{lum} simultaneously to maximize energy saving and cut down on the lighting inside buildings.

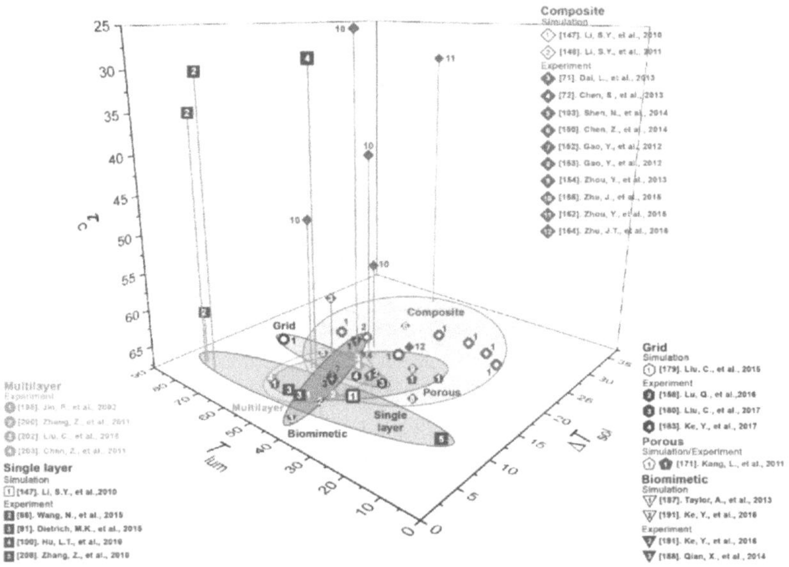

Figure 12.2 Summary of thermochromic performance (τ_c, ΔT_{sol}, and T_{lum}) in some of the best-reported and selected works. Reproduced from [1], Copyright © 2018 Elsevier.

VO_2 thermochromic smart windows are promising in terms of applications, and the research is in the beginning stages. Several issues still need to be addressed to further widen the scope of their applications.

12.1 Reduce Transition Temperature; Enhance Visible Transmittance and Solar Modulation Ability

As shown in Fig. 12.2, the simultaneous reduction of τ_c and enhancement of ΔT_{sol} and T_{lum} is a daunting task due to the intrinsic physical limitations of VO_2. Low VO_2 loadings are generally favorable for increasing the T_{lum}, but this is often at the cost of a degraded ΔT_{sol}. Doping is an effective strategy to decrease the τ_c, but typically the other two parameters are also compromised. Currently, doping is limited to continuous films and nanocomposites; but investigation can be extended to other categories, such as biomimetic, gridding, controlled porosity, and multilayered structure engineering (Fig. 12.1). More dopants and co- or multidoping need to be examined at the atomic level by combining simulations and experiments with more structure designs at the nano- or microlevel.

VO_2 coatings for glazing can combine spectral selectivity with angular-dependent properties, but these coatings are usually optimized only for light at normal incidence. Multilayer designs with angular performance tailored to specific building sites and orientations are options for the future. Angular selectivity remains a possibility that needs further exploration, including for angular-selective thermal emission/reflection. For the first time, the variable solar elevation angle has been taken into consideration and the 3D printing technique is employed to fabricate tilted microstructures to dynamically modulate solar transmission in relation to the solar elevation angle by Long's group [2].

Measurements with the incident light perpendicular to the solar modulation devices at low and high temperatures may not be able to reflect the actual solar modulation ability precisely, especially for a location with four distinct seasons. Meanwhile, by taking advantage of the dynamic feature of solar radiation, it may be possible to achieve enhanced solar modulation ability as well as reasonable transmittance of visible light (Fig. 12.3a,b). A custom-designed tilted grating-shape composite structure of polymer and VO_2/W-doped VO_2 nanoparticles (NPs) on glass substrates will target applications in real scenarios in cities situated at different latitudes. By using the 3D printing technology (Digital Light Processing 3D printing) (Fig. 12.3b), the tilted angle of the composite structures can be manipulated according to the solar elevation angle variation in different seasons. As demonstrated in Fig. 12.3a, if we consider that the windows are placed parallel to the ground surface, in summertime, solar radiation will be perpendicular to the tilted structure so that the majority of the solar transmission will be absorbed due to the near-infrared (NIR) blocking effect of metallic VO_2 at high temperatures and the high effective blocking volume. However, in wintertime, solar transmission will be largely enhanced due to the solar irradiation parallel to the tilted surface, which will effectively reduce the blocking volume together with the NIR transparent characteristic of insulating VO_2.

For flat films, the standard practice is to place the incident light perpendicular to the windows (Fig. 12.3c); the transmittance spectra of these films at both temperatures, mimicking the incident angles in different seasons, are shown in Fig. 12.3d. By fixing the concentration at 1.5 wt.%, the width at 400 μm, and the tilted angle at 45°, a structured VO_2-polymer composite structure with 200 μm spacing and a 200 μm width was designed and printed. As shown in Fig. 12.3e, while determining the solar modulation ability of the tilted composite structure, the incident light source was placed at 0° and 45° with respect to the tilted surface to simulate wintertime and summertime, respectively. The transmittance spectra of the above-mentioned composite structures with 0° incident light at 20°C and 45° incident light at 90°C are shown in Fig. 12.3f. In contrast to a flat VO_2-polymer composite film with the same concentration and thickness, a large contrast occurs in both visible and infrared (IR) range. Thereby, the ΔT_{sol} has been improved significantly from 8.4% to 21.3%. Meanwhile, because of the gaps between the two structures, the proportion of the visible light that can pass through, especially in the wintertime, has been remarkably enhanced, where the $T_{lum(winter)}$ is increased from 36.1% to 51%, the $T_{lum(summer)}$ is increased from 32.8% to 33.1%, and the $T_{lum(average)}$ is increased from 34.5% to 42%. This first attempt to adaptively regulate solar modulation by taking the solar elevation angle variation into consideration together with one of the best-reported thermochromic properties (τ_c = 40°C, $T_{lum(average)}$ = 40.8%, and ΔT_{sol} = 23.3%) could further enhance the thermochromic performance.

12.2 Color

VO_2 film coatings with an unattractive yellow or brown color, or even a dark-blue color, on glass is another big barrier to the commercialization due to the strong absorption of such coatings. Different cultures may require certain modifications to tune the color. Although some simulations have suggested that doping may

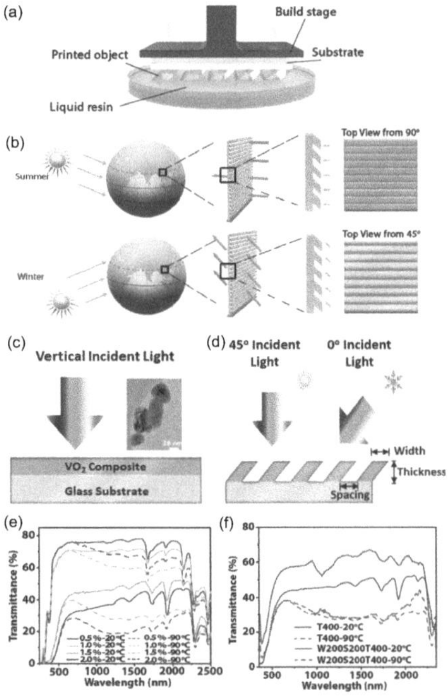

Figure 12.3 (a) Schematic diagram of the solar modulation mechanism for 3D-printed composite structures, together with the side-view picture and the top-view picture. In wintertime, some of the solar radiation transits through the gaps between the two composite structures, leading to a relatively high T_{lum}. In summertime, before being transmitted indoors, the solar radiation passes through the composite structures with the longest path length, resulting in a superior solar modulation ability. (b) Schematic diagram of the 3D printing process for the structured composite targeting solar modulation. (c) Schematic illustration of the conventional approach, with the vertical incident light for transmittance measurement on flat films; the insert is the TEM picture of the VO_2 NPs. (d) UV-vis-NIR transmittance spectrum of the flat polymer-VO_2 composite films with a 200 μm thickness and various VO_2 concentrations, varying from 0.5 wt.% to 2 wt.% at both high and low temperatures. (e) Schematic illustration of the transmittance measurement approach for the structured polymer-VO_2 composite. The angle of the incident light is labeled according to the angle between light source and the tilted surface of the structures as a simulation of various solar elevations at midday for different seasons. The dimensional parameters of the structured polymer-VO_2 composite are defined as labeled, including width, thickness, and spacing. (f) UV-vis-NIR transmittance spectrum of 3D-printed composite structures at a tilted angle of 45°, with a 200 μm width (W), 200 μm spacing (S), and a 400 μm thickness (W), with 0° incident light at 20°C and 45° incident light at 90°C, together with the flat polymer-VO_2 composite film of identical thickness and with the same VO_2 concentration at high and low temperatures, respectively. Reproduced from [2], Copyright © 2020 John Wiley and Sons.

be an effective strategy to alter the color by shifting the adsorption edges, experimental results have shown that only limited changes can be achieved. Complexion with dyes or responsive matrixes, either organic or inorganic, can modulate the original color of VO_2 at the expense of the thermochromic properties of VO_2 films or transparency, which may not be a good solution in some applications. Structural coloration as discussed in Chapter 6 has proved effective in altering the color, but some degradation of the thermochromic performance and the complicated fabrication approach need to be investigated further. To address this concern, more creative ideas are needed.

12.3 Emissivity

Solar heating results from the visible and NIR region of the solar spectrum, whereas blackbody emissivity of room temperature objects occurs in the mid-IR region. The former contributes to heat gains from the sun, while the latter is an important way to transmit heat from the hot side to the cold side. To achieve smart energy-saving functionality, a coating should have switchable transmittance to control the heat gain from sun as well as the emissivity to tune the heat exchange between the indoors and the outdoors favorably in different seasons. VO_2 films usually have emissivity as large as 0.83, which is much larger than that of double-layered silver-based low-emission (low-E) glass (0.3–0.4). This implies that glass coated with only VO_2 layers may not be effective in environmental heat management.

Vanadium dioxide thin films were deposited on transparent conducting fluorine-doped tin oxide (FTO) substrates using a lower-temperature solution process (Fig. 12.4a). The introduction of this type of low-E buffer layer gives reduced emissivity in both states compared with the normal VO_2/SiO_2 (Fig. 12.4b). For a 42 nm thick VO_2 film deposited on FTO glass, an averaged visible transmittance of 38% was obtained. The emissivity for the samples was quite low (0.15 and 0.19, respectively, for high- and low-temperature phase), whereas a solar transmittance modulation of 3.5% was achieved [2].

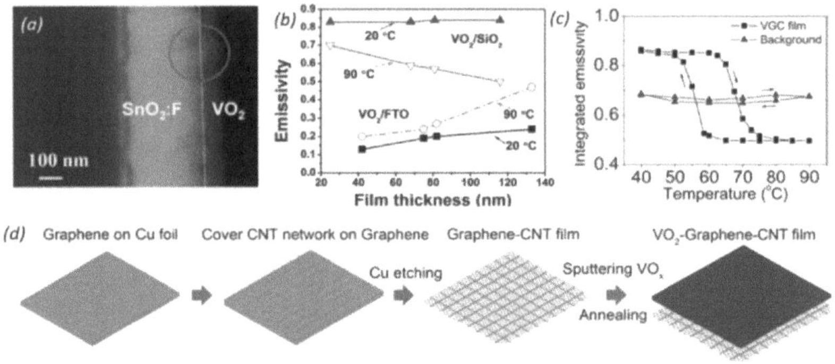

Figure 12.4 (a) Backscattered electron image for the VO_2/FTO/substrate double-layered film. (b) Thickness dependence of emissivity for VO_2/FTO/substrate films. The emissivity of VO_2 on fused-silica substrates was adopted for comparison. Lines with solid and open circles are the emissivity measured at 20°C and 90°C, respectively. (c) Temperature-dependent emissivity of the VGC film on black tape and the background coating. (d) Fabrication and characterizations of a VGC film. Reproduced from [3, 4], Copyright © 2011 Royal Society of Chemistry and Copyright © 2015 American Chemical Society, respectively.

A flexible foil designated as VO_2/graphene/CNT (VGC) consists of a VO_2 layer, with a negative differential thermal emissivity, coated on a graphene/carbon nanotube thin film. A slight joule heating drastically changes the emissivity of the device, achieving rapid switchable with temperature (Fig. 12.4c). Figure 12.4d shows the fabrication process of such a foil [3].

Simulation of heat and energy transfer through windows is needed to understand the optical response of coatings of different wavelengths to achieve better energy saving performance in thermochromic smart windows.

12.4 Stability

Vanadium, a transition metal, is a multivalence element that exhibits +3, +4, +5, and mixed valences. In ambient atmosphere, VO_2(M) can be readily oxidized to vanadium oxides or their hydroxides, and the presence of moisture accelerates this process. This instability results in serious durability problems in practical applications. Therefore, Chapter 11 is dedicated to this issue. The formation of VO_2 in an inert oxide shell, which is used to isolate VO_2 from oxygen, may alleviate the instability issue but often at the cost of the aggregation of well-dispersed NPs and the complexity of process control. More fundamental studies are needed in this field as instability is smart windows' Achilles' heel and architecture has a lot of influence on durability.

12.5 Toxicity

The fate of NPs in a variety of environmental and manufacturing settings has attracted immense attention. Toxicity is a major obstacle against their practical application, especially in applications that involve close contact with people and pets. Although some VO_2 products are commercially available (such as VO_2 foils in China), thorough research into the toxicological impact of VO_2 materials, especially in the form of NPs, on human health and the environment and the possible hazards is still in its infancy and must be urgently addressed to achieve large-scale applications of VO_2. Toxicity is one of the biggest concerns in VO_2-based smart windows since it is well known that V_2O_5 is an extremely toxic substance. However, the toxicity of vanadium is determined by its valence state and solubility. A higher oxidation state and high solubility would lead to higher toxicity [5, 6]. VO_2 would have less toxicity than V_2O_5. Actually, as an essential trace element in the human body, vanadium and its toxic mechanism must be treated differently from that of some heavy metal ions (such as Hg, Pb, and Cd). The human body can easily get rid of excess vanadium by metabolism. It should be noted that vanadium is a beneficial element in low concentrations and only harms human health if there is excessive intake in the short term or exposure to high concentrations in the long term.

Toxicity studies of VO_2 are far fewer than those of V_2O_5, especially studies related to the VO_2 NPs used for smart windows. The current works show that VO_2 NPs are low in toxicity to bacterial [7] and human lung cells (A549) [8] and their toxicity is dependent

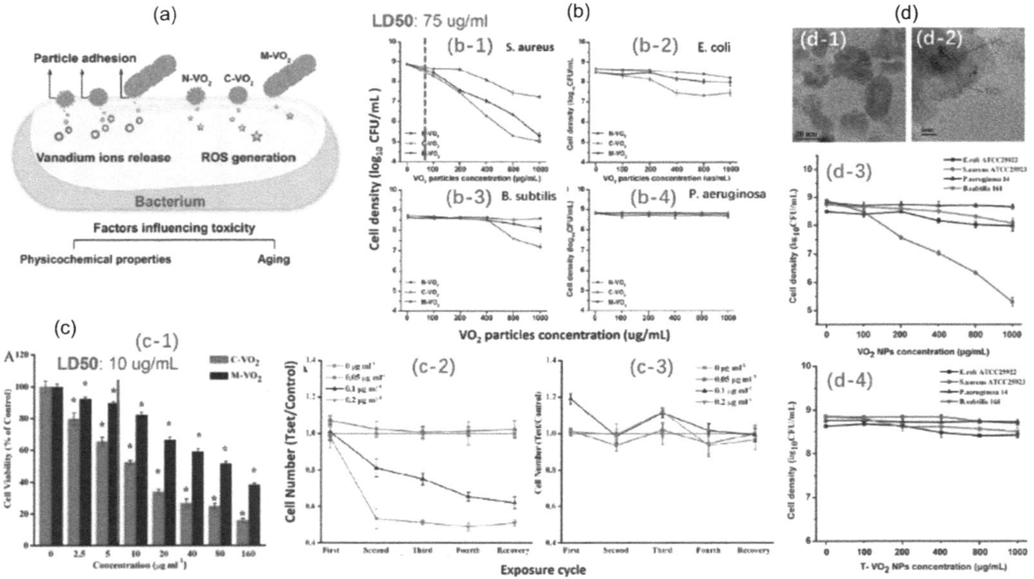

Figure 12.5 (a) Schematic illustration of the toxicity mechanisms of different VO_2 particles: VO_2 microparticles (M-VO_2), highly crystalline VO_2 (C-VO_2), and low-crystalline VO_2 (N-VO_2). (b) Antibacterial effects of different VO_2 particles against *S. aureus* (b-1), *E. coli* (b-2), *B. subtilis* (b-3), and *P. aeruginosa* (b-4); VO_2 particles are low in toxicity toward bacteria, and the toxicity is species dependent. (c) Toxicity toward A549 cells, (c-1) MMP levels of A549 cells after exposure to VO_2 particles for 24 h, and proliferation profiles of A549 cells during long-term exposure to C-VO_2 (c-2) and M-VO_2 (c-3), which shows that the toxicity of VO_2 particles is size dependent. (d) A comparison of toxicity of VO_2 NPs before (d-1 and d-3) and after (d-2 and d-4) they are coated with TiO_2. Reproduced from [7, 8], Copyright © 2018 and 2019 Royal Society of Chemistry, respectively.

on particle size, crystallinity, dose, and species (Figs. 12.5b and 12.5c). Toxicity of VO_2 NPs is derived from the released vanadium ions and the produced reactive oxygen species (Fig. 12.5a) [7]. Being coated with other materials would significantly reduce their toxicity (Fig. 12.5d).

As to the usage of VO_2-based smart windows, the toxicity of VO_2 should be ignored since VO_2 is used in trace amounts in smart windows: for VO_2 thin films, the optimized thickness is about 50–100 nm (0.22–0.43 g/m^2), and for VO_2-based nanocomposite coatings, about 0.5g/m^2 of VO_2 NPs is used. Even if all the vanadium in 1 m^2 of smart films/coatings dissolved and oxidized to V^{5+}, the dose will not be enough to kill an adult, not to mention the release of VO_2 is a long process and the VO_2 is usually covered by or wrapped in other materials, such as antireflective films (SiO_2 [9], TiO_2 [10], ZrO_2 [11], etc.) and polyurethane matrix [12], which would further extend the releasing process. On the other hand, the antibacterial effect of vanadium would endow the VO_2 smart films/coatings with antibacterial property [6].

The only problem would be in the production section, where high concentrations of VO_2 exist. The producer must pay attention to how to control the dust of the VO_2 nanopowder and how to deal with the waste containing vanadium. However, these are not a challenge in technology in the current chemical industry.

12.6 Scaled-Up Production of VO_2 Nanoparticles

VO_2 coatings are a potential energy-efficient option for buildings, and their production needs to be scaled up scientifically and technologically. Almost all the synthesis strategies/methods discussed in this book, either gas based (chemical vapor deposition [CVD] and physical vapor deposition [PVD]) or solution based (hydrothermal and sol-gel process), are currently limited to the lab scale. Among them, the CVD method developed so far is one of the most suitable ones for industrial processing, similar to coating silver or tin oxide on float glass. However, the cost issue in operation and small-scale production and the difficulty in controlling the pristine VO_2(M) phase as well as the elemental-doped ones while using gas-based methods have limited their large-scale production.

In the ongoing and future development of thermochromic VO_2 smart coatings, development of facile but efficient scale-up strategies is critical. This requires the integration of a few of the technologies into the glass production lines, contrary to the direct deposition of VO_2 on substrates using the CVD or PVD method. The combination of chemical and physical methods may achieve dramatic effects. For instance, a two-step method has been recently developed for generating glass coatings for smart windows, which may be a feasible solution for carrying out industrial-scale productions. By this method, VO_2 NPs with a uniform size distribution could be first prepared by the hydrothermal method (chemical method) in acid (oxalic acid) or base (ammonia) systems and then dispersed in an aqueous solution as a nanoink for the preparation of film coatings by physical techniques, such as spraying, rolling, casting or screen printing, which could significantly reduce costs and readily achieve thermochromic coatings to satisfy the huge market needs. In such a strategy, the preparation of uniform-sized VO_2 NPs (ranging from 30 nm to 100 nm) is the most crucial step, while the pertinent parameters (temperature, pressure, reaction time, concentration, pH, drying conditions, etc.) need to be carefully optimized.

12.7 Process

The harsh preparation requirements of VO_2 and the complicated fabrication procedure of devices are a great challenge in current approaches, such as solution-based methods, CVD, and PVD, which have a limited ability to produce VO_2-based devices in mild conditions with the simple methods summarized in this book. Development of manufacturing technologies like roll-to-roll coating of polymer foil may open paths toward the production of cheap products. However, the ultimate goal to come up with a simple, moderate, and highly productive technique is one of the important research directions to bring the VO_2-based smart device out from the laboratory and into our daily lives.

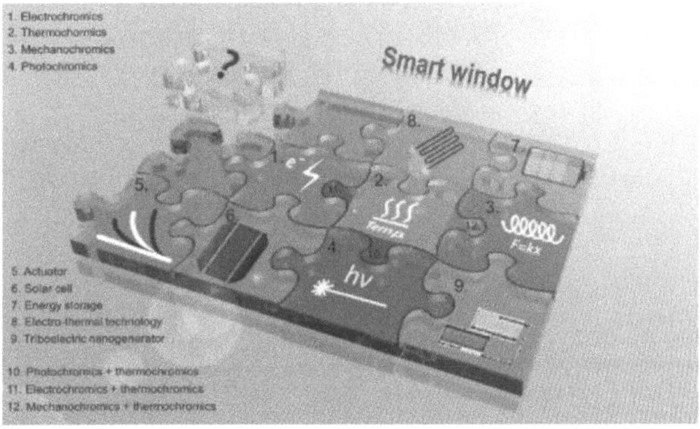

Figure 12.6 Illustration of the four optical modulation modes, the multi-functional technologies, and other potential technologies. Reproduced from [13], Copyright © 2019 John Wiley and Sons.

12.8 Combination of Thermochromic VO_2 with Other Energy-Saving System and Other Functionalities

To fully utilize the large surface areas available in the urban windows, in one recent review paper published by Long et al. [13], future developments were suggested to continue to break the boundaries among different modes of stimuli to integrate functionalities like energy production and photocatalytic self-cleaning (Fig. 12.6). One of the recent trends is to rationally combine the different stimuli to further enhance chromic window performance. For example, the combination of thermochromic VO_2 with mechanical kirigami structures renders a much-enhanced performance by passively modulating the indoor solar energy in response to the environmental temperature together with actively controlling the kirigami structure (Fig. 12.7a) [1, 14]. The adaptive thermal-responsive system is applied in a house demo to demonstrate the potential for the thermochromic smart window (Fig. 12.7b). It was observed the film changes from a noncompact (strain of 100%) to a compact (strain of 0%) state through a gradual close up of kirigami cuts with increasing temperature, accompanied by a less clear image of the indoor clamp (Fig. 12.7c).

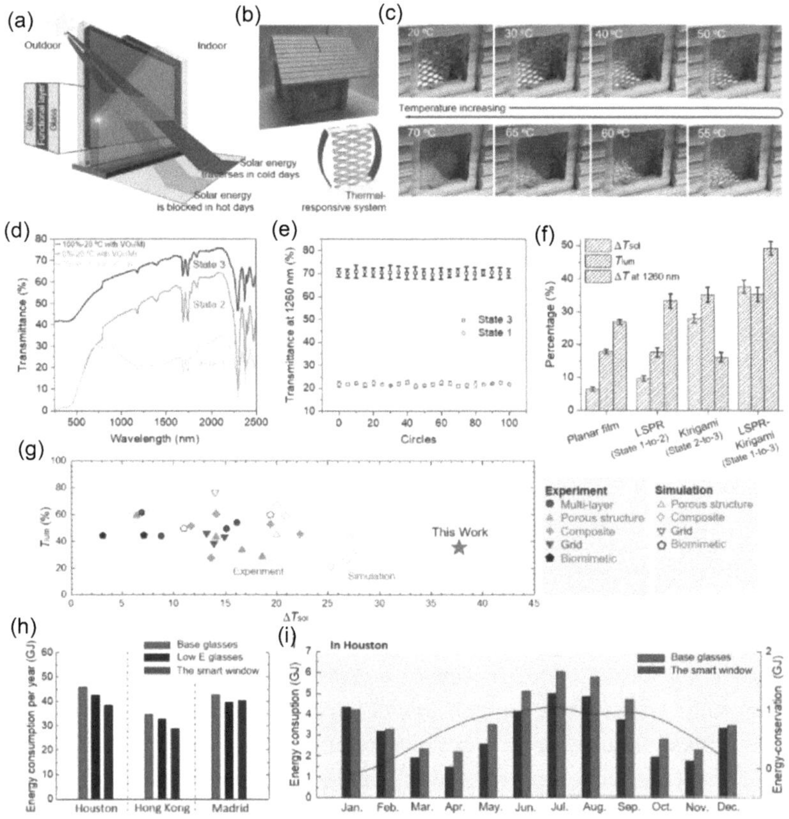

Figure 12.7 Smart, energy-efficient windows. (a) Illustration of an energy-saving smart window that adaptively reduces the indoor solar energy irradiation under elevated temperatures. (b) The photograph of a house model and a schematic of the thermal-responsive system for the smart window demo. (c) The configuration transition of the smart window model under increasing temperatures. The fully thermal-responsive system is applied. (d) Transmittance spectra of the film at (state 1) 90°C with a compact configuration, (state 2) 20°C with a compact configuration, and (state 3) 20°C with a noncompact configuration (strain of 100%). (e) Cycling test of the film switching between states 1 and 3. (f) Calculated ΔT_{sol}, T_{lum}, and ΔT_{1260nm} of the film in states 1–3 and the reference film based on planar VO_2 and PDMS, respectively. Reproduced from [15], Copyright © 2019 Elsevier.

In Fig. 12.7d, the respective contributions of the active local surface plasmonic resonance (LSPR) induced by VO_2 particles and the kirigami structure are identified by comparing three states: state 1 (0% strain @ 95°C); state 2 (0% strain @ 20°C); and state 3 (100% strain @ 20°C). More specifically, the contributions of active LSPR, kirigami, and their synergistic effect are calculated on the basis of the contrasts of states 1-to-2, 2-to-3, and 1-to-3 in Fig. 12.7d, respectively. The results of ΔT_{sol}, average visible transmittance (T_{lum}), and the transmittance contrast at 1260 nm (ΔT_{1260nm}) are plotted in Fig. 12.7f. The sole active LSPR and kirigami structure effects contribute to ΔT_{sol} of 9.7% and 28%; however, the combined effect of the LSPR and kirigami structure gives a record-high ΔT_{sol} (up to 37.7%), as well as a significantly enhanced T_{lum} (from 17.6% [plasmonic sample] to 35.2% ["plasmonic-kirigami sample"]) as shown in Fig. 12.7f. Meanwhile, the introduction of active LSPR largely enhances the NIR controllability and the contrast at 1260 nm (ΔT_{1260}) increases from 16.1% on the kirigami sample to 49.3% on the plasmonic kirigami sample (Fig. 12.7f). Both ΔT_{sol} and T_{lum} were simultaneously enhanced by this simple approach.

Energy harvesting, superhydrophobicity for self-cleaning windows [16, 17] energy storage [18], water harvesting from the environment and its storage [19, 20], and anti-icing [21–23] are highly desired functionalities as these will make such windows more competitive in the market and very economical in terms of the energy bill. The inevitable challenge lies in drawing a balance between the luminous transmission, transparency, energy saving, device performance, and durability.

12.9 Energy Calculation

Last but not the least, the actual energy consumption needs to be simulated in different climates to offer customized designs and maximize the energy saving performance [2, 13, 14]. In the recently developed 3D-printed tilted structure, as shown in Fig. 12.3, W-doped VO_2, with a $\Delta T_{sol} = 23.3\%$ and $T_{lum} = 40.8\%$ [2], was further added to test the real-world applications, outperforming all the previous reported results, as summarized in one of the

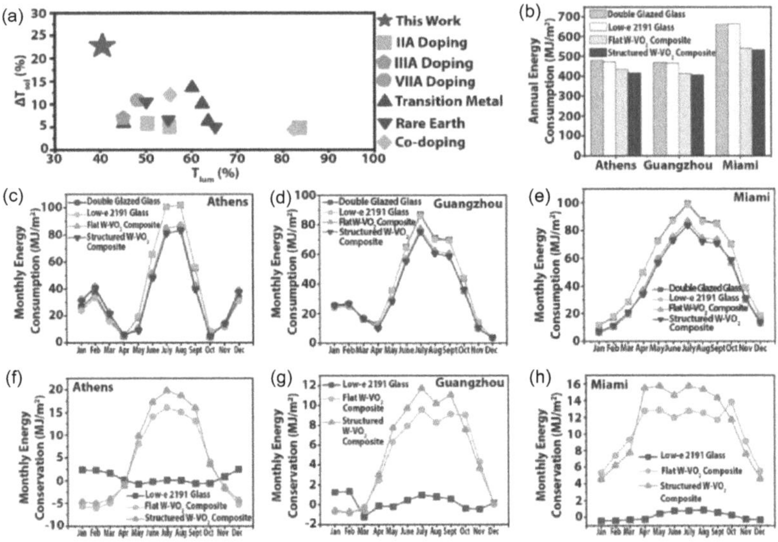

Figure 12.8 (a) Comparison of τ_c, T_{lum}, and ΔT_{sol} between the best-reported and selected works and the tilted VO_2 and W-doped VO_2 composite structures. (b) Annual energy consumption of double clear glass, low-E 2191 glass, a flat W-doped VO_2 composite, and a structured W-doped VO_2 composite in Athens, Guangzhou, and Miami. (c) Monthly energy consumption of double clear glass, low-E 2191 glass, a flat W-doped VO_2 composite, and a structured W-doped VO_2 composite in Athens. (d) Monthly energy consumption of double clear glass, low-E 2191 glass, a flat W-doped VO_2 composite, and a structured W-doped VO_2 composite in Guangzhou. (e) Monthly energy consumption of double clear glass, low-E 2191 glass, a flat W-doped VO_2 composite, and a structured W-doped VO_2 composite in Miami. (f) Monthly energy conservation of low-E 2191 glass, a flat W-doped VO_2 composite, and a structured W-doped VO_2 composite in Athens while using double clear glass as the baseline. (g) Monthly energy conservation of low-E 2191 glass, a flat W-doped VO_2 composite, and a structured W-doped VO_2 composite in Guangzhou while using double clear glass as the baseline. (h) Monthly energy conservation of low-E 2191 glass, a flat W-doped VO_2 composite, and a structured W-doped VO_2 composite in Miami while using double clear glass as the baseline. Reproduced from [2], Copyright © 2020 John Wiley and Sons.

reviews (Fig. 12.8a) [1]. Using Energyplus to demonstrate the actual energy-saving performance of such a structured W-doped VO_2 thermochromic smart window, a simulation of the annual energy consumption was performed at various cities, including Athens, Guangzhou, and Miami, as they have similar solar incident angle differences between winter and summer. Four windows were used to gauge the performance: low-E glass (IGDB 2191), double-glazed glass, flat W-doped VO_2 composite film (1.5 wt.% and 200 μm thickness), and structured W-doped VO_2 composite (1.5 wt.%, 200 μm thickness, 200 μm spacing, and 200 μm width) (Fig. 12.8b). In all three cities, compared with the benchmark of the double-glazed glass and low-E glass, both thermochromic smart windows provided better energy saving while the structured window gave the best performance among all four windows. This phenomenon is more pronounced in Miami, where energy consumption can be reduced by over 20% annually. According to the monthly energy consumption diagram of all three cities (Fig. 12.8c–e), the gap between the double-glazed glass and the structure W-doped VO_2 composite reaches the maximum in summer, which indicates that the energy consumption for cooling can be largely reduced by as-designed smart window structures. While using the double clear glass as the baseline, the energy conservation abilities of the low-E 2191 glass, the flat W-VO_2 composite, and the structured W-VO_2 composite at the three cities are shown in Fig. 12.8f–h. The structured W-VO_2 composite surpassed the conventional energy-saving low-E glass in terms of energy conservation ability by over 10 times in all three cities, especially in the summertime. Moreover, comparisons have been done between the flat W-doped VO_2 composite smart window and the structure W-doped VO_2 composite smart window. Regarding the energy conservation ability, an average performance enhancement of 23.4%, 22.5%, and 22.7% could be achieved during summertime in Athens, Guangzhou, and Miami, respectively, which indicates the advantage of such uniquely designed structures.

In conclusion, the VO_2-based smart window is a highly interdisciplinary technology. By reviewing the recent progress in materials science, we expect that this summarized work and first VO_2-based thermochromic book may accelerate the development of smart windows that can regulate heat favorably. More innovative ideas are needed to custom-design specific structures to cater to different geographic and weather conditions while taking real-world sceneries into consideration, such as different solar incident angles and actual energy consumption. More research could potentially integrate multifunctional devices to collect, store, and convert energy, while ensuring privacy control, entertainment, and sensing to fully utilize the large glass surface areas available in urban cities. More fundamental studies on reducing the transition temperature, increasing the durability, and eliminating toxicity are urgently needed before this technology can come into the market. This highly interdisciplinary research is predicted to have a high social impact in the near future as it addresses the most urgent global warming issues.

Acknowledgment

Y. F. Gao drafted Section 12.5, X. C. Jiang drafted Sections 12.6 and 12.7, and Y. Long drafted the rest and edited the whole chapter. For the funding support, Y. Long wishes to thank the National Research Foundation, Prime Minister's Office, Singapore, under its Campus for Research Excellence and Technological Enterprise (CREATE) program.

References

1. Cui, Y., et al. (2018). Thermochromic VO_2 for energy-efficient smart windows, *Joule*, **2**(9), pp. 1707–1746.
2. Zhou, C., et al. (2020). 3D printed smart windows for adaptive solar modulations, *Adv. Opt. Mater.*, **8**(11), p. 2000013.
3. Zhang, Z., et al. (2011). Solution-based fabrication of vanadium dioxide on F:SnO_2 substrates with largely enhanced thermochromism and low-emissivity for energy-saving applications, *Energy Environ. Sci.*, **4**(10), pp. 4290–4297.
4. Xiao, L., et al. (2015). Fast adaptive thermal camouflage based on flexible VO_2/graphene/CNT thin films, *Nano Lett.*, **15**(12), pp. 8365–8370.
5. Ghosh, S. K., Saha, R. and Saha, B. (2015). Toxicity of inorganic vanadium compounds, *Res. Chem. Intermed.*, **41**(7), pp. 4873–4897.

6. Li, J., et al. (2016). Oxidative stress-mediated selective antimicrobial ability of nano-VO_2 against gram-positive bacteria for environmental and biomedical applications, *Nanoscale*, **8**(23), pp. 11907–11923.
7. Wu, D., et al. (2018). Toxicity assessment and mechanistic investigation of engineered monoclinic VO_2 nanoparticles, *Nanoscale*, **10**(20), pp. 9736–9746.
8. Xi, W.-S., et al. (2019). Short-term and long-term toxicological effects of vanadium dioxide nanoparticles on A549 cells, *Environ. Sci.: Nano*, **6**(2), pp. 565–579.
9. Chen, Z., et al. (2011). VO_2-based double-layered films for smart windows: optical design, all-solution preparation and improved properties, *Sol. Energy Mater. Sol. Cells*, **95**(9), pp. 2677–2684.
10. Jin, P., et al. (2003). Design, formation and characterization of a novel multifunctional window with VO_2 and TiO_2 coatings, *Appl. Phys. A*, **77**(3), pp. 455–459.
11. Xu, G., et al. (2004). Optimization of antireflection coating for VO_2-based energy efficient window, *Sol. Energy Mater. Sol. Cells*, **83**(1), pp. 29–37.
12. Gao, Y., et al. (2012). Enhanced chemical stability of VO_2 nanoparticles by the formation of SiO_2/VO_2 core/shell structures and the application to transparent and flexible VO_2-based composite foils with excellent thermochromic properties for solar heat control, *Energy Environ. Sci.*, **5**(3), pp. 6104–6110.
13. Ke, Y., et al. (2019). Smart windows: electro-, thermo-, mechano-, photochromics, and beyond, *Adv. Energy Mater.*, **9**(39), p. 1902066.
14. Ke, Y., et al. (2018). Emerging thermal-responsive materials and integrated techniques targeting the energy-efficient smart window application, *Adv. Funct. Mater.*, **28**(22), p. 1800113.
15. Ke, Y., et al. (2019). Adaptive thermochromic windows from active plasmonic elastomers, *Joule*, **3**(3), pp. 858–871.
16. Zheng, J., Bao, S. and Jin, P. (2015). TiO_2(R)/VO_2(M)/TiO_2(A) multilayer film as smart window: combination of energy-saving, antifogging and self-cleaning functions, *Nano Energy*, **11**, pp. 136–145.
17. Qian, X., et al. (2014). Bioinspired multifunctional vanadium dioxide: improved thermochromism and hydrophobicity, *Langmuir*, **30**(35), pp. 10766–10771.
18. Cai, G., Wang, J. and Lee, P. S. (2016). Next-generation multifunctional electrochromic devices, *Acc. Chem. Res.*, **49**(8), pp. 1469–1476.

19. Helseth, L. E. and Guo, X. D. (2016). Fluorinated ethylene propylene thin film for water droplet energy harvesting, *Renewable Energy*, **99**, pp. 845–851.
20. Nandakumar, D. K., et al. (2018). A super hygroscopic hydrogel for harnessing ambient humidity for energy conservation and harvesting, *Energy Environ. Sci.*, **11**(8), pp. 2179–2187.
21. Mazumder, P., et al. (2014). Superomniphobic, transparent, and anti-reflection surfaces based on hierarchical nanostructures, *Nano Lett.*, **14**(8), pp. 4677–4681.
22. Mitridis, E., et al. (2018). Metasurfaces leveraging solar energy for icephobicity, *ACS Nano*, **12**(7), pp. 7009–7017.
23. Yamazaki, T., et al. (2019). Antifreeze liquid-infused surface with high transparency, low ice adhesion strength, and antifrosting properties fabricated through a spray layer-by-layer method, *Ind. Eng. Chem. Res.*, **58**(6), pp. 2225–2234.

Index